CAMBRIDGE STUDIES IN MODERN OPTICS: 1

Interferometry

Interferometry

W. H. STEEL
CSIRO Division of Applied Physics, Sydney, Australia

SECOND EDITION

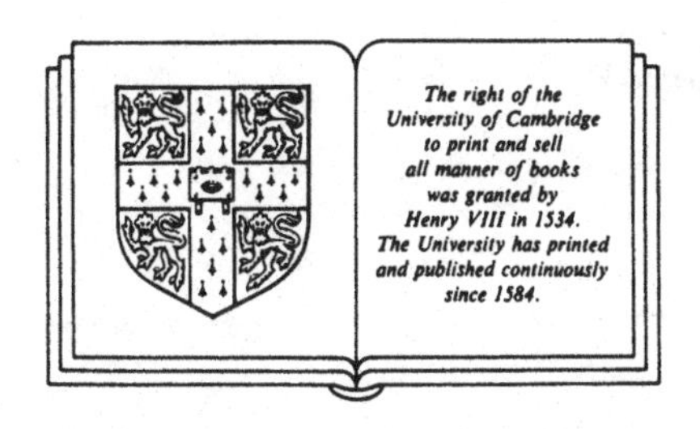
The right of the
University of Cambridge
to print and sell
all manner of books
was granted by
Henry VIII in 1534.
The University has printed
and published continuously
since 1584.

CAMBRIDGE UNIVERSITY PRESS

CAMBRIDGE

NEW YORK NEW ROCHELLE MELBOURNE SYDNEY

CAMBRIDGE UNIVERSITY PRESS
Cambridge, New York, Melbourne, Madrid, Cape Town, Singapore, São Paulo, Delhi

Cambridge University Press
The Edinburgh Building, Cambridge CB2 8RU, UK

Published in the United States of America by Cambridge University Press, New York

www.cambridge.org
Information on this title: www.cambridge.org/9780521311625

First published 1967
Second edition 1983
Reprinted 1985, 1987
First paperback edition 1986
Reprinted 1987
Re-issued in this digitally printed version 2009

A catalogue record for this publication is available from the British Library

Library of Congress Catalogue Card Number: 67-12140

ISBN 978-0-521-25320-8 hardback
ISBN 978-0-521-31162-5 paperback

Contents

Preface to the first edition

To the astronomer, interferometry suggests the measurement of stellar diameters or the techniques of radio astronomy, both implying studies of the spatial distribution of radiation. For the spectroscopist, it measures spectral distributions and represents a new technique offering high resolving power and increased sensitivity. To the optical designer, it is a method of testing lenses and, to the biologist, it is a new branch of microscopy. To the metrologist and the engineer, it is the means of converting the international standard of length into a practical scale. Each user is familiar with the techniques and applications of his own field but not often those of other fields.

My aim in writing this monograph is to present a theory of interferometry and a description of its techniques that are valid for all applications and in all regions of the spectrum where interferometers are used. The treatment naturally reflects my own interests and prejudices, and is biased towards applications with visible light; in justification, however, it can be said that it is here that the greatest development of the interferometer has occurred, microwave and radio interferometers being simple instruments by comparison. These last instruments are included as well as the new field of infra-red spectroscopy by Fourier transform methods. Most of the mathematics is given in the earlier chapters, so that the later chapters on the various instruments and the different applications are mainly descriptive; those not interested in the details of the theory can thus omit Chapter 4 [Chapter 6 in the second edition]. I have aimed to include most techniques under the description of instruments, rather than under the particular application for which they were developed, with the hope of suggesting their wider use in other fields.

This book owes much to discussions with colleagues, chiefly at the Air Force Cambridge Research Laboratories, Bedford, the National Standards and Radiophysics Laboratories, Sydney, and the University of Sydney. A special debt is due to Professor A. Maréchal, under whom I first worked in Physical Optics, and I am grateful to Professor P. Rouard for giving me the opportunity of undergoing the important discipline of presenting these ideas as a short course at Marseilles. Finally, I acknowledge with thanks the permission of those authors, publishers, and learned societies whose figures I have reproduced here.

Sydney, March 1967 W.H.S.

Preface to the second edition

When the first edition of *Interferometry* was written, most interferometers used light sources that were far from coherent. To emphasize this, I avoided the traditional treatment in terms of coherent sources and simple waves and gave only a theory in terms of coherence functions. But the rapid increase in the use of laser sources has given practical relevance to the wave treatment, which has now been added.

At the same time the great expansion of the subject has made it impossible to treat in detail the methods and techniques used in all spectral regions. To include new chapters on hologram and speckle interferometry, the sections on microwave and radio interferometry have been cut down, except where these fields provide the best illustration of some principle or technique. Since the first edition was published, radio astronomy has been treated by the monograph by Christiansen and Högbom. It has also been necessary to make drastic cuts to the number of references given.

I again acknowledge with thanks the permission of those authors, publishers, and learned societies whose figures I have reproduced, and the debt I owe to my colleagues here for their expert knowledge of interferometry, which has helped me so often. Finally, I have had the good fortune to have worked near two of the most interesting interferometers of recent times, the intensity interferometer of R. Hanbury Brown and J.P. Wild's radioheliograph.

Sydney (1983) W. H. STEEL

1

Introduction

1.1. Uses of interferometry

Interference occurs when radiation follows more than one path from its source to the point of detection. It may be described as the local departures of the resultant intensity from the law of addition, for, as the point of detection is moved, the intensity oscillates about the sum of the separate intensities from each path. Light and dark bands are observed, called interference fringes. The phenomenon of interference is a striking illustration of the wave nature of light and it has had a considerable influence on the development of physics. Young's observation and explanation of the interference of the beams through two holes provided the basis for Fresnel's wave theory of light and the same experiment has been used as the foundation of modern coherence theory. Einstein's special theory of relativity is supported by the negative result of the Michelson–Morley experiment [Shankland, 1973].

Derived from interference is the technique of interferometry, now one of the important methods of experimental physics, with applications extending into other branches of science. The father of visible-light interferometry was Michelson [1902, 1927], who was awarded in 1907 the Nobel prize in physics for 'his optical instruments of precision and the spectroscopic and metrological investigations he has executed with them'. Applications to other spectral regions are more recent: the first use of interferometry in radio astronomy was reported in 1947, and infrared interference spectroscopy is even more modern.

A list of the fields of application of interferometry is given in Table 1.1. These applications range through the electromagnetic spectrum from X-rays to radio waves and apply also to acoustical waves [Magome *et al.*, 1981], electrons [Endo, Matsuda & Tonomura, 1979], and neutrons [Bonse & Rausch, 1979]. Techniques are furthest advanced, however, for visible light. The table follows from the theory given in Chapter 7, which provides a general description of two-beam interferometers. This theory is the unifying theme for this study of interferometry, while convenient boundaries to the subject are provided by those that traditionally separate interference from the other main branches of physical optics: diffraction and polarization. In some respects this boundary is artificial: a diffraction grating is probably more logically treated as a multiple-beam interferometer, as is the equivalent system in radio astronomy. Further, it is often customary to treat certain examples of diffraction, such as the holograms of Gabor [1949], as due to the interference between the *direct* and the *diffracted* radiation from an object. All such phenomena will not be classed here as interference; phase-contrast microscopy, for example, is not included. Traditionally, the effects of diffraction caused by apertures within the interferometer are ignored in interference theory, but they must be taken into account when the wavelength is large or when very great precision is required.

Table 1.1. *Fields of interferometry*

Measurement		Applications
Direct	Derived	
(1) Fringe position	(*a*) Mean phase difference	(i) Length standard and wavelengths (ii) Length comparisons and machine control (iii) Refractometry (iv) Velocity of light
	(*b*) Phase variations	(i) Interference microscopy (ii) Microtopography (iii) Optical testing (iv) Hologram interference
(2) Fringe visibility	(*a*) Spectrum of source	Profiles of symmetrical spectral lines
	(*b*) Spatial distribution at source	Stellar diameters
(3) Full intensity distribution (position and visibility)	(*a*) Spectrum of source	(i) Direct interference spectroscopy (ii) Fourier spectroscopy
	(*b*) Spatial distribution at source	(i) Optical transfer function (ii) Radio astronomy

In Table 1.1, the spectral and spatial distributions are those of intensity and apply to a source outside the interferometer. The measurements of phase apply to a sample inside it and, in fact, the amplitude is also given in this case. An alternative classification of the applications of interferometry is therefore possible: for an external sample it measures the spatial and spectral distributions of the intensity; for a sample inside it measures the spatial distribution of the complex amplitude transmission factor, provided the radiation used is sufficiently monochromatic.

General introductions to interferometry have been given by Williams [1930] and Tolansky [1955] and more details of the instruments by Candler [1951]. More recent treatments are given by Françon [1956, 1966], and Dyson [1970].

1.2. Interference and coherence

Interference and coherence are the experimental and theoretical aspects of the same phenomenon. While radiation from two separate sources produces no interference (except over short time intervals), that from the same source travelling by several different paths may do so. Initially the words 'coherent' and 'incoherent' were used to describe the addition of radiation in these two cases. When two beams came from the same source, they were coherent (provided the source was sufficiently small and of a sufficiently narrow bandwidth) and would interfere when added together. If from different sources, they were incoherent and no interference was produced; their intensities added directly.

Later the idea of partial coherence was introduced and coherence was given a scale of values, derived from the amount of interference produced. In recent years, coherence theory has developed considerably beyond the field of simple interferometry, but the earlier results are of direct application. It is on this theory that most of this study of interferometers is based.

Recently, however, the development of lasers has made available a light source that produces pairs of beams that are highly coherent with each other for almost all interferometer arrangements. Their interference can be treated by a simpler theory, expressed in terms of trains of waves. This is the historical treatment of interference, which was very much an idealization before lasers.

1.3. Classifications of interferometers

There are two traditional methods of classifying interferometers: by the number of interfering beams and by the method used to separate these beams. Modern advances in interferometry have introduced new instruments and new applications that give rise to further classifications.

1.3.1. Laser or thermal source

One of these is in terms of the source of radiation that is used. Of the applications listed in Table 1.1, those listed under measurements of fringe position are measurements that can be made with any suitable light source. For these a laser is being increasingly used. But for the other applications the source cannot be chosen at will, since these are measurements of the properties of the source itself.

1.3.2. Number of beams

Interferometers are classed as *two beam* or *multiple beam* according to the number of beams that interfere. As the typical multiple-beam interferometer usually has many interfering beams, there remains a group of interferometers that are not directly included in these classes: those with three or four interfering beams. These may be considered as variants of the two-beam interferometer, to which they are more closely allied.

Two-beam interferometers produce interference fringes with a sinusoidal variation of intensity. In a multiple-beam interferometer, each pair of beams contributes a Fourier component to the fringe pattern. The fringe pattern is still periodic and, in principle, any periodic profile should be possible. However, most interferometers have profiles that approximate to a series of narrow spikes, the Dirac comb.

1.3.3. Separation of beams

Another traditional basis for classifying interferometers is by the method used to divide the light into separate beams. If the radiation from the source may pass through one of several apertures in the same plane, it is said to be separated into beams by *division of wavefront*. These beams are made up of radiation that has left the source in different directions. If the beams consist of radiation that has left the source in the same direction but is then separated by a beam splitter, there is said to be *division of amplitude*.

Interferometers with division of amplitude can be further subdivided according to the type of beam splitter used or the number of passes made through it. A simple *partially reflecting* beam splitter consists of a thin film, either metal or dielectric, usually on a transparent support: the classical part-silvered mirror is an example. One beam is the reflected light, the other the transmitted. Although such films affect somewhat the polarization of the beams, this effect is relatively unimportant.

But if a stack of layers is used with each surface at the Brewster angle, the two beams leave with orthogonal linear polarizations. Some of the modern variants of the Nicol prism [Thompson, 1905] will produce the same result, provided they are made to pass both the transmitted and reflected beams. A Rochon or Wollaston prism transmits the two beams, but they leave in different directions. Any of these can be used as a *polarizing beam splitter*. The two beams are then labelled by their states of polarization and will be affected differently by further polarizing components. Thus, a second beam splitter can bring the two beams together without the further splitting that would occur at a partial reflector. In certain interferometers, a birefringent material such as calcite is used, not only for the beam splitter, but also to provide the different paths for the two beams, one of which acts as an ordinary ray, the other as an extraordinary ray. This is then a *polarization interferometer*.

The third beam splitter is the *diffracting beam splitter* which has either a random or a periodic variation of (complex) transmission across its surface; its use is reviewed by Lohmann [1962*a*]. The incident light is then divided into a direct beam and one or more diffracted beams. When the transmission of the beam splitter is periodic, the fringes obtained with an interferometer consisting of two beam splitters in series have the special name of *moiré fringes*, and a new branch of engineering metrology has been built around their use [Guild, 1960; Takasaki, 1979].

Any inteferometer with division of amplitude can appear in three very different guises, depending on which of the three beam splitters is used. An example is shown in fig. 1. Each is a *lateral-shearing interferometer* which, from each ray from the source, produces two rays leaving some point in different directions. The first is the form due to Bates [1947]. It is completely adjustable, and both the angle between the rays and the position of their intersection can be varied. The polarization form is a Wollaston prism which, when placed between two polarizers oriented at $\pi/4$ to it, will give two interfering beams at a fixed shear. The last form is the Ronchi grating with one direct and several diffracted beams, any pair of which represents a shearing interferometer with a fixed shear. Although the last two forms are not adjustable, they can be made adjustable if two are used in series. In some circumstances, however, lack of adjustment is an advantage. Usually interferometers based on these last beam splitters are more compact and more stable than those based on the first type.

The change of shear with wavelength differs for each form. For the form with mirrors and partial reflectors, the shear is independent of wavelength. The Wollaston prism has a shear that increases as the wavelength decreases while the grating has a shear that is proportional to wavelength. The fringe spacing for this last type of interferometer is independent of wavelength

[Leith & Swanson, 1980] and the fringes appear as images of the rulings of the grating: moiré fringes are black and white, not coloured.

1.3.4. Number of passes

Each of these lateral-shearing interferometers may be described as a *single-pass* or *straight-through* interferometer. It will be seen later that there are often advantages in returning the radiation back through the same system to give a *reflex* system. If a sample being measured is inside such an interferometer, the radiation goes through it twice and the sensitivity is doubled by such a *double-pass* measurement. At the same time the second passage can be used to cancel out effects, such as shears, that would otherwise reduce or even destroy the visibility of the interference fringes.

1.3.5. Order of correlation

It will be shown later that a simple interferometer measures second-order correlations of the radiation field. Hanbury Brown & Twiss [1954, 1956] have developed a new class of interferometers, *intensity interferometers* or *correlation interferometers*, that measure fourth-order correlations. In principle, higher-order correlations can also be measured, but these have more limited application.

Fig. 1 Three forms of a lateral-shearing interferometer (*a*) form due to Bates [1947] with partially reflecting beam splitter; (*b*) birefringent form – a Wollaston prism; (*c*) diffraction form – a Ronchi grating.

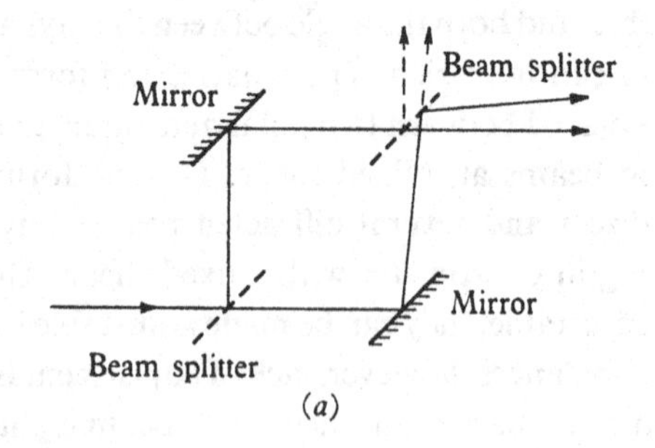

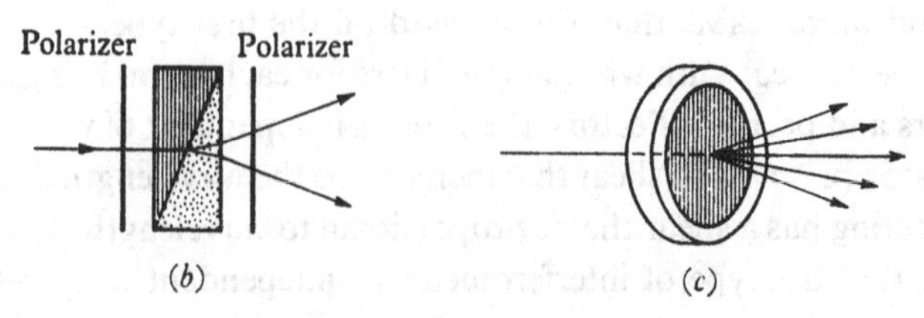

1.3.6. Hologram interference

The new field of hologram interferometry has introduced a new class of interferometers and a new parameter, the time at which a wave was produced. Holograms can store waves so that they can be reconstructed later to interfere with either another stored wave or a wave produced directly. These two types of hologram interferometry are known, respectively, as *frozen-fringe* and *live-fringe* interferometry. A third method, *time-averaged* interferometry, is used to study vibrating specimens.

1.4. Basic interferometers

Some of the more important interferometers are introduced in figs. 2 to 5. Those in fig. 2 are two-beam interferometers with division of wavefront. The first is the classical experiment of Young where light from a source passes through two pinholes in a screen and the interference is observed on a second screen placed beyond the first. The second example is Lloyd's mirror where the interfering beams come from the source and from its image in the mirror. This is the first interferometer used in radio astronomy, the mirror there being the sea. The third example is the Rayleigh interferometer used for refractometry.

The more important two-beam interferometers with division of amplitude are shown in fig. 3. The first is the Mach–Zehnder interferometer,

Fig. 2 Two-beam interferometers with division of wavefront: (*a*) Young's experiment; (*b*) Lloyd's mirror; (*c*) Rayleigh interferometer.

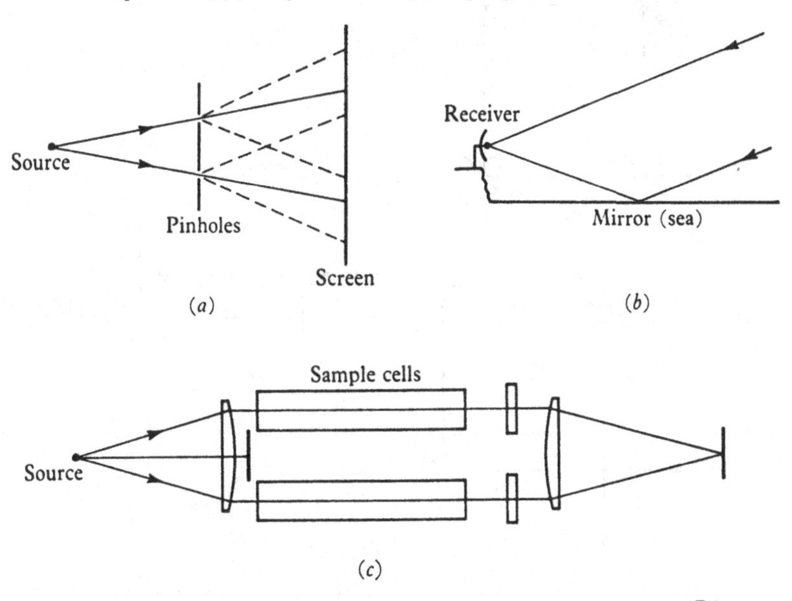

where the differences between the two beams are fairly generally adjustable. The beams are separated by one beam splitter and recombine at another. Fewer adjustments are possible with the Michelson interferometer, in which only one beam splitter is used, the two beams being reflected back to recombine where they separated. In the final interferometer they also come back to the same beam splitter, but on the other side, having each traced a loop. This is the cyclic interferometer, usually named after Sagnac, although it also was used earlier by Michelson [Hariharan, 1975*b*].

Two modifications of these interferometers are shown in fig. 4. The first is a Mach–Zehnder interferometer, in which the two mirrors have been replaced by mirror pairs [Hariharan, 1969]. This makes it possible to vary the path difference between the two beams, an adjustment that is not readily made to the simpler form. The second instrument is a Michelson interferometer, in which the mirrors have been replaced by cube corners.

Two multiple-beam interferometers are shown in fig. 5. The first is the Fabry–Perot interferometer with division of amplitude, first used by Boulouch [1893]. The second is the Christiansen cross [Christiansen *et al.*, 1961].

Fig. 3 Two-beam interferometers with division of amplitude: (*a*) Mach–Zehnder; (*b*) Michelson; (*c*) cyclic or Sagnac.

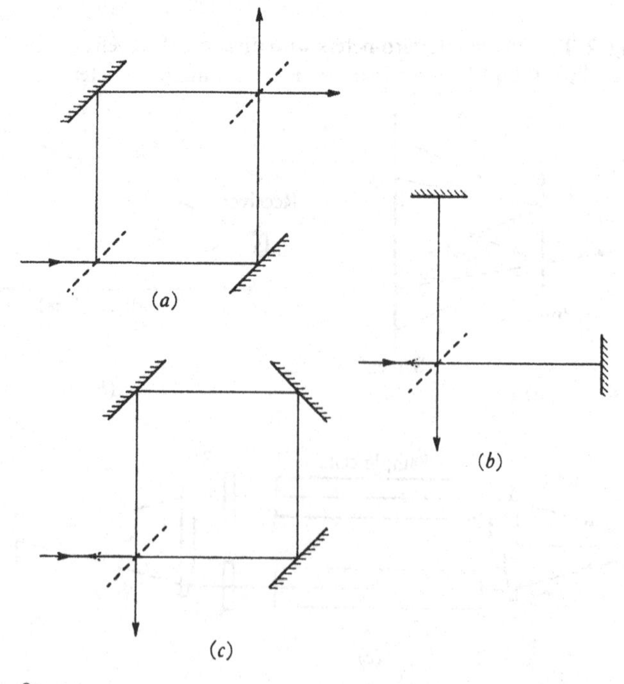

1.5. Description of an interferometer: complementary fringes

The Michelson interferometer of fig. 3(*b*) is probably the most widely used instrument in interferometry and it provides a suitable example for a general description of the properties of an interferometer and also for the theory given later. Let us consider it illuminated by a single spectral line from a small source that is effectively at infinity, that is, at the focus of a suitable collimating lens (fig. 6). Provided the two arms have about the same length, an observer looking into the interferometer and focusing on the

Fig. 4 Modified forms of two-beam interferometers with division of amplitude: (*a*) Mach–Zehnder with mirror pairs [Hariharan, 1969]; (*b*) Michelson with retroreflectors.

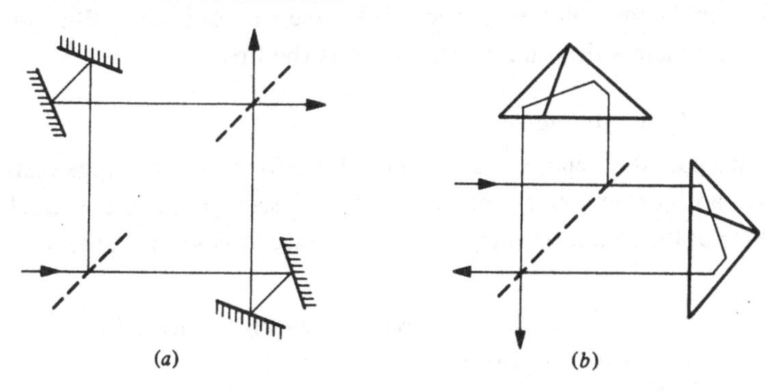

Fig. 5 Multiple-beam interferometers: (*a*) Fabry–Perot interferometer; (*b*) Christiansen cross.

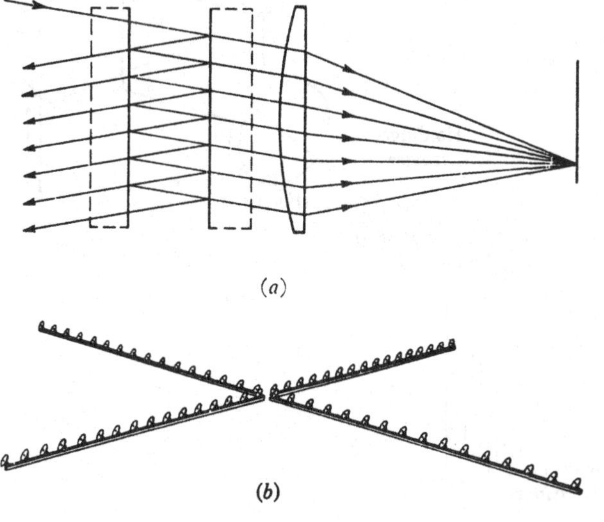

mirrors at the ends of the arms will see straight fringes there. These are contours of the wedge between one mirror and the image of the other reflected in the beam splitter, and are called *fringes of equal thickness* or *Fizeau fringes*. If one mirror is moved back parallel to itself to introduce a difference in the optical paths in the two arms, the fringes lose contrast. The narrower the bandwidth of the light, the further the mirror can be moved while still retaining visible fringes. If either mirror is tilted to remove the wedge, the fringes *spread out* to give a field of uniform intensity.

A second *complementary* set of fringes is produced by the other pair of beams that leave the beam splitter. In interferometers in which the beams are reflected back to recombine at the place where they separated, such as the Michelson, the second interferogram goes back to the source and another beam splitter is needed to see these fringes. But in other interferometers they are as accessible as the first set.

1.5.1. Alternative fringes

With a path difference present, not too large to reduce the fringe visibility to zero, an independent set of fringes may be seen. These are located at the image of the source, that is, infinity. They are only visible if the first set has

Fig. 6 The Michelson interferometer with collimated light, showing the two alternative sets of fringes.

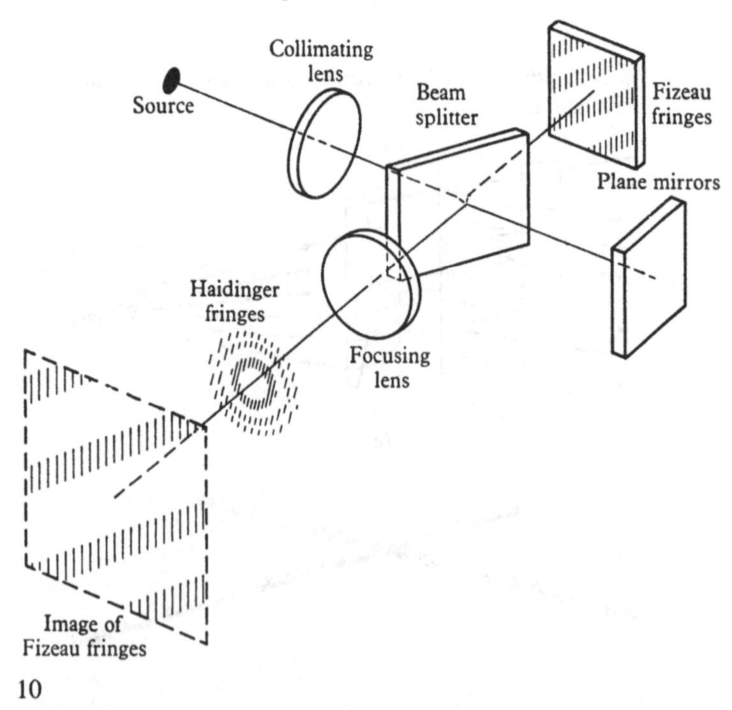

been spread out completely or if the apertures of the mirrors have been stopped down so as to be smaller than the fringe spacing. These fringes are circular and are known as *fringes of equal inclination* or *Haidinger fringes.* If the optical path difference is increased by moving the mirror further, these fringes lose contrast at the same rate as the Fizeau fringes; at the same time their spacing decreases.

This historic distinction between the two sets of fringes is meaningful for the Michelson interferometer with the source effectively at infinity, because one set of fringes consists always of straight lines and the other of concentric circles. But other interferometers may not have these differences, while the distinction between equal inclination and equal thickness depends on the position of the point of reference. In a general treatment, I prefer to call the fringes that the interferometer was designed to record the *observed* or *test fringes*, whether they are Fizeau or Haidinger fringes. The effective source for these fringes is then crossed by an independent set of fringes, the *source fringes*, which may be made visible in suitable circumstances.

Any interferometer has this basic pair of sets of *alternative* fringes, which are the limits between which a whole range of fringe shapes can be obtained. To see an intermediate shape it is necessary to limit the size of the beams in some other suitable plane, which then becomes the effective source.

1.5.2. Fringes of equal chromatic order

There remains another type of fringes. If the Haidinger fringes of a Michelson interferometer are focused on the slit of a spectroscope and the monochromatic spectral-line source replaced by white light, the spectrum obtained is crossed by fringes. These show the variation of fringe position with wavelength and are known as *fringes of equal chromatic order* or a *channelled spectrum.*

2

Mathematical foundations

The theory of optics makes considerable use of Fourier transforms [Papoulis, 1968], particularly since the work of Duffieux [1946]. In interferometry they relate the interferograms obtained to either the spectral or spatial properties of the radiation source. These and other mathematical methods are summarized in this chapter, the treatment following the descriptive method used by Bracewell [1965].

2.1. Fourier transforms

The notation used for the Fourier transform of $f(x)$ is

$$F(u)=\int_{-\infty}^{+\infty} f(x)e^{-2\pi iux}\,dx, \tag{2.1}$$

while

$$f(x)=\int_{-\infty}^{+\infty} F(u)e^{2\pi iux}\,du. \tag{2.2}$$

Usually the two functions are called Fourier transforms of each other without regard to the sign of the exponent. In the rare case when this is of importance, (2.2) is said to define the inverse transform.

An alternative name for the transform is the *spectrum* of the function and it is said to be a function of *reciprocal space* while the function itself is in *real space.* As established notation in physics does not always allow this change between small and capital letters to denote transforms, the transform of $f(x)$ may also be denoted by $\bar{f}(u)$ or by $\mathscr{F}\{f(x); u\}$.

The conditions for $f(x)$ to have a transform are usually stated as:

(i) $\int_{-\infty}^{+\infty} |f(x)|dx$ should be finite and definite, and

(ii) $f(x)$ should have only a finite number of discontinuities, none of which is infinite.

These conditions are satisfied by the functions found in physics, and the idea of a function and its transform are familiar concepts: a waveform and its spectrum or an aperture and its far-field diffraction pattern. However, some idealizations that are very useful in physics, such as impulses with infinite discontinuities or sine waves of infinite duration, do not satisfy these conditions. Nor are they, strictly speaking, physically realizable, but should be regarded as the limit of a sequence of realizable functions. Although such functions have not a transform in the strict sense, a transform in the limit can be defined.

Other definitions of Fourier transforms are currently used. In the Bateman tables [1954] the unsymmetrical relations

$$F(u)=\int_{-\infty}^{+\infty} f(x)e^{-iux}\,dx,$$

$$f(x)=\frac{1}{2\pi}\int_{-\infty}^{+\infty} F(u)e^{iux}\,du, \tag{2.3}$$

are given. If the forms (2.1) and (2.2) are used for the transform between time and frequency with the variables now t and v, equation (2.3) follows when v is replaced by the angular frequency $\omega=2\pi v$. A third pair of definitions is the symmetrical one,

$$F(u)=(2\pi)^{-1/2}\int_{-\infty}^{+\infty} f(x)e^{-iux}\,dx,$$

$$f(x)=(2\pi)^{-1/2}\int_{-\infty}^{+\infty} F(u)e^{iux}\,du. \tag{2.4}$$

2.1.1. Sine and cosine transforms

If a real function is symmetrical, its transform is real and is equal to the cosine transform, which is defined by

$$F_c(u)=2\int_0^{\infty} f(x)\cos 2\pi ux\,dx, \tag{2.5}$$

and the inverse holds for positive x, namely

$$f(x)=2\int_0^{\infty} F_c(u)\cos 2\pi ux\,du. \tag{2.6}$$

Sine transforms may be defined in the same way,

$$F_s(u)=2\int_0^{\infty} f(x)\sin 2\pi ux\,dx, \tag{2.7}$$

$$f(x)=2\int_0^{\infty} F_s(u)\sin 2\pi ux\,du. \tag{2.8}$$

The exponential Fourier transform of any function can thus be expressed in terms of the sine and cosine transforms of its odd and even parts,

$$\mathscr{F}\{f(x);u\}=\mathscr{F}_c\{\tfrac{1}{2}[f(x)+f(-x)];u\}-\mathrm{i}\mathscr{F}_s\{\tfrac{1}{2}[f(x)-f(-x)];u\}. \tag{2.9}$$

2.1.2. More than one variable

Functions of more than one variable also have Fourier transforms. If f is a function of the variables $x_1, x_2, \ldots, x_n$, it is most conveniently written as $f(\mathbf{x})$, where $\mathbf{x}$ is an n-dimensional vector. Its transform is then the function of the n-dimensional vector $\mathbf{u}$ and is given by

$$F(\mathbf{u})=\int_{-\infty}^{+\infty} f(\mathbf{x})\mathrm{e}^{-2\pi\mathrm{i}\mathbf{u}.\mathbf{x}}\,\mathrm{d}\mathbf{x}, \tag{2.10}$$

while

$$f(\mathbf{x})=\int_{-\infty}^{+\infty} F(\mathbf{u})\mathrm{e}^{2\pi\mathrm{i}\mathbf{u}.\mathbf{x}}\,\mathrm{d}\mathbf{u}, \tag{2.11}$$

where the integrations are n-fold.

Both one- and two-dimensional transforms are common in optics, the first in spectroscopy and the second in the theory of image formation. In the latter case, if the function is radially symmetrical, a change to polar coordinates, (r, θ) for $\mathbf{x}$ and (ρ, ϕ) for $\mathbf{u}$, reduces the two-dimensional Fourier transform to a one-dimensional Hankel transform,

$$\begin{aligned} F(\rho) &= \int_0^{2\pi}\int_0^{\infty} f(r)\mathrm{e}^{-2\pi\mathrm{i}r\rho\cos(\theta-\phi)}\,\mathrm{d}\theta r\,\mathrm{d}r, \\ &= 2\pi\int_0^{\infty} f(r)\mathrm{J}_0(2\pi\rho r)r\,\mathrm{d}r, \end{aligned} \tag{2.12}$$

while

$$f(r)=2\pi\int_0^{\infty} F(\rho)\mathrm{J}_0(2\pi\rho r)\rho\,\mathrm{d}\rho. \tag{2.13}$$

2.2. Linear systems

Many physical systems are linear or approximately so: an increase in the input leads to a proportional increase in the output. But the output is not a simple copy of the input, for the properties of the system also enter. No practical system can copy perfectly, the finer variations of the input being always smoothed to some extent.

At some value x of the independent variable, the output does not depend only on the value of the input $i(x)$ at x but also on its values for a range of values around x. The influence of the input is spread through a range of outputs and this spread is described by an *instrument function* or *Green's function* $k(x)$. The amount by which the input $i(x-y)$ at a distance y before x affects the output at x is weighted by the value of $k(y)$. The total output at x is then

$$o(x)=\int_{-\infty}^{+\infty} k(y)i(x-y)\mathrm{d}y. \tag{2.14}$$

It is assumed here that the instrument function $k(y)$ does not vary with x: that, as well as being linear, the system is *invariant*. In different branches of physics, different names are used for the instrument function: *scanning function* in spectroscopy, *spread function* in imagery, and *impulse response* in electronics.

Equation (2.14) is a *convolution* integral, so named because, as a function of the dummy variable y, the sign of the argument of i is reversed, or i is folded over with respect to k. A simple change of variable shows that convolution is symmetrical and

$$o(x)=\int_{-\infty}^{+\infty} k(y)i(x-y)\mathrm{d}y=\int_{-\infty}^{+\infty} k(x-y)i(y)\mathrm{d}y, \tag{2.15}$$

and the symmetrical notation $k * i$ is used so that (2.15) may be written as

$$o=k * i. \tag{2.16}$$

Since the Fourier transform of the convolution of two functions is the product of the two transforms,

$$O(u)=K(u)I(u), \tag{2.17}$$

the spectrum of the output is obtained by multiplying that of the input by a *transfer function* $K(u)$. In the theory of electrical filters, transfer functions are functions of frequency. But in spectroscopy, where the scanning function is a function of frequency, the transfer function is a function of a time, the time difference within the instrument.

A linear, invariant system is thus described either by a convolution or a linear filtering (2.17) [O'Neill, 1963]. Although non-linear systems are met in practice, a linear treatment is usually the first approximation used to study them. If a system is not invariant over its full range, it can be made approximately invariant by dividing up this range and studying each part with a different instrument function.

Closely related to the convolution integral, but without the folding, is the cross-correlation integral

$$\int_{-\infty}^{+\infty} f^*(y)g(y+x)\mathrm{d}y, \tag{2.18}$$

where f^* is the complex conjugate of f, with the special case of autocorrelation

$$\int_{-\infty}^{+\infty} f^*(y)f(y+x)\mathrm{d}y. \tag{2.19}$$

The Fourier transform of the autocorrelation of $f(x)$ is

$$|F(u)|^2. \tag{2.20}$$

the *Wiener spectrum* or *power spectrum* of $f(x)$. This last result is the Wiener–Khinchin theorem.

2.3. Generalized functions

An ideal system would give an output identical with the input. Equation (2.14) can be used to define the instrument function for such a system. Such an equation defines a *generalized function* or *distribution*, as discussed by Lighthill [1958], the one defined here being the Dirac δ-function, $\delta(x)$. For any function chosen as $f(x)$, $\delta(x)$ is defined as satisfying the integral equation

$$f(x)=\int_{-\infty}^{+\infty} \delta(y)f(x-y)\mathrm{d}y. \tag{2.21}$$

Any attempt to find particular values of $\delta(x)$ shows that

$$\delta(x)=0 \quad (x\neq 0),$$

$$\delta(x)=\infty \quad (x=0).$$

However, the integral of $\delta(x)$ from $-\infty$ to ∞ is unity, as can be seen when $f(x)=1$ in (2.21). Thus the δ-function is not a normal function. It usually occurs in integrals and then can be treated as the limit of an ordinary function $f(\varepsilon, x)$,

$$\delta(x)=\lim_{\varepsilon\to 0} f(\varepsilon, x)$$

where $f(\varepsilon, x)$ satisfies the conditions

$$f(\varepsilon, x)\to 0 \quad (x\neq 0),$$

and

$$\int_{-\infty}^{+\infty} f(\varepsilon, x)\mathrm{d}x=1 \quad \text{for all } \varepsilon. \tag{2.22}$$

The actual form chosen for $f(\varepsilon, x)$ depends on the particular problem. By the use of a suitable $f(\varepsilon, x)$, such as $(1/\varepsilon)\mathrm{rect}(x/\varepsilon)$, it can be shown that, in the limit, $\delta(x)$ is the Fourier transform of unity. The inverse result follows from (2.21).

Equation (2.21) has another interpretation, as comparison with (2.15) shows. It asks the question 'what input gives the instrument function as its output', the answer being obviously, 'a δ-function'. Hence in electronics the instrument function is called the impulse response.

A second generalized function is the Dirac comb, represented by Bracewell's symbol III(x), the Cyrillic letter *shah*. It is the infinite series of δ-functions at unit spacing,

$$\mathrm{III}(x) = \sum_{n=-\infty}^{\infty} \delta(x-n) \tag{2.23}$$

and its transform (in the limit) is III(u), also a Dirac comb.

Both the δ-function and the Dirac comb can be extended to more than one dimension. Thus, in two dimensions, III(x) represents a form of grating and III(x) III(y) a two-dimensional network of spikes, a 'bed of nails'.

2.4. Some useful functions

In addition to these generalized functions, there are some ordinary functions commonly used that are simple but discontinuous functions. Some of these are illustrated in fig. 7.

The rectangular function

$$\begin{aligned} \mathrm{rect}\,x &= 1, \quad |x| < \tfrac{1}{2}, \\ &= \tfrac{1}{2}, \quad |x| = \tfrac{1}{2}, \\ &= 0, \quad |x| > \tfrac{1}{2}. \end{aligned} \tag{2.24}$$

It is a centred square of unit height and width. The function $a\,\mathrm{rect}\,(x/b+c)$ is thus a rectangle of height a and width b, centred at $x = -bc$.

The triangular function

$$\begin{aligned} \mathrm{tri}\,x &= 1-|x|, \quad |x| < 1, \\ &= 0, \qquad |x| \geqslant 1. \end{aligned} \tag{2.25}$$

It satisfies the relation $\mathrm{tri}\,x = \mathrm{rect}\,x * \mathrm{rect}\,x$.

The sign (signum) function

$$\begin{aligned} \mathrm{sgn}\,x &= -1, \quad x < 0, \\ &= 0, \quad x = 0, \\ &= 1, \quad x > 0, \end{aligned} \tag{2.26}$$

The Heaviside function or *unit step*

$$\varepsilon(x) = \tfrac{1}{2} + \tfrac{1}{2}\operatorname{sgn} x. \tag{2.27}$$

Bessel functions occur frequently in the study of radially symmetrical optical systems, usually in the normalized form

$$\Lambda_n(x) = \frac{2^n n!}{x^n}\mathbf{J}_n(x), \tag{2.28}$$

where $\Lambda_n(0) = 1$ for all n, and the factorial function $n!$ is sometimes called the Γ-function $\Gamma(n+1)$ when n is not integral. Often it is preferable to use a different normalization so that the transform has unit value at the origin. Such a function is

Fig. 7 Some useful functions, including the generalized functions $\delta(x)$ and III(x).

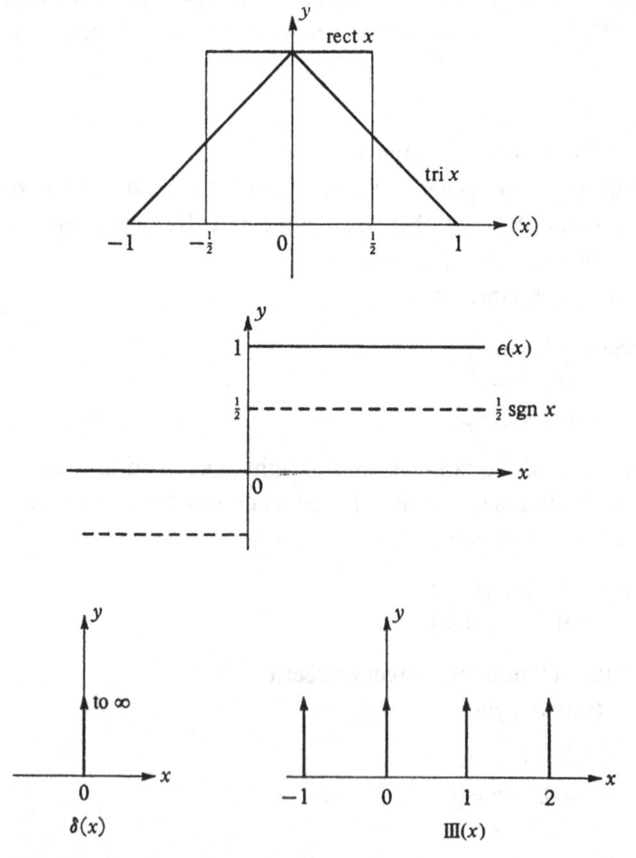

$$\frac{\sqrt{\pi}}{2}\frac{(n-\frac{1}{2})!}{n!}\Lambda_n(\pi x)=\frac{2^{n-1}(n-\frac{1}{2})!}{\pi^{n-1/2}x^n}\mathbf{J}_n(\pi x), \tag{2.29}$$

which is shown in fig. 8 with its transform $(1-4u^2)^{n-1/2}$ rect u for different values of n. For optical systems with rectangular apertures, instead of Bessel functions of integral order, the function that occurs is

$$\Lambda_{1/2}(\pi x)=\sin \pi x/\pi x=\operatorname{sinc} x. \tag{2.30}$$

2.5. Table of Fourier transforms

Some general theorems for Fourier transforms follow.

Area $$\int_{-\infty}^{+\infty} f(x)\mathrm{d}x=F(0), \tag{2.31}$$

the area under a function is the centre value of its transform.

Power $$\int_{-\infty}^{+\infty} f(x)g^*(x)\mathrm{d}x=\int_{-\infty}^{+\infty} F(u)G^*(u)\mathrm{d}u, \tag{2.32}$$

with the special result, due to Rayleigh,

$$\int_{-\infty}^{+\infty} |f(x)|^2\,\mathrm{d}x=\int_{-\infty}^{+\infty} |F(u)|^2\,\mathrm{d}u. \tag{2.33}$$

Other important results are given in the table of transforms, Table 2.1.

Fig. 8 Normalized Bessel functions $f_n(x)=\frac{2^{n-1}(n-\frac{1}{2})!}{\pi^{n-1/2}x^n}\mathbf{J}_n(\pi x)$ and their Fourier transforms.

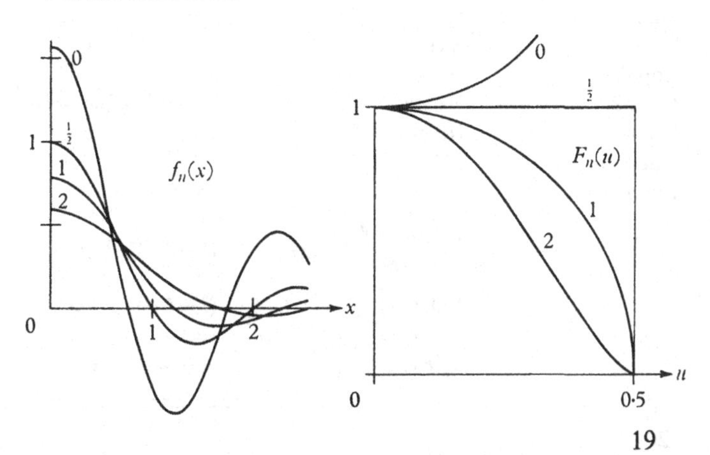

2.5.1. Uncertainty relation

Many of the functions listed above have a maximum at their centre and decrease symmetrically to either side. They may, therefore, be characterized by their height and their width. Several measures of the width are possible, for example,

equivalent width $\int_{-\infty}^{+\infty} f(x)\mathrm{d}x \Big/ f(0)$,

mean-square width $\int_{-\infty}^{+\infty} x^2 f(x)\mathrm{d}x \Big/ \int_{-\infty}^{+\infty} f(x)\mathrm{d}x$,

or the half-intensity width. The latter is most often used in optics and is the separation of the two points at which the function has dropped to half its maximum value.

The widths of a function and its transform are reciprocally related. The

Table 2.1. *Fourier transforms*: $F(u)=\int_{-\infty}^{+\infty} f(x)\mathrm{e}^{-2\pi i u x}\,\mathrm{d}x$

	$f(x)$	$F(u)$
Scale and shift	$f\left(\frac{x}{a}+b\right)$	$\lvert a\rvert \mathrm{e}^{2\pi i a b u} F(au)$
Complex conjugate	$f^*(x)$	$F^*(-u)$
n-th derivative	$f^{(n)}(x)$	$(2\pi i u)^n F(u)$
Convolution	$f(x) * g(x)$	$F(u)G(u)$
Wiener–Khinchin	$\int_{-\infty}^{+\infty} f^*(y)f(y+x)\mathrm{d}y$	$\lvert F(u)\rvert^2$
Rectangle	rect x	sinc u
Triangle	tri x	$\mathrm{sinc}^2\, u$
Sign	sgn x	$1/\pi i u$
Step	$\varepsilon(x)$	$\frac{1}{2}\delta(u)+1/2\pi i u$
Impulse	$\delta(x)$	1
Cosine	$\cos \pi x$	$\frac{1}{2}\delta(u+\frac{1}{2})+\frac{1}{2}\delta(u-\frac{1}{2})$
Comb	III(x)	III(u)
Gaussian	$\mathrm{e}^{-\pi x^2}$	$\mathrm{e}^{-\pi u^2}$
Lorentzian	$(1+x^2)^{-1}$	$\pi \exp(-\lvert 2\pi u\rvert)$
Bessel	$\frac{\sqrt{\pi}}{2}\frac{(n-\frac{1}{2})!}{n!}\Lambda_n(\pi x)$	$(1-4u^2)^{n-1/2}$ rect u
	$\frac{1}{2}\pi \mathrm{J}_0(\pi x)$	$(1-4u^2)^{-1/2}$ rect u
	sinc x	rect u
	$\mathrm{J}_1(\pi x)/2x$	$(1-4u^2)^{1/2}$ rect u

constant of proportionality depends on the measures used for the width and on the actual function. For the gaussian function $e^{-\pi x^2}$ this constant is least and, in terms of half-intensity widths,

$$\Delta x \Delta u = 0.88.$$

For other functions, this product is larger. As the exact value is usually unimportant and only the fact that it is of the order of unity, this *uncertainty relation* is normally stated as

$$\Delta x \Delta u \gtrsim 1, \tag{2.34}$$

the product of the widths of a function and its transform is of the order of unity, or greater.

2.6. Series, sampling, and Gibbs phenomenon

The integrated modulus of a periodic function is infinite and such a function does not satisfy the conditions for possessing a Fourier transform. As is well known, however, it has a Fourier series. Instead of treating this as a separate concept, it can be regarded as a special case of a Fourier transform in the limit.

Any function $f(x)$ that vanishes outside the region $|x| < \frac{1}{2}a$ can be made into a periodic function of period a by convolution with $\text{III}(x/a)$. This is shown in fig. 9. The transform of the periodic function $f(x) * \text{III}(x/a)$ is the product $F(u)\text{III}(au)$. This is a series of impulses, spaced at intervals of $1/a$ and each of weight (integral value) equal to the corresponding value of $F(u)$, the transform of the non-periodic function $f(x)$. This series of impulses is the Fourier transform equivalent of a Fourier series, and the Dirac comb makes possible their treatment by the same theory.

Fig. 9 Transforms of non-periodic and periodic functions.

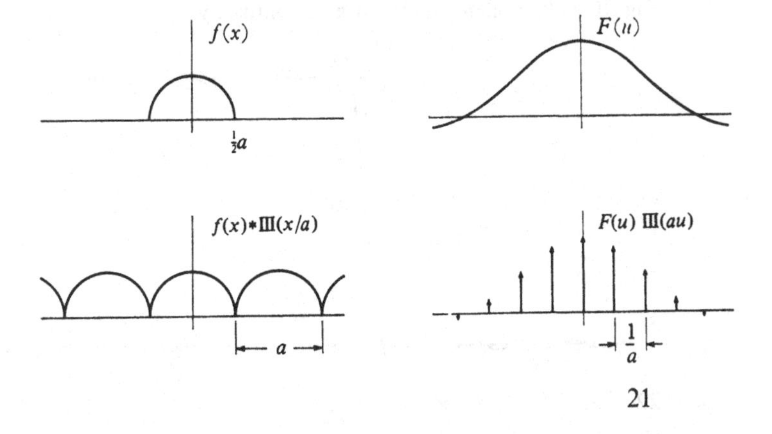

2.6.1. Sampling theorem

If the periodic function above is multiplied by rect (x/a), the original non-periodic function is obtained again:

$$f(x) = [f(x) * \mathrm{III}(x/a)]\,\mathrm{rect}\,(x/a). \tag{2.35}$$

The transform of this expression gives the *sampling theorem*

$$F(u) = [F(u)\mathrm{III}(au)] * \mathrm{sinc}\,au. \tag{2.36}$$

Since the spectrum of $F(u)$ vanishes for $|x| > \frac{1}{2}a$, $F(u)$ is said to be *band limited.* The sampling theorem (2.36) states that such a function is completely specified by the discrete set of its values at an interval $1/a$; these are represented by $F(u)\mathrm{III}(au)$. All other values may be obtained from these by convolution with sinc au.

The converse of this result is that a function that is sampled at an interval $\Delta x = b$ has a spectrum which is made periodic by the sampling, the period being $1/b$ in u.

2.6.2. Gibbs phenomenon

Another different application of convolution with a sinc function is the Gibbs phenomenon of Fourier series. If the series representing a periodic function is terminated at some finite value, the effect on the function is to introduce ripples around any discontinuity. A similar effect occurs when the transform of a non-periodic function is truncated.

If the transform $F(u)$ of $f(x)$ is truncated at $u = \frac{1}{2}a$ this is equivalent to multiplication of $F(u)$ by rect (u/a). Inverse transformation then gives, not $f(x)$, but its convolution with sinc ax. Where $f(x)$ varies slowly, this convolution has little effect but, at discontinuities, gives the result of fig. 10. The value overshoots by a factor of 0.09 on each side of the discontinuity and then shows decreasing oscillations about the original function. In

Fig. 10 Gibbs phenomenon at a discontinuity.

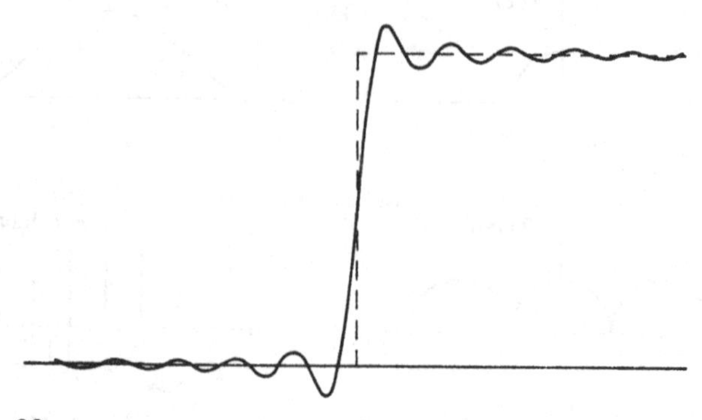

electronics, this effect is called *ringing*, in optics *diffraction bands* or *diffraction rings*, and in radio astronomy *side lobes*.

The cause of this phenomenon is the large oscillations of the sinc function. Convolution with a transfer function with smaller oscillations reduces the effect: convolution with sinc^2 gives no oscillations at finite discontinuities. In reciprocal space, this is equivalent to tapering $F(u)$ continuously to zero rather than truncating it abruptly. In optics, the removal or reduction of diffraction bands by tapering the transform is known as *apodization*, and fig. 8 shows some possible tapering functions and their transforms. Others can be found in Table 2.1.

2.7. Analytic signal

In the study of wave motions we are concerned with real functions of time that can be expressed as a Fourier integral

$$V^{(\mathrm{r})}(t)=2\int_0^{\infty} a(\nu)\cos\left[\phi(\nu)-2\pi\nu t\right]\mathrm{d}\nu. \tag{2.37}$$

There are well-known advantages in using complex exponential functions in place of real trigonometric ones. If we artificially bring in negative frequencies by defining for them $a(-\nu)=a(\nu)$, $\phi(-\nu)=-\phi(\nu)$, and write $a(\nu)^{\mathrm{i}\phi(\nu)}$ as $v(\nu)$, then

$$V^{(\mathrm{r})}(t)=\int_{-\infty}^{+\infty} v(\nu)\,\mathrm{e}^{-2\pi\mathrm{i}\nu t}\,\mathrm{d}\nu. \tag{2.38}$$

Alternatively, we can consider only positive frequencies and define the complex function

$$V(t)=\int_0^{\infty} v(\nu)\mathrm{e}^{-2\pi\mathrm{i}\nu t}\,\mathrm{d}\nu. \tag{2.39}$$

Expansion of (2.39) shows that the real part of V is half the real signal $V^{(\mathrm{r})}$ and the imaginary part is $\frac{1}{2}V^{(\mathrm{i})}$, where

$$V^{(\mathrm{i})}(t)=2\int_0^{\infty} a(\nu)\sin\left[\phi(\nu)-2\pi\nu t\right]\mathrm{d}\nu. \tag{2.40}$$

The imaginary part is thus not arbitrary but is completely specified by the real function: the Fourier transform of $V^{(\mathrm{i})}$ is found by changing the phase of the Fourier transform of $V^{(\mathrm{r})}$ by $-\frac{1}{2}\pi$. The function $V(t)$ is called the *analytic signal* associated with $V^{(\mathrm{r})}$; it is a complex function that is analytic in the lower half-plane. As such, if it is known over part of this region, it can be evaluated for the rest by analytic continuation.

The two parts $V^{(r)}$ and $V^{(i)}$ are *Hilbert transforms* of each other. The transform relation is given by a comparison of (2.39) with (2.38) when it is seen that the transform of V, since it contains no negative frequencies, is given by multiplication of $v(\nu)$ with a step function

$$\tilde{V}(\nu) = v(\nu)\varepsilon(\nu).$$

If transforms are taken

$$V(t) = V^{(r)}(t) * \tfrac{1}{2}[\delta(t) + 1/\pi i t] \tag{2.41}$$

so that

$$V^{(i)}(t) = \frac{1}{\pi} P \int_{-\infty}^{+\infty} \frac{V^{(r)}(t')}{t'-t} \mathrm{d}t'. \tag{2.42}$$

while

$$V^{(r)}(t) = -\frac{1}{\pi} P \int_{-\infty}^{+\infty} \frac{V^{(i)}(t')}{t'-t} \mathrm{d}t', \tag{2.43}$$

where P denotes that the Cauchy principal value of the integral is taken.

2.8. Random functions

Many of the quantities encountered in physics are represented by randomly varying functions and can be treated only by statistical methods. The functions are examined in terms of correlations, of which autocorrelation and average power are special cases. Since an integral such as (2.18) would be infinite if f and g are random functions, the cross correlation of two stationary random functions is defined as

$$\phi_{fg}(x) = \lim_{Y\to\infty} \frac{1}{2Y} \int_{-Y}^{+Y} f^*(y) g(y+x)\, \mathrm{d}y. \tag{2.44}$$

Usually these are functions of time and the time average is denoted by

$$\phi_{fg}(\tau) = \lim_{T\to\infty} \frac{1}{2T} \int_{-T}^{+T} f^*(t) g(t+\tau)\, \mathrm{d}t,$$

the correlation for a time interval τ, which is written as

$$\phi_{fg}(\tau) = \langle f^*(t) g(t+\tau) \rangle. \tag{2.45}$$

The autocorrelation function is then

$$\phi_{ff}(\tau) = \langle f^*(t) f(t+\tau) \rangle, \tag{2.46}$$

and the average power is $\phi_{ff}(0)$.

The power spectrum is, by the Wiener–Khinchin theorem, the transform of the autocorrelation function

$$\Phi_{ff}(\nu) = \int_{+\infty}^{+\infty} \phi(\tau)\, e^{2\pi i\nu t}\, d\tau. \tag{2.47}$$

It is related to F_T, the Fourier transform of the random function taken over a finite interval $-T$ to T, by the expression

$$\Phi_{ff}(\nu) = \lim_{T \to \infty} \frac{1}{2T} |F_T(\nu)|^2$$

where an average must be taken over the ensemble of functions $f(t)$.

When a randomly varying function is being measured, it may have a mean value or a mean-square value that is of interest. On top of this is superimposed the fluctuations which appear as an unwanted signal or *noise*. A typical example is the power received from a radiation field. The spectrum of this power is limited to some frequency band $\Delta\nu$ by the filtering action of the detector and any associated equipment. If it is detected for a total time T, the mean energy received is proportional to $T\Delta\nu$. The mean square of the fluctuations, the noise power, is also proportional to the same quantity, so its square root is proportional to $(T\Delta\nu)^{1/2}$. The final *signal-to-noise ratio* is also proportional to $(T\Delta\nu)^{1/2}$.

3

Optical foundations

Interferometry is a branch of optics and this, in its widest sense, means the study of electromagnetic radiation. In this chapter are summarized some of the concepts and methods of optics, which, although not strictly a part of interferometry, are frequently used in later chapters. They apply particularly to two branches of optics, spectroscopy and image formation, which are respectively studies of the spectral and spatial distributions of radiation.

The components that make up an interferometer, lenses, mirrors, and prisms, are those of any optical system. Their action can be described by geometrical optics, that is, by light rays.

3.1. Reflexion at plane mirrors

Many interferometers contain plane mirrors and their analysis depends on locating the images given by these. When a complicated instrument is to be designed, it is often useful to represent a beam of light by a strip of paper. A fold in this represents reflexion at a mirror.

To give the final location of the image, vector or matrix methods are most useful, yet the laws of reflexion are not usually given in this form. If the ray incident on a plane mirror is in the direction $\mathbf{r}$ and the mirror normal is in the direction $\mathbf{n}$, the direction of the reflected ray is given by

$$\mathbf{r}' = \mathbf{r} - 2(\mathbf{r} \cdot \mathbf{n})\mathbf{n}, \tag{3.1}$$

all being unit vectors. This relation contains the two laws of reflexion, for

(i) $\mathbf{r}'$ lies in the plane of incidence containing $\mathbf{r}$ and $\mathbf{n}$, and
(ii) $\mathbf{r}'$ and $\mathbf{r}$ make equal angles to the normal on opposite sides ($\mathbf{r}' \cdot \mathbf{n} = -\mathbf{r} \cdot \mathbf{n}$).

This vector relation can be written in matrix form.

The same relation (3.1) holds for the reflexion of other vectors. If $\mathbf{o}$ is a vector denoting the orientation of an object or the direction of polarization, the same property in the image is given by

$$\mathbf{o}' = \mathbf{o} - 2(\mathbf{o} \cdot \mathbf{n})\mathbf{n}. \tag{3.2}$$

3.1. *Reflexion at plane mirrors*

Methods for finding the arrangement of mirrors that will convert some pair of vectors **r** and **o** into **r**′ and **o**′ are given by Rosendahl [1960] and Sivtsov [1980].

The position of the image of a point is given by a modification of (3.1),

$$\mathbf{x}' = \mathbf{x} - 2[(\mathbf{x} - \mathbf{a}) \cdot \mathbf{n}]\mathbf{n}, \tag{3.3}$$

where **x** and **x**′ are position vectors of the object point and its image and **a** the position of a point on the mirror. The same result is given in matrix form by Brouwer [1964]. The change in this position, produced by changes $\delta\mathbf{x}$ and $\delta\mathbf{a}$ of the object and mirror positions and a change $\delta\mathbf{n}$ of the mirror orientation, is given by

$$\begin{aligned}\delta\mathbf{x}' = \delta\mathbf{x} - 2(\delta\mathbf{x} \cdot \mathbf{n})\mathbf{n} + 2(\delta\mathbf{a} \cdot \mathbf{n})\mathbf{n} - 2[(\mathbf{x} - \mathbf{a}) \cdot \delta\mathbf{n}]\mathbf{n} \\ - 2[(\mathbf{x} - \mathbf{a}) \cdot \mathbf{n}]\, \delta\mathbf{n}.\end{aligned} \tag{3.4}$$

Reflexion at an odd number of mirrors produces a *reversion*: the right-handed set of vectors **i**, **j**, **k** is imaged into the left-handed set **i**′, **j**′, **k**′, for which

$$\mathbf{i}' \times \mathbf{j}' = -\mathbf{k}', \text{ etc.}$$

3.1.1. Retroreflectors

A plane mirror is a *specular reflector* and the reflected ray leaves from the point of arrival of the incident ray but at an equal angle on the opposite side of the normal. The complementary system is also of importance in optics. It is the *retroreflector* in which the reflected ray leaves at the same angle as the incident ray but from a point an equal distance on the opposite side of the reflector centre to the point of incidence.

Some common retroreflectors are shown in fig. 11. The first is the *trihedral mirror*, three mutually perpendicular plane mirrors. It is shown made as a prism, the *cube corner*. The next two examples consist of a reflector at the focus of an imaging system, which may be either a refracting surface or a concave mirror. These last two are often called *cat's eyes*.

Fig. 11 Three forms of retroreflectors.

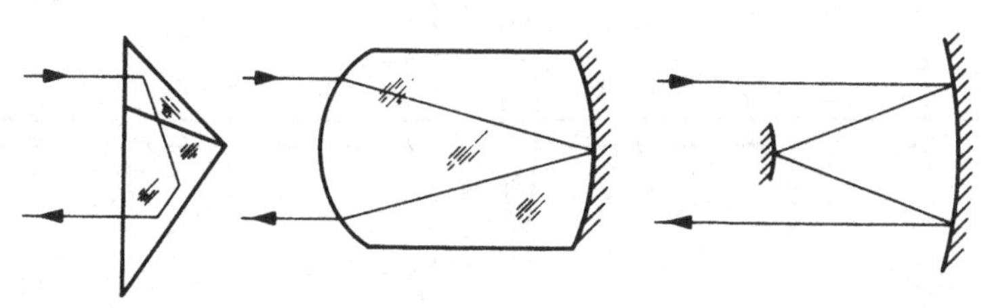

The ray reflected from a plane mirror is unaffected by a displacement of the mirror in its own plane, but its direction is changed if the mirror is tilted. Only an image at the mirror surface is not affected. The ray reflected from a retroreflector is unaffected by a tilt of the retroreflector about its centre, but it is displaced if the retroreflector is displaced laterally. Only an image at infinity is not affected.

The two can be combined, as shown by Terrien [1959], to provide a reflecting system that is insensitive to both tilts and lateral displacements. This system is discussed in §8.3.2. It is still sensitive however to longitudinal displacements. The process which returns a beam back on its path independently of all rotations and displacements is *phase conjugation* [Hopf, 1980]. This is not an effect that can be obtained with simple optical components.

A *roof mirror*, two plane mirrors meeting at a right angle, acts like a specular reflector for rays in the plane containing the roof edge, and like a retroreflector for rays in the plane perpendicular to this.

3.2 Imaging systems

The general properties of an image-forming system are shown in fig. 12. This forms an image B', said to be in the *image space* following the system, of a point B in the *object space* that precedes it. These spaces have refractive indices n and n'. The object and image are said to be *conjugates* or to be in *conjugate planes* and their area is limited by an aperture or stop, the *field stop*. This is shown at the image in fig. 12 but it may equally well be at the object. If there is no field stop, there is no sharp edge to the image.

The amount of light arriving at each image point B' depends on the size of another stop, the *aperture stop*. If this is in the object space it is called the *entrance pupil*, while its image is the *exit pupil*. Field and aperture stops form independent systems, each with a set of images through the optical

Fig. 12 The properties of an image-forming system.

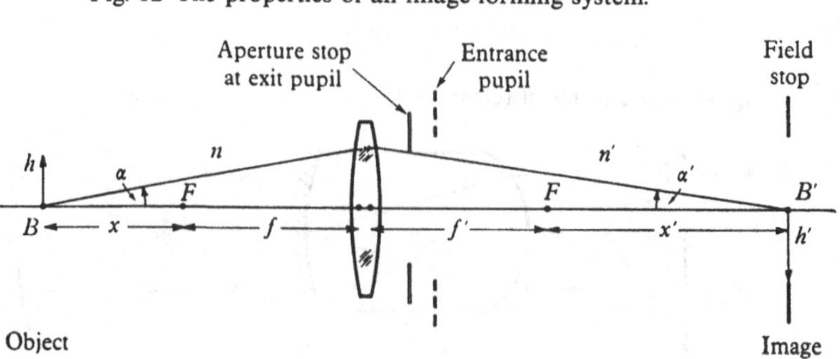

instrument. In theory, the stops are interchangeable and the system forms an image of the aperture stop with the field stop now determining the amount of light to each point. But practical optical systems must be corrected for aberrations and this correction is not equally good for all object positions.

When each stop has a small angular size as seen from a stop of the other set, the position and size of the image are given by the simple lens formulae. Thus, if the object and image focal points are F and F' and the focal lengths f and f', where

$$f'/n' = -f/n, \tag{3.5}$$

the distances of the object and image from their corresponding focal points are related by

$$xx' = ff'. \tag{3.6}$$

These relations are given in matrix form by O'Neill [1963] or Brouwer [1964].

The lateral sizes of the object and image are related by the magnification m, where

$$m = h'/h = -f/x = -x'/f'. \tag{3.7}$$

The longitudinal magnification is the square of this;

$$\delta x'/\delta x = m^2. \tag{3.8}$$

If a ray leaving the object at an angle α to the optical axis reaches the image at α',

$$n\alpha/n'\alpha' = m,$$

so that

$$nh\alpha = n'h'\alpha' = \text{invariant}, \tag{3.9}$$

and

$$n\alpha^2\,\delta x = n'\alpha'^2\,\delta x' = \text{invariant}. \tag{3.10}$$

For larger angles, the angles should be replaced by their sines, and $n \sin \alpha$ and $n' \sin \alpha'$ are the *numerical apertures* of the system on the object and image side, when the ray considered is one to the edge of the aperture stop. The invariant (3.9) can also be written in the two-dimensional form

$$E = n^2 A\Omega = \text{invariant}, \tag{3.11}$$

where A is the area of the object and Ω the solid angle that the entrance pupil subtends at its centre. (In most cases, we are concerned with systems in air and the factor n^2 is omitted.) This invariant E provides a most useful geometrical description of an optical instrument and the associated source or detector and it can also be applied to the abstraction of a beam of light. Since there are several contending names for it in English, I shall use the French name, *étendue*, the 'extent' of the instrument or beam.

3.3. Units of radiometry

Wavelength and frequency

Different parts of the electromagnetic spectrum are distinguished by their different wavelengths or different frequencies. Since diffraction gratings give spectra that are approximately linear in wavelength, wavelengths have been widely used in spectroscopy. Unless specified otherwise, they are those in air [CIE, 1957], that is

$$\lambda = \lambda_v / n, \tag{3.12}$$

the vacuum wavelength λ_v divided by the refractive index of air. Although used here, this convention is not always followed; λ is often used for vacuum wavelength.

Wavelength, however, is an unfortunate choice as atomic energy levels have separations proportional to the frequency of the radiation emitted or absorbed. For frequencies above those in the microwave region, it is still wavelength that is usually measured and spectroscopists use the wavenumber σ, the reciprocal of the vacuum wavelength ($\sigma = 1/\lambda_v$). But with the extension of frequency measurements to visible radiation [Evenson *et al.*, 1972; Baird *et al.*, 1979] and the change to base the metre on a standardized value of c, the speed of light, of 299 792 498 ms^{-1}, the frequency $\nu = c/\lambda_v$ is being increasingly used. For visible radiation the convenient unit is the terahertz, 1 THz $= 10^{12}$ Hz; for the ultraviolet it is the petahertz, 1 PHz $= 10^{15}$ Hz.

Power or flux

The detectors used for microwaves provide an electrical signal that follows in amplitude and phase the electric field of the radiation. At frequencies higher than about 0.5 THz, detectors can no longer follow the fluctuations of the field and respond only to the power of the radiation.

The quantity of radiation, the *radiant flux*, is measured in the unit of power, the watt, W. In addition to the total flux Φ, the flux per unit frequency interval or spectral flux, Φ_ν or $\phi(\nu)$, is used, where

$$\Phi = \int_0^\infty \phi(\nu)\,d\nu.$$

All the other radiometric units given have these two forms: a total radiation unit and a spectral unit. To conform with the notation used for Fourier transforms, spectral quantities will be denoted by small letters.

In radio astronomy, a different measure, the *brightness temperature* in kelvins, is used. This is the temperature of a black body which, when subtending the same angle as the observed object, would produce the power observed.

Irradiance

The *irradiance* on a receiving surface is the flux falling on unit area, from whatever angle it arrives. Allied to this is the *optical intensity* I of the radiation which is the radiant flux crossing unit area of a plane normal to the direction of propagation. Intensity in this sense is a term of theoretical optics; in radiometry and photometry it has the different meaning of the flux emitted by a source per unit solid angle.

Radiance

An emitting surface is characterized by its radiance, L, the flux emitted per unit area per unit solid angle. In air, the flux through an optical system is

$$\Phi = LA\Omega = LE, \tag{3.13}$$

the unit of L being W m^{-2} sr^{-1}. For a system with no losses due to reflexion or absorption, the flux leaving is equal to the flux entering. Since the étendue is invariant, (3.13) shows that, neglecting these losses, the radiance is also invariant through an optical system.

Numerical factors

Passive components may be described by the dimensionless factors of reflectance $\mathscr{R}$, absorptance $\mathscr{A}$, and transmittance $\mathscr{T}$, and these also have both total and spectral forms. These factors apply to the optical intensity. Those for the amplitude of a light wave, the *amplitude transmission factors k* are the square roots of these.

To each radiometric unit there is a corresponding photometric unit in which the power is weighted by a visibility function denoting the spectral sensitivity of the normal human eye. This is now an internationally standardized function, the *luminous efficiency* or *luminosity* of radiation of that frequency.

3.4. Wave propagation and diffraction

Although electromagnetic radiation implies a vector field satisfying Maxwell's equations, in many optical problems an approximate description by a scalar wave function is completely adequate. In fact, cases for which Maxwell's equations have been solved rigorously are rare.

The approximate treatment is due to Huygens and Fresnel and is given in terms of a scalar U, the complex amplitude. The actual scalar components of the field vector at a point $\mathbf{x}$ and a time t are rapidly fluctuating real functions $V^{(r)}(\mathbf{x}, t)$. Detectors of power measure $[V^{(r)}(\mathbf{x}, t)]^2$ averaged over a time long compared with these fluctuations. This is the optimal intensity I, which, in terms of the analytical signal V associated with $V^{(r)}$ by (2.39), is, to a factor of 2,

$$I(\mathbf{x}) = \langle V^*(\mathbf{x}, t)V(\mathbf{x}, t)\rangle. \tag{3.14}$$

If the radiation has a bandwidth small compared with its mean frequency, so that

$$\Delta\nu \ll \nu, \tag{3.15}$$

it is said to be *quasi-monochromatic*. If further it comes from a sufficiently small source and has an intensity I, it can be represented by a train of waves of amplitude $I^{1/2}$. At a distance R measured in the direction of propagation, the wave is given by

$$I^{1/2}(R)\cos(2\pi\nu t - \kappa R),$$

where κ is the *propagation constant* $2\pi/\lambda$ or $2\pi n\nu/c$. This description can be interpreted in two ways. At any fixed point at this distance R there is a periodic disturbance of this form. Alternatively each value of the phase, for example $\kappa R = 2N\pi$, defines a surface, the *wavefront*, that advances with the wave. If the time-dependent term is omitted and the cosine replaced by a complex exponential, we have the complex amplitude

$$U(R) = I^{1/2}(R)\exp i\kappa R. \tag{3.16}$$

3.4.1. Plane waves

If the wavefront is a plane normal to the unit vector $\mathbf{r}$, the ray vector, and $\mathbf{x}$ is a position vector of some point in it, the distance travelled, R, is equal to $\mathbf{r}\cdot\mathbf{x}$ for all points on the wavefront. The complex amplitude of a uniform plane wave then has the form

$$U(\mathbf{x}) = a\,e^{i\kappa\mathbf{r}\cdot\mathbf{x}} \tag{3.17}$$

where a is a constant.

3.4.2. Spherical waves

The amplitude of a spherical wave is not constant but decreases as the distance from the source increases. It is necessary to introduce the concept of the *strength* of the source; a source of unit strength produces a wave of unit amplitude at unit distance. For a source of strength A, the complex amplitude of the spherical wave it emits is

$$U(\mathbf{x}) = A\, \mathrm{e}^{\mathrm{i}\kappa R}/R, \tag{3.18}$$

since $|\mathbf{x}| = R$ if the source is at the origin of $\mathbf{x}$.

3.4.3. General wave

The complex amplitude for any wave for which the amplitude and phase are both functions of $\mathbf{x}$ can be written as

$$U(\mathbf{x}) = a(\mathbf{x}) \exp \mathrm{i}[\kappa \mathbf{r} \cdot \mathbf{x} + \psi(\mathbf{x})] \tag{3.19}$$

where the phase has been written as the sum of two terms, one representing the phase across a best-fit plane, which has a normal $\mathbf{r}$, and the other giving the phase variations from this plane.

3.4.4. Diffraction at an aperture

The propagation of a spherical wave from a point source is given by (3.18). Diffraction of this wave through an aperture $\mathscr{A}$ is then given by the Fresnel–Kirchhoff formula for which the notation is given in fig. 13. If the angles between rays are small so that the cosines of the angles between them can be taken as unity the obliquity factors usually included in this formula may be omitted. If R and R' are the lengths of the rays from the source to the aperture and from the aperture to the point P, the complex amplitude at P is given by the Fresnel–Kirchhoff formula

Fig. 13 Notation for the Fresnel–Kirchhoff formula for diffraction at an aperture.

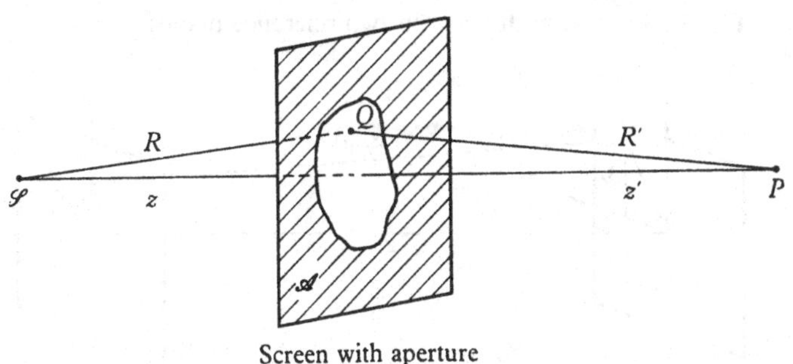

$$U(P) = -\frac{\mathrm{i}A}{\lambda}\iint \frac{\mathrm{e}^{\mathrm{i}\kappa(R+R')}}{RR'}\,\mathrm{d}\mathscr{A}, \tag{3.20}$$

the integration being taken over some surface $\mathscr{A}$ across the aperture. This expression can be written as

$$U(P) = -\frac{\mathrm{i}}{\lambda}\iint_{\mathscr{A}} U(Q)\frac{\mathrm{e}^{\mathrm{i}\kappa R'}}{R'}\,\mathrm{d}\mathscr{A} \tag{3.21}$$

where

$$U(Q) = A\exp(\mathrm{i}\kappa R)/R$$

is the amplitude at a point Q on $\mathscr{A}$. For small angles, the variations of R and R' over $\mathscr{A}$ may also be neglected in the product RR' (but not in the exponential term). Then

$$U(P) \approx -\frac{\mathrm{i}A\,\mathrm{e}^{\mathrm{i}\kappa(z+z')}}{\lambda zz'}\iint_{\mathscr{A}} \mathrm{e}^{\mathrm{i}\kappa(\Delta R'+\Delta R)}\,\mathrm{d}\mathscr{A}, \tag{3.22}$$

where z and z' are the lengths of the central rays and $\Delta R = R - z$, $\Delta R' = R' - z'$.

3.5. Diffraction in optical systems

The amplitude given by (3.22) is evaluated by expanding the optical path in the exponent as a function of the coordinates of the object, image, and pupil. These approximate expansions will be used frequently and are given in full.

The distance between two points $(\mathbf{x})$ and $(\mathbf{u})$ on parallel planes, a distance z apart (fig. 14), is, to the second order in $|\mathbf{x}|$ and $|\mathbf{u}|$,

$$R(\mathbf{x}, \mathbf{u}, z) = z - \mathbf{u}\cdot\mathbf{x}/z + |\mathbf{x}|^2/2z + |\mathbf{u}|^2/2z\ldots,$$

and

$$R(\mathbf{x}, 0, z) = z + |\mathbf{x}|^2/2x\ \ldots$$

Fig. 14 Optical path between two reference planes.

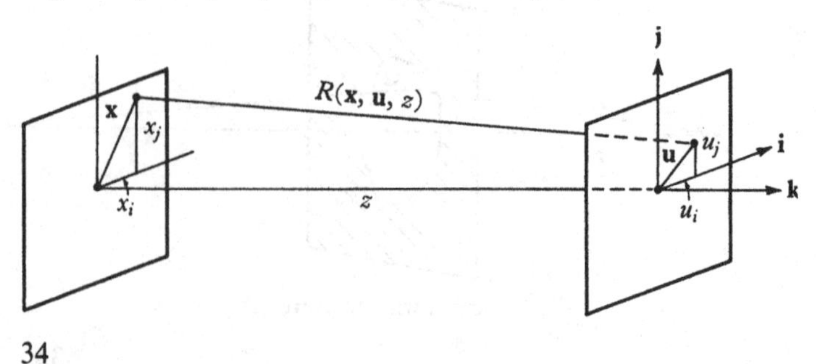

Then

$$\Delta R_u = -\mathbf{u}\cdot\mathbf{x}/z + |\mathbf{u}|^2/2z\ldots, \tag{3.23}$$

and similarly

$$\Delta R_x = -\mathbf{u}\cdot\mathbf{x}/z + |\mathbf{x}|^2/2z\ldots$$

are the variations of path difference caused by changes of either **u** or **x** from zero.

A perfect imaging system converts a spherical wave at the entrance pupil to a spherical wave of different curvature at its exit pupil. Over these spherical surfaces the term in $|\mathbf{u}|^2$ in ΔR_u vanishes. Then

$$\Delta R' + \Delta R = \mathbf{u}\cdot(\mathbf{x}' - \mathbf{x})/z - \tfrac{1}{2}\Delta z'|\mathbf{u}|^2/2z'^2\ldots, \tag{3.24}$$

the last term having been added to take account of any defocusing $\Delta z'$.

It is convenient to introduce new coordinates, defined in terms of the numerical apertures, $\sin\alpha$ and $\sin\alpha'$, in object and image space. These are

$$\begin{aligned} &\xi = (\mathbf{x}\sin\alpha)/\lambda, \quad \xi' = (\mathbf{x}'\sin\alpha')/\lambda, \\ &\boldsymbol{\eta} = \mathbf{u}/(z\sin\alpha) = \mathbf{u}'/(z'\sin\alpha'), \\ &\zeta = (\Delta z'\sin^2\alpha')/\lambda. \end{aligned} \tag{3.25}$$

Equation (3.22) with a constant phase term omitted can now be written as

$$U(P) = (A/\lambda)\sin\alpha\sin\alpha' \iint_{\mathscr{A}} \exp 2\pi i\{\boldsymbol{\eta}\cdot(\xi' - \xi) - \tfrac{1}{2}|\boldsymbol{\eta}|^2\zeta\}\, d\boldsymbol{\eta}. \tag{3.26}$$

If the optical system is not perfect, it introduces an extra term $w(\boldsymbol{\eta})$ into the optical path, where w is the wave aberration of the system. The defocusing term $\tfrac{1}{2}|\boldsymbol{\eta}|^2\zeta$ can be included as part of this. The wave aberration of a symmetrical optical system can be expressed in the series given by Zernike [1934]

$$w(\eta) = \sum_{l,m,n} b_{lnm}\eta^n\sigma^{2l+m}\cos m\phi, \tag{3.27}$$

where b_{lnm} is an aberration coefficient, $\eta = |\boldsymbol{\eta}|$, ϕ is the azimuth of $\boldsymbol{\eta}$ (or **u**), and σ is the angle off-axis at which the aberration is measured. From the wave aberration we define the *pupil function*

$$\begin{aligned} f(\boldsymbol{\eta}) &= \exp\{i\kappa w(\boldsymbol{\eta})\}, \quad \text{over } \mathscr{A}, \\ &= 0, \text{ elsewhere.} \end{aligned} \tag{3.28}$$

This definition enables (3.26) to be written as an integral to infinity,

$$U(P)=(A/\lambda)\sin\alpha\sin\alpha'\int_{-\infty}^{+\infty} f(\boldsymbol{\eta})\exp 2\pi i\boldsymbol{\eta}\cdot(\boldsymbol{\xi}'-\boldsymbol{\xi})\,d\boldsymbol{\eta}, \tag{3.29}$$

$$=(A/\lambda)(\sin\alpha\sin\alpha')F(\boldsymbol{\xi}'-\boldsymbol{\xi}); \tag{3.30}$$

to a constant factor, the amplitude distribution at the image is the two-dimensional Fourier transform of the pupil function.

Equation (3.30) can be generalized to state that any amplitude distribution $f(\boldsymbol{\eta})$ produces, at infinity, an amplitude that is its Fourier transform, this transform being in terms of the direction cosines of the rays to infinity. The approximation is good provided the term in $|\mathbf{u}|^2$ in (3.23) is small compared with λ, that is,

$$z \gg |\mathbf{u}|^2_{max}/\lambda. \tag{3.31}$$

This is the *far-field condition.*

Fourier transforms are thus more than a convenient mathematical tool in optics, they are also an approximate representation of the physics of diffraction at infinity or near a focus, historically called *Fraunhofer diffraction.*

3.5.1. Coherent object

If the object plane contains a screen of variable transmission, illuminated by a small source of quasi-monochromatic light, the complex amplitudes from each object point add in the image plane. The illumination is then said to be coherent and the system is linear in complex amplitude. The total image is obtained as the convolution of the amplitude distribution over the object $U(\boldsymbol{\xi})$, and the function $F(\boldsymbol{\xi})$ of (3.30); this is the *spread function* for coherent imagery. Its transform, the pupil function $f(\boldsymbol{\eta})$, is then the transfer function for coherent illumination.

An optical system commonly used with coherent illumination is that for optical information processing [Goodman, 1968; Françon, 1979], shown in fig. 15. To satisfy fully the conditions under which a Fourier transform relates the complex amplitudes at the object and pupil, these are placed at the two principal foci of a lens and so each appears at infinity when viewed

Fig. 15 Coherent optical system for spatial filtering.

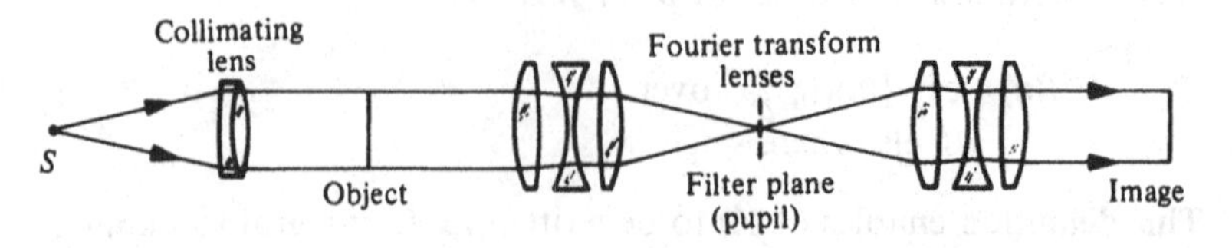

from the other. The same applies for the second lens between the pupil and the image. These lenses are specially corrected for both principal foci, so the transfer function $f(\eta)$ is determined by any mask placed at the pupil plane. This acts as a filter of spatial frequencies and the final filtered image has an amplitude

$$U'(\xi')=U(\xi')*F(\xi'). \tag{3.32}$$

3.5.2. Incoherent object

When the optical system is linear in intensities, the illumination is said to be incoherent. This holds closely for self-luminous objects or for objects illuminated by radiation incident over a large angle. The spread function is now the intensity image of a point source, that is, $|F(\xi)|^2$. Its transform, the *optical transfer function*, is, by (2.20), the autocorrelation function of the pupil function $f(\eta)$. As a function of a vector *spatial frequency* $\mathbf{s}$, the transfer function is given by

$$e(\mathbf{s})=\int_{-\infty}^{+\infty} f^*(\boldsymbol{\eta})f(\boldsymbol{\eta}+\mathbf{s})\,\mathrm{d}\boldsymbol{\eta}. \tag{3.33}$$

For a perfect optical system with a circular aperture, the amplitude spread function is

$$F(\xi)=\mathrm{J}_1(2\pi\rho)/\pi\rho, \quad \rho=|\xi|, \tag{3.34}$$

and the intensity spread function, the Airy disk, is its square. The corresponding optical transfer function is

$$e(\mathbf{s})=\frac{2}{\pi}\,[\arccos\tfrac{1}{2}s-\tfrac{1}{2}s(1-\tfrac{1}{4}s^2)], \quad s=|\mathbf{s}|. \tag{3.35}$$

These functions are shown in fig. 16.

Fig. 16 Spread function $E(x)$ and transfer function $e(s)$ for a perfect, diffraction-limited system.

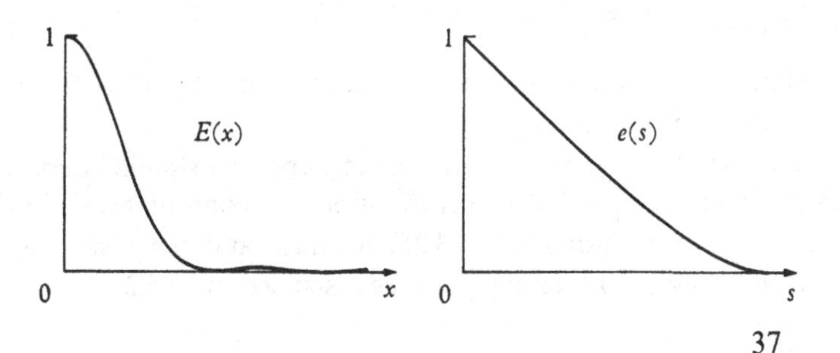

3.5.3. Aberration tolerances

An image-forming system is said to be *diffraction limited* when its aberrations do not noticeably affect the image quality. Following Maréchal [1947], the convention is used that a reduction of the central intensity of a point image to 0.8 of the value due to diffraction only is not noticeable.

The relative central intensity is given by (3.28) and (3.29) as

$$I(0)=\frac{1}{\pi^2}\left|\int_{-\infty}^{+\infty}[\exp 2\pi i w(\eta)/\lambda]\,d\eta\right|^2,$$
$$=1-(2\pi^2/\lambda^2)\langle w^2(\eta)\rangle, \qquad (3.36)$$

where $\langle w^2\rangle$ is the mean square variation of the wave aberration across the pupil. From this expression tolerances can be derived for the different aberrations b_{lnm} of (3.27); for example, the tolerance on defocusing is the well-known Rayleigh limit, $\lambda/4$.

3.5.4. Amplitudes around focus

For a system without aberration, but possibly with a defocusing, the distribution of complex amplitude around the geometrical focus has been evaluated for both circular and rectangular apertures.

For circular apertures, the distribution is a radially symmetrical function of the lateral displacement $\rho=|\xi|$ and also depends on the defocusing ζ. In polar coordinates, (3.26) with normalization becomes

$$U(\rho,\zeta)=2\int_0^1 \mathrm{J}_0(2\pi\rho r)\,e^{-\pi i\zeta r^2} r\,dr. \qquad (3.37)$$

This integral has been evaluated by Lommel in terms of functions now named after him. In the focal plane, it reduces to the Airy disk, in which the amplitude is

$$U(\rho,0)=\mathrm{J}_1(2\pi\rho)/\pi\rho, \qquad (3.38)$$

while along the optical axis

$$U(0,\zeta)=e^{-(1/2)\pi i\zeta}\,\mathrm{sinc}\,\tfrac{1}{2}\zeta. \qquad (3.39)$$

The full function is plotted later in fig. 46 (a component $\kappa\,\Delta z'$ of the phase does not appear in this figure).

When the pupil is rectangular, its numerical aperture should be specified in two planes as α_i, α_j and two sets of diffraction coordinates defined in terms of these. The variables in (3.26) separate and the result can be obtained in terms of Fresnel integrals, as discussed in §7.5.2.

3.5.5. Diffraction gratings

A diffraction grating, used to disperse light into a spectrum, has a transmittance or reflectance that varies periodically in one direction. A simple grating with a period a in the x direction has a transmittance

$$K(x)=K_0+K_1\cos 2\pi ax/\lambda.$$

When it is illuminated normally by a plane wave, the far-field amplitude is

$$k(u)=\tfrac{1}{2}K_1\delta(u+\lambda/a)+K_0\delta(u)+\tfrac{1}{2}K_1\delta(u-\lambda/a),$$

a direct beam at $u=0$ and two diffracted beams at $u=\pm\lambda/a$. For the more usual grating, which has higher harmonics present, there are higher-order diffracted beams at $\pm N\lambda/a$, where N is an integer.

When the diffraction is not in the principal plane of the grating, the diffracted direction **u** can be expressed in terms of the sines of the angles β_i and β_j that the diffracted ray makes with the normal to the grating. When the illumination is at the angles α_i and α_j to the normal, the diffracted rays are given by [Guild, 1956],

$$\sin\beta_{Ni}=\sin\alpha_i+N\lambda/a,$$
$$\sin\beta_{Nj}=\sin\alpha_j. \qquad (3.40)$$

3.6. Polarization

So far we have ignored the fact that light is a transverse wave and can exhibit polarization. For a wave propagated in the z-direction, separate treatments are needed for the two components of the electric vector and hence for the two complex analytic signals E_x and E_y associated with them.

When the light is fully polarized, these analytic signals can be combined into a polarization vector

$$\varepsilon=\begin{pmatrix}E_x\\E_y\end{pmatrix}, \qquad (3.41)$$

and calculations made by the Jones [1941] calculus, reviewed by Shurcliff [1962]. The action of optical components that affect polarization is obtained by multiplying this by a matrix that characterizes the component. Thus, a birefringent plate that introduces a retardation 2γ between the two orthogonal linear polarizations, a *linear retarder*, is represented, to a constant phase factor, by the matrix

$$\mathbf{R}(\gamma)=\begin{pmatrix}e^{i\gamma} & 0\\0 & e^{-i\gamma}\end{pmatrix}. \qquad (3.42)$$

Rotation matrices are also required if the optic axis of the retarder is at an azimuth ϕ to the x-axis. These are given by Shurcliff as

$$\mathbf{S}(\phi)=\begin{pmatrix} \cos\phi & \sin\phi \\ -\sin\phi & \cos\phi \end{pmatrix}. \tag{3.43}$$

Jones [1941], however, has used the opposite sign for ϕ. The final polarization, after passage through the retarder, is given by

$$\boldsymbol{\varepsilon}'=\mathbf{S}(-\phi)\mathbf{R}(\gamma)\mathbf{S}(\phi)\boldsymbol{\varepsilon}. \tag{3.44}$$

An important retarder in interferometry is the reflecting surface of a mirror or beam splitter, for which the complex amplitude reflexion factors $r_{\parallel}$ and $r_{\perp}$ (or r_p and r_s) for light polarized parallel or perpendicular to the plane of incidence may differ in both amplitude and phase. A mirror is thus represented by the matrix

$$\mathbf{M}=\begin{pmatrix} r_{\parallel} & 0 \\ 0 & r_{\perp} \end{pmatrix}. \tag{3.45}$$

It should be remembered that, when determining the change of azimuth ϕ between one plane of incidence and the next, account should be taken of the reversion of the frame of reference at each reflexion, as this changes the sign of ϕ.

Final intensities are given by

$$I=\langle E_x^* E_x + E_y^* E_y \rangle = \langle \varepsilon^{\dagger} \varepsilon \rangle, \tag{3.46}$$

the time-average of the scalar product of $\boldsymbol{\varepsilon}$ and its transpose or Hermitian conjugate $\varepsilon^{\dagger}$. Further, two beams of unit intensity with polarizations ε_1 and ε_2 are said to have polarizations differing in state by an angle c, where

$$\cos c = \langle |\varepsilon_1^{\dagger}\, \varepsilon_2| \rangle. \tag{3.47}$$

When $\cos c = 0$, the polarizations are *orthogonal.*

3.7. Reflexion and transmission at surfaces

Reflexion and transmission coefficients, such as those used in (3.45), are given for a boundary separating media of refractive indices n_1 and n_2 by the Fresnel formulae:

$$t_{\parallel}=\frac{2n_1\cos\theta_1}{n_2\cos\theta_1+n_1\cos\theta_2}, \quad t_{\perp}=\frac{2n_1\cos\theta_1}{n_1\cos\theta_1+n_2\cos\theta_2}.$$

$$r_{\parallel}=\frac{n_2\cos\theta_1-n_1\cos\theta_2}{n_2\cos\theta_1+n_1\cos\theta_2}, \quad r_{\perp}=\frac{n_1\cos\theta_1-n_2\cos\theta_2}{n_1\cos\theta_1+n_2\cos\theta_2}, \tag{3.48}$$

where θ_1 is the angle of incidence of the light at the boundary and θ_2 the angle at which the transmitted light is refracted. These coefficients give the ratios of the complex amplitude of the transmitted and reflected waves to that of the incident wave. If the direction of the light is reversed so that it reaches the boundary from medium 2, the coefficients become

$$t' = tn_2 \cos\theta_2/n_1 \cos\theta_1, \qquad (3.49)$$
$$r' = -r,$$

for both polarizations. The transmittance $\mathscr{T}$ and reflectance $\mathscr{R}$ for intensities are given by

$$\mathscr{T} = |t|^2 n_2 \cos\theta_2/n_1 \cos\theta_1$$
$$= |t'|^2 n_1 \cos\theta_1/n_2 \cos\theta_2 = |tt'|,$$
$$\mathscr{R} = |r|^2, \qquad (3.50)$$

again for both polarizations.

Typical curves of the reflectance of a boundary between two transparent media are shown in fig. 17 as a function of the angle of incidence θ_1. It is seen that, while the reflectance of the component polarized perpendicular to the plane of incidence increases steadily with angle, that for the parallel component falls to zero at the *Brewster angle*, which is given by

$$\tan\theta_1 = n_2/n_1. \qquad (3.51)$$

If unpolarized light is incident at the boundary at this angle, the reflected

Fig. 17 Reflectances at a boundary between media for which the ratio of refractive indices is 3, as a function of the angle of incidence, θ_1.

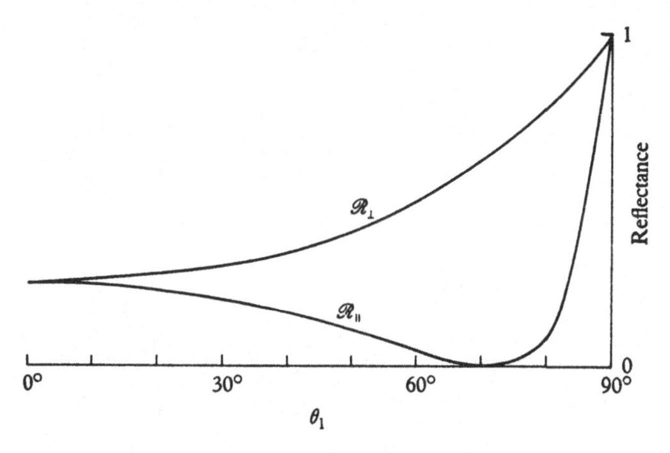

light is linearly polarized in the perpendicular direction.

The formulae (3.48) may also be used when the media are absorbing. Then the complex index $\hat{n}$ is used, where

$$\hat{n}=n+ik. \tag{3.52}$$

3.7.1. Total internal reflexion

When light reaches the boundary between a denser and a less dense medium, where $n_2 < n_1$, the angle of refraction θ_2 becomes $\frac{1}{2}\pi$ when $\sin\theta_1 = n_2/n_1$. This equation defines the *critical angle* θ_1. For greater angles of incidence there is total reflexion and no light is transmitted through the boundary.

Totally reflecting prisms are widely used in optical systems as alternatives to mirrors. When used with linearly polarized lasers, they can have entrance and exit faces at the Brewster angle and are then, in theory, loss free. Like a mirror they are a linear retarder with a phase difference between the two linear polarizations of

$$\tan\psi=\cos\theta_1(n_1^2\sin^2\theta_1-n_2^2)^{1/2}/n_1\sin^2\theta_1, \tag{3.53}$$

but no change of amplitude.

3.7.2. Optical thin films

To produce the desired reflectance and transmittance at the surface of an optical component, thin layers of dielectrics or metals are used, usually deposited by evaporation. They can give surfaces with either a high or a low reflectance or beam splitting surfaces that divide a beam into two beams of approximately equal intensity. The study and design of such coatings is a special application of multiple-beam interference. The methods used are given, for example, by Macleod [1969].

4

Wave interference

When the source is a laser, radiation from it can be treated as a simple wave motion (§3.4). When two waves cross, their effects add and the complex amplitude at any point is the sum of those of the separate waves. If both waves have the same frequency, a pattern of *standing waves* is formed, made up of regions where the wave amplitudes always add, separated by regions where they always subtract. The amplitudes at the latter regions are zero when the two waves have equal amplitude. This effect is familiar to anyone who has watched waves on water, when an approaching train of waves is crossed by waves reflected by a steep bank. The regions of maximum and minimum disturbance follow lines that remain fixed. In spite of the name 'interference' for this effect, one wave does not interfere with the other in the common meaning of the word. After they have crossed, each wave continues on unaffected by its encounter with the other.

The same effect for light is shown by Young's demonstration of interference, which he described in 1802. Monochromatic light from a small pinhole S (fig. 18) illuminates two closely spaced pinholes S_1 and S_2. From each of these, spherical waves, spread out by diffraction, illuminate a plane $\mathcal{O}$. The illuminated region of this plane is crossed by light and dark bands, the interference fringes, shown in fig. 19. These are the lines where the *plane of observation* or *interference plane* $\mathcal{O}$ intersects the maxima and minima of the standing waves. If a three-dimensional object is placed in the standing waves, these fringes contour its surface.

In water waves, the wave amplitude can be detected directly and the standing-wave pattern can be seen. But with light, the standing waves are only made evident when they are received by some detector of intensity, such as a scanning photocell or a screen viewed by eye. It is possible to make a distinction between standing waves when there is no detector and the interference pattern when there is, but this is not yet an established convention.

4.1. Plane waves

The simplest example of interference is that between plane waves (fig. 20).

Wave interference

For two waves that have complex amplitudes of the form (3.17), the total amplitude at a position $\mathbf{u}$ in the plane of observation is

$$U(\mathbf{u}) = a_1\, e^{i\kappa \mathbf{r}_1 \cdot \mathbf{u}} + a_2\, e^{i\kappa \mathbf{r}_2 \cdot \mathbf{u}}, \tag{4.1}$$

where the position vector $\mathbf{x}$ has been replaced by the vector $\mathbf{u} = \mathbf{x} - \mathbf{x}_0$ in the plane through $\mathbf{x} = \mathbf{x}_0$. The intensity at $\mathbf{u}$ is

$$I(\mathbf{u}) = a_1^2 + a_2^2 + 2a_1 a_2 \cos \kappa[(\mathbf{r}_2 - \mathbf{r}_1)\cdot\mathbf{u}]. \tag{4.2}$$

This intensity is made up of a uniform background $a_1^2 + a_2^2$ on which is superimposed a cosinusoidal variation

$$G(\mathbf{u}) = 2a_1 a_2 \cos \kappa[(\mathbf{r}_2 - \mathbf{r}_1)\cdot\mathbf{u}]. \tag{4.3}$$

It is this variation $G(\mathbf{u})$ that is called the *interferogram*.

The vector $\mathbf{r}_2 - \mathbf{r}_1$ is a measure of the angle between the two interfering waves. It is the angular *tilt* $\boldsymbol{\theta}$ between the interfering beams, where

$$\boldsymbol{\theta} = \mathbf{r}_2 - \mathbf{r}_1 \tag{4.4}$$

In terms of this vector,

$$\begin{aligned} G(\mathbf{u}) &= 2a_1 a_2 \cos \kappa\boldsymbol{\theta}\cdot\mathbf{u}, \\ &= 2a_1 a_2 \cos(2\pi\boldsymbol{\theta}\cdot\mathbf{u}/\lambda). \end{aligned} \tag{4.5}$$

Fig. 18 Formation of Young's fringes by light waves.

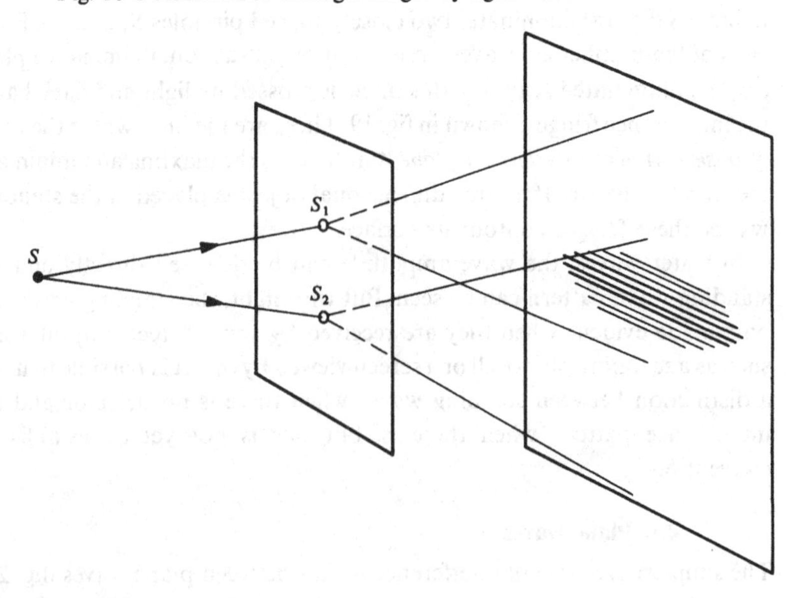

The maxima of the standing waves occur where

$$\boldsymbol{\theta}\cdot\mathbf{x}=N\lambda, \tag{4.6}$$

the constant N, the *order of interference*, being an integer. The equation (4.6) is that for a set of equally spaced planes normal to the tilt vector, planes that bisect the external angle between two wave surfaces.

Although it is natural to express the tilt as an angle, it is convenient for the theory later to express it also as a distance, the lateral separation of two images of the light source, S_1S_2 in fig. 18. I shall denote this measure of the tilt by the vector $\mathbf{t}$ and its length by t.

In fig. 18, S_1 and S_2 are the sources of two spherical waves. At a large distance z, the fringes produced are approximately straight lines perpendicular to $\mathbf{t}$ with the spacing $z\lambda/t$.

Fig. 19 Young's fringes.

Their contrast is given by the *visibility*, defined by Michelson as

$$\mathscr{V}=(I_{max}-I_{min})/(I_{max}+I_{min}). \tag{4.7}$$

For the fringes described by (4.2),

$$\mathscr{V}=\frac{2a_1a_2}{a_1^2+a_2^2},$$
$$=\frac{2I_1^{1/2}I_2^{1/2}}{I_1+I_2}, \tag{4.8}$$

where I_1 and I_2 are the intensities due to each beam alone. The visibility is a maximum, with a value of unity, when $I_1=I_2$. Then the centres of the dark fringes have zero intensity. Even beams that have a considerable difference in intensity give fringes of reasonable visibility: when $I_1=4I_2$, $\mathscr{V}=0.8$.

When the two interfering waves are travelling in the same direction, there is no tilt, and the fringes are said to be *spread out* to give a uniform intensity across the interference plane. The value of this intensity depends on R, the difference between the two distances R_1 and R_2 travelled by the wave, and is

$$a_1^2+a_2^2+2a_1a_2\cos 2\pi R/\lambda.$$

Fig. 20 Interference between two plane waves.

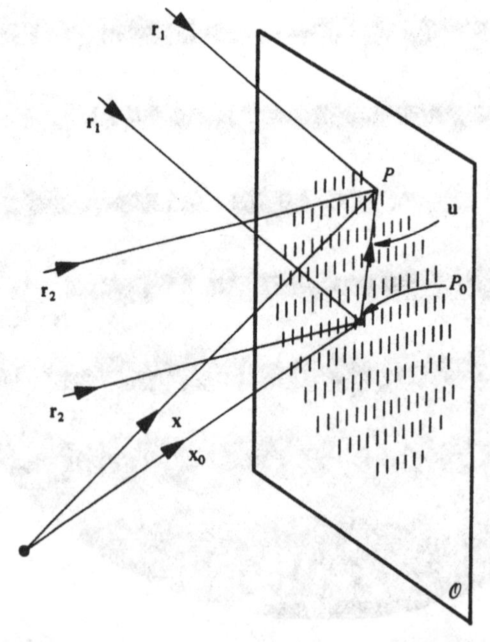

The part that varies with the path difference is

$$G(R) = 2a_1a_2 \cos 2\pi R/\lambda$$
$$= 2a_1a_2 \cos 2\pi\nu\tau, \tag{4.9}$$

where τ is the *delay*, the *optical path difference* $p = nR$ divided by the speed of light,

$$\tau = p/c. \tag{4.10}$$

This is still called an interferogram but it is now a function of the path difference, or the delay, rather than of the position **u** in a plane.

4.2. Spherical waves

The interference shown in fig. 18 is actually the interference between two spherical waves, the complex amplitudes of the two waves from S_1 and S_2 having the form (3.18). When the two pinholes have the same size, the two effective sources have the same strength A, and the intensity where the beams overlap is

$$I(\mathbf{u}) = A^2/R_1^2 + A^2/R_2^2 + 2(A^2/R_1R_2)\cos\kappa(R_2 - R_1). \tag{4.11}$$

The equation for the surfaces of the maxima of the standing waves is

$$R_2 - R_1 = N\lambda, \tag{4.12}$$

the bipolar equation for confocal hyperboloids of revolution of two sheets. A section of these is shown in fig. 21.

The intersection of these surfaces with an interference plane will always

Fig. 21 Section of standing-wave pattern from two point sources.

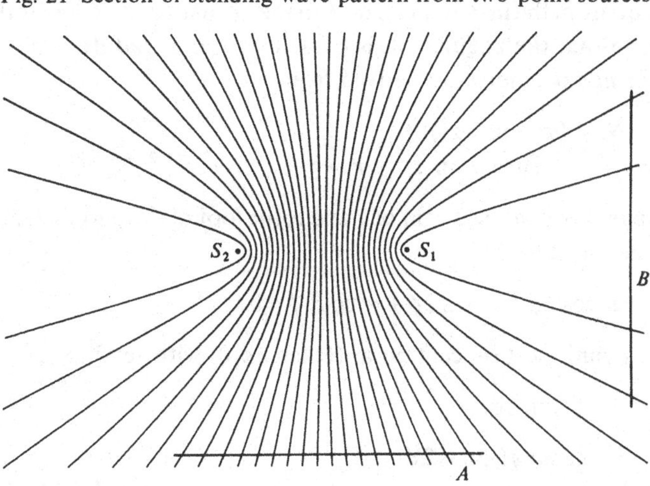

give fringes having the shape of conic sections (which are no longer confocal). A plane in the position A will show hyperbolic fringes, which are almost straight lines when the distance to A is large compared with the separation S_1S_2 of the sources. The plane sees this separation as a tilt. A plane at B will see no tilt but a longitudinal separation of the sources. One wave arrives having travelled a distance S_1S_2 further than the other: it has a *lead* $l = S_1S_2$ and the fringes are concentric circles. They are not uniformly spaced, but have a constant increment of area from one circle to the next. For inclined positions between A and B, the fringes are ellipses, parabolas, or hyperbolas.

Fig. 21 represents all cases of interference when the two waves come from real sources. When one of the sources is virtual, so that a wave converging to it interferes with a wave diverging from the other, the equation for the standing waves becomes

$$R_2 + R_1 = N\lambda, \tag{4.13}$$

and the surfaces are spheroids, that is, ellipsoids of revolution. When the two sources appear in line, the fringes are again concentric circles; otherwise they are ellipses.

4.2.1. Derivation of the source separation

From the shape of the fringes it is possible to find the separation of the two sources. A simple analysis shows how this can be derived from the shape of one fringe and its separation from the next, but in practice it would be better to use a best-fit solution that involved the positions of all the fringes. The notation is shown in fig. 22. The interference is observed in a plane $\mathcal{O}$, with coordinates u and v, at a distance z from the source S_1. The other source is a distance d from the first at an angle ϕ to the normal to $\mathcal{O}$. There is thus a tilt $d \sin \phi$, so that the angular tilt is $(d/z) \sin \phi$, and a lead $d \cos \phi$.

The fringe positions are derived from

$$N\lambda = (u^2 + v^2 + z^2)^{1/2} \\ - [(u - d \sin \phi)^2 + v^2 + (z - d \cos \phi)^2]^{1/2} \tag{4.14}$$

The fringe through the origin of u and v, the foot of the perpendicular from S_1 to the plane, is given by

$$u^2 \cos 2\phi - uz \sin 2\phi + v^2 \cos^2 \phi = 0.$$

This is a conic section centred at $(\frac{1}{2}z \tan 2\phi, 0)$ with semi-axes

$$a = \tfrac{1}{2}z \tan 2\phi,$$

$$b = \tfrac{1}{2}z \sin \phi (\pm \cos 2\phi)^{-1/2},$$

where the positive sign denotes an ellipse with $\phi < \pi/4$, and the negative sign a hyperbola with $\phi > \pi/4$. For either, the eccentricity is

$$e = \tan \phi, \tag{4.15}$$

and the radius of curvature at the origin is

$$\rho = z \tan \phi. \tag{4.16}$$

Either of these expressions gives the angle ϕ, the direction of the separation between the two sources.

The distance d is found from the spacing Δu, along the axis $v=0$, of two fringes differing in order by ΔN. For $v=0$, equation (4.14) gives

$$N\lambda = d\cos\phi + (ud/z)\sin\phi - (u^2 d/2z^2)\cos\phi + \ldots \tag{4.17}$$

Near $u=0$, $\lambda\,\Delta N \approx (d\,\Delta u/z)\sin\phi$, that is

$$d = \frac{\lambda z}{\sin\phi}\frac{\Delta N}{\Delta u}, \tag{4.18}$$

Fig. 22 Notation for the derivation of the source separation.

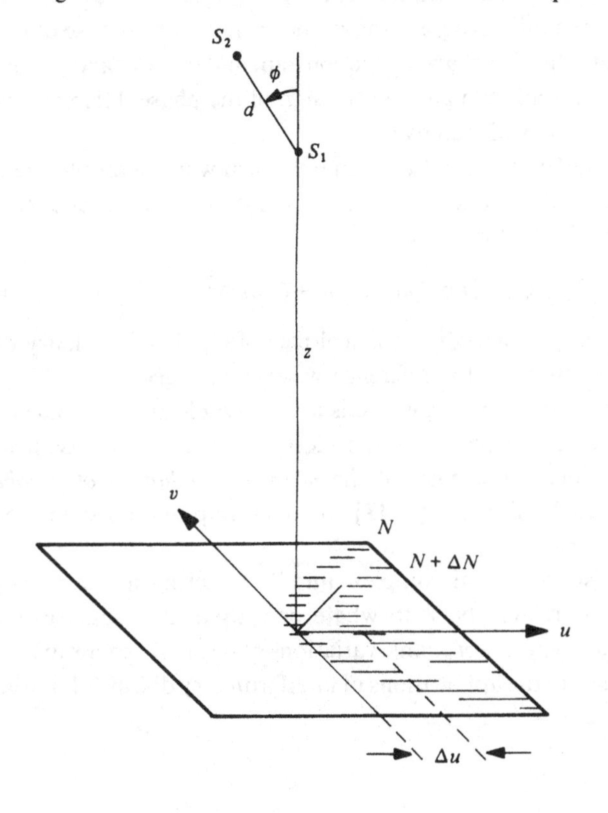

unless $\phi=0$. This shows that the fringe spacing depends only on the tilt, $d\sin\phi$, and not on the lead, $-d\cos\phi$.

When $\phi=0$, equation (4.17) gives for two fringes of order N and $N+\Delta N$ at u_1 and u_2

$$d=\frac{2\lambda z^2\,\Delta N}{u_1^2-u_2^2}. \tag{4.19}$$

4.3. General wave

When two waves represented by the general equation (3.19)

$$U(\mathbf{u})=a(\mathbf{u})\exp\mathrm{i}[\kappa\mathbf{r}\cdot\mathbf{u}-\psi(\mathbf{u})]$$

interfere, the resulting intensity is

$$I(\mathbf{u})=a_1^2(\mathbf{u})+a_2^2(\mathbf{u})+2a_1(\mathbf{u})a_2(\mathbf{u})\cos[\kappa\boldsymbol{\theta}\cdot\mathbf{u}+\psi_2(\mathbf{u})-\psi_1(\mathbf{u})]. \tag{4.20}$$

The interferogram has the basic pattern of the straight fringes corresponding to the mean angular tilt, $\boldsymbol{\theta}$, between the two waves. This pattern is distorted by the variations of the phase difference, $\psi_2-\psi_1$. When there is no tilt, the interferogram shows the cosine of this phase difference, provided the amplitudes a_1 and a_2 are constant and do not vary with $\mathbf{u}$. This representation has an ambiguity in the sign of the phase difference, which the introduction of a tilt removes.

When the interference is between an unknown wave of amplitude $U_2(\mathbf{u})$ and a uniform plane reference wave of complex amplitude U_1, the interferogram has the form

$$G(\mathbf{u})=2U_1|U_2(\mathbf{u})|\cos[\arg U_2(\mathbf{u})+\text{constant}], \tag{4.21}$$

that is, to a constant factor, it is the real part of U_2. The imaginary part is obtained if the phase of the reference wave is changed by $\frac{1}{2}\pi$. Thus the complex amplitude of a wave, which is unobservable at the frequencies of visible light, is made observable by the addition of a reference wave from the same source. This is the basis of the *method of addition of a coherent background* used by Zernike [1948] to study amplitude distributions in diffraction patterns.

Once the distribution of complex amplitude across a wave has been found, it can be related back to whatever caused it: irregularities on a reflecting surface, lens aberrations, variations of refractive index in the light path, etc. These are the applications of interferometry discussed in Chapter 11.

4.4. Shearing interferometers

The term *shear* was introduced into interferometry by Bates [1947]. Instead of measuring a wavefront against a substantially error-free reference wave, he compared it with 'a sheared image of itself'. He thus introduced the concept of a *common-path interferometer*, in which the two interfering waves follow the same path, with a small separation, resulting in an instrument that is less affected by outside disturbances than one in which they follow widely separated paths.

Bates's shearing interferometer is shown in fig. 23. The interferogram obtained is a picture, not of the wave itself, but of finite differences across the wavefront. If one wave is sheared a vector distance $\mathbf{s}'$ with respect to the other, this distance being measured between the two images of the wave in the final interference plane, the interferogram, instead of having the form (4.21), is

$$G(\mathbf{u}) = 2|U_1(\mathbf{u})|\,|U_2(\mathbf{u}-\mathbf{s}')|\cos[\arg U_2(\mathbf{u}-\mathbf{s}') - \arg U_1(\mathbf{u})]. \tag{4.22}$$

A tilt may also be included to add straight fringes to the interferogram.

The properties of a shearing interferometer are shown in fig. 24. In observation space, where the fringes are formed, there are again two sheared wavefronts and, if a tilt is present, two images of the source are seen.

Fig. 23 Bates' wave-shearing interferometer.

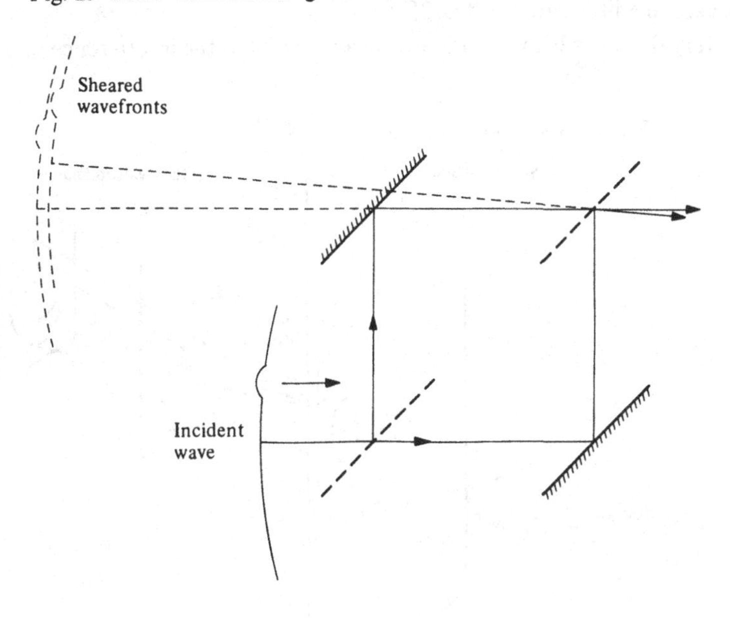

From the specimen, where the object being studied is located, a single wavefront is seen against two separated images of the interference plane. These have a vector separation **s**, the image of **s**′ back through the interferometer. (The quantities **s** and **s**′ have the same magnitude for an interferometer made up of plane reflectors only, with no focusing components.) From the light source *S*, the same separation of these images is seen, and I shall extend the use of the term shear to the separation of the two images of the plane of observation as seen from the source. While this extension has no significance for the interference of simple waves, it becomes an important parameter when the interferometer has an extended source. It is not necessarily the same as the shear seen from the specimen, since there may be a further shear before the specimen that can cancel out the shear that follows it.

The distinguishing feature of Bates's interferometer was a shear that is constant across the wavefront. It was later realized that other shears were possible that varied across the wave: *radial*, *rotational*, and *folding shears*, and so Bates's shear is now called a *lateral shear*, although the variable shears are also lateral. For a radial shear, one image of the wavefront is magnified with respect to the other so that corresponding points can be regarded as sheared apart radially. In a rotational shear, one image is rotated with respect to the other, a rotation of π giving an *inverted shear*. In a folding shear, one wavefront is reverted about some line to give a shear that is normal to this line and increasing as the distance from it. The four shears are illustrated in fig. 25.

It is also possible to imagine a shear normal to the interference plane or a

Fig. 24 Two views of shear.

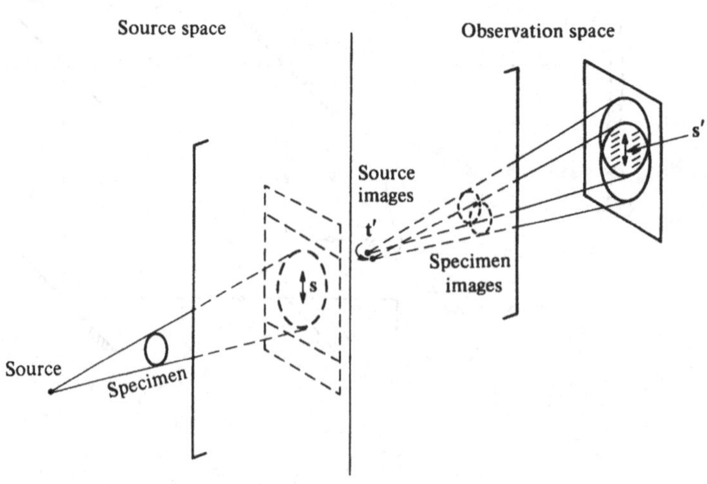

combination of an in-plane shear with this. Such an interferometer would compare a wave with an out-of-focus view of itself. This could be called a longitudinal shear but I prefer to describe it as a *shift* between the two wavefronts. Schulz [1964] has given a theory of interferometers in which the shift and shear are combined into a screw displacement between the wavefronts.

4.5. Heterodyne interferometry

If the two interfering waves have different frequencies ν and $\nu+\Delta\nu$, the time dependent term $2\pi\nu t$ must be retained in the complex amplitude, which is then

$$U(\mathbf{r}, t)=a \exp \mathrm{i}(\kappa \mathbf{r} \cdot \mathbf{x}-2\pi\nu t)$$

for a uniform plane wave. When two waves of different frequency interfere, the interferogram is

$$G(\mathbf{u})=2a_1 a_2 \cos 2\pi[\boldsymbol{\theta} \cdot \mathbf{u}/\lambda - n\,\Delta\nu(\mathbf{r}_1+\mathbf{r}_2) \cdot \mathbf{u}/c - t\,\Delta\nu]. \tag{4.23}$$

This represents a travelling intensity wave with a frequency $\Delta\nu$. The fringes move past any point $\mathbf{u}$ at this frequency and a detector there will convert the interferogram into an alternating electrical signal. The fringes are no longer

Fig. 25 Types of shear: (*a*) lateral; (*b*) radial; (*c*) rotational; (*d*) folding.

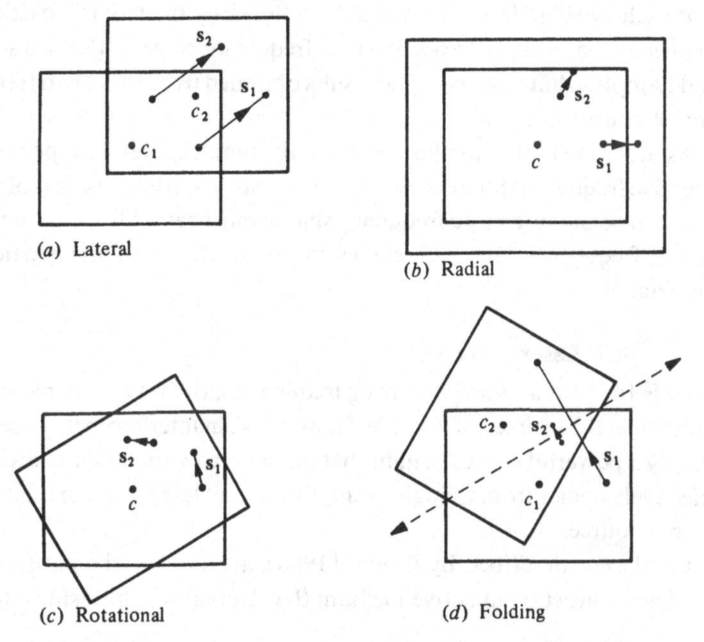

exactly normal to the angular tilt θ nor spaced at exactly λ/θ, but the departures are usually negligible, since practical frequency differences $\Delta\nu$ are very small compared with the frequency ν of light.

While interference effects can be obtained between light from two different lasers, they are of short duration since the frequency difference does not remain constant. Practical heterodyne interferometry requires the use of two beams from the same source and some method of producing a constant frequency difference between them. Another way of looking at this latter is that it is a method that gives a linearly increasing phase difference between the beams. Methods of producing a frequency difference are given later in §4.7.2.

4.6. Laser Doppler

A technique allied to interferometry, but usually treated as a separate subject [Durst, Melling & Whitelaw, 1976], is the *laser Doppler* method of measuring the velocity of small particles. If a particle moves through the standing waves produced by two intersecting beams from the same laser, the amount of light that it scatters varies sinusoidally as it passes through the maxima and minima. The frequency of this variation of the scattered light multiplied by the spacing of the pattern, λ/θ, gives the component of the particle velocity normal to the standing waves, that is normal to a direction midway between those of the two beams.

The more usual treatment of laser Doppler considers the Doppler shifts that each wave undergoes when it is scattered by the moving particle. The frequency that is measured is the beat frequency between the two scattered and Doppler-shifted waves. The results obtained from these two treatments are, of course, the same.

As described, the method cannot distinguish between positive and negative frequency differences and so does not give the direction of motion. This can be obtained by a frequency shift of one beam. The resulting change in the frequency difference shows in which direction the particles are moving.

4.7. Laser sources

For visible light and neighbouring frequencies, the wave treatment applies only when the source is a laser. As far as its use in interferometry goes, this is simply a powerful source of light that has a narrow spectral bandwidth and effectively comes from a single point. But not all lasers approximate well to such a source.

Gas lasers, described by Bloom [1968], are shown schematically in fig. 26. They consist of an active medium that amplifies light inside a resonant cavity, a Fabry–Perot interferometer. The active medium shown is a gas in a narrow capillary through which an electrical discharge is passing. The cavity is a resonator for radiation along the axis for which the separation of the reflecting surfaces is an integral number of half wavelengths. This radiation constitutes the *axial* or TEM_{00} *modes* of the laser. Lasers can oscillate also in off-axis modes, but these are not usually suitable sources for interferometry.

When a mode occurs within the spectrum over which the active medium amplifies, its *gain curve*, the laser radiates in this mode. The spectrum then looks like that in fig. 27, a number of narrow modes a few kilohertz wide, spaced across a gain curve that may be some gigahertz wide. For a cavity of length d and refractive index n, the spacing between the modes is

$$\Delta\nu = c/2nd. \tag{4.24}$$

When there are Brewster windows at the ends of the tube containing the active medium, the laser oscillates only for radiation in the linear

Fig. 26 Form of a gas laser: (*a*) with external mirrors and windows at the Brewster angle; (*b*) with internal mirrors.

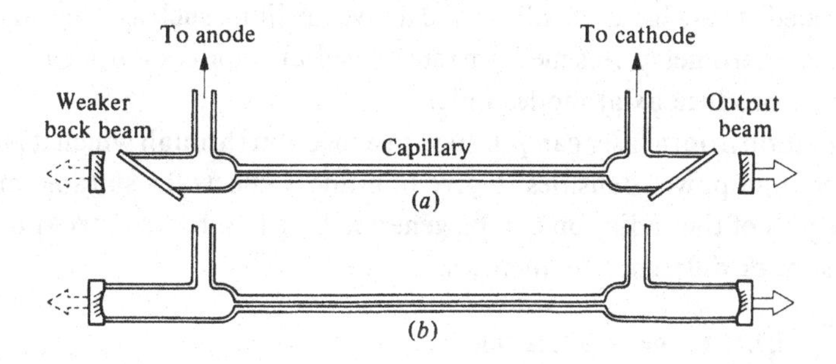

Fig. 27 Spectrum of laser radiation.

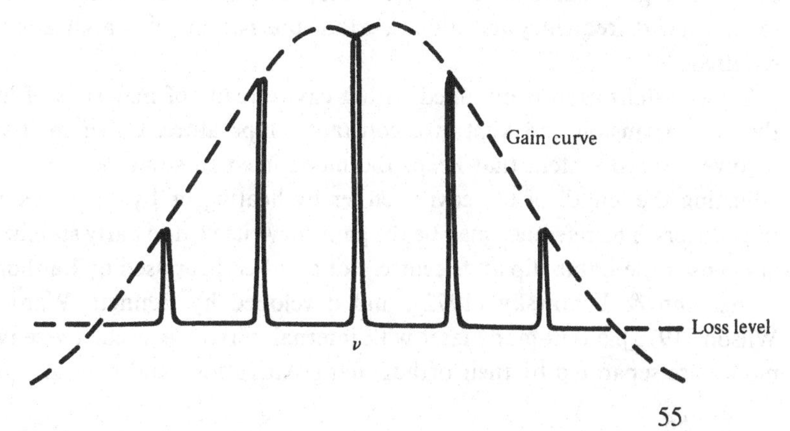

polarization for which there is no reflexion loss. But when the cavity mirrors are fixed directly to the ends of the tube, there is no preferred direction of polarization. In practice each mode is linearly polarized, alternate modes being polarized usually horizontally and vertically.

For small delays the laser acts as a light source of spectral bandwidth equal to that of the gain curve, and the modulation of an interferogram falls off as the delay approaches the reciprocal of the bandwidth. But, because of the mode structure, there are repetitions of the region of high modulation whenever the interferometer delay is close to an integral multiple of the delay of the laser cavity. A nearly monochromatic wave is obtained only when the laser oscillates in a single mode. Either the laser must be short, so that only one mode occurs under the gain curve, or a mode selector must be used. This is commonly a Fabry–Perot étalon, placed inside or outside the cavity. It usually has plane surfaces, but Hariharan [1982] has shown that, for higher-powered lasers, an étalon with spherical surfaces gives better selection and more power. An alternative to this is the Fox–Smith selector [Smith, 1965].

The lasers most often used for interferometry are gas lasers with a continuous output. But for measurements on moving bodies, pulsed lasers are needed. These have usually a solid active medium, such as ruby, and, for use in interferometry, this medium must be selected for good optical quality if it is to produce axial modes only.

Radiation from a laser can produce, in a medium through which it passes, high enough power densities to give non-linear effects. In suitable media, harmonics of the radiation can be generated, or two beams mixed to give their sum or difference frequencies.

4.7.1. Laser stabilization

As the length of a laser cavity changes with temperature, the modes move across the gain curve and the frequency obtained varies. To obtain radiation with frequency variations less than one part in 10^6, stabilization is required.

Some stabilization is obtained with a cavity made of materials of low thermal expansion and kept at a constant temperature. Other methods involve a servo system that keeps the mode fixed to some reference, by adjusting the length of the cavity, either by heating or by piezo-electric transducers. The reference may be the gain curve itself, and early stabilized lasers used the Lamb dip at its centre. For another, proposed by Balthorn, Kunzmann & Lebowsky [1972], and developed by Bennett, Ward & Wilson [1973], a two-mode laser with internal mirrors is used. These two modes are separated by their orthogonal polarizations and kept at equal power. Alternatively, a single linearly polarized mode can be split into two circularly polarized modes by the Zeeman effect when an axial magnetic field is applied to the laser. The powers of these can be kept equal, or their frequency difference kept to a minimum [Baer, Kowalski & Hall, 1980]. This minimum occurs when the two components are equally spaced about the centre of the gain curve.

The most precise stabilization, to a few parts in 10^{11}, is obtained when a separate reference material is used and the laser stabilized, by saturated absorption, to a line in its spectrum. Lasers stabilized in this way are used as reference standards of length [Baird & Hanes, 1974; Wallard, 1973; Chebotayev, 1980].

Lasers also have fluctuations of power. There are separate stabilization methods for these, but the fluctuations are reduced in any case when the frequency is stabilized.

4.7.2. Frequency changing

The two circularly polarized components from the longitudinal Zeeman laser produce heterodyne interference, and such a laser is used in a length-measuring interferometer [Liu & Klinger, 1979]. Other methods of obtaining heterodyne interference involve starting with two beams of the same frequency and changing the frequency of one or both of them [Dändliker, 1980].

Light diffracted by a moving grating has its frequency changed. A grating of spacing a moving at a speed v changes the frequency of the N-th order diffracted beam by

$$\Delta\nu = Nv/a. \tag{4.25}$$

This method is illustrated in fig. 28(a). If the grating is used as a beam splitter, the two first-order beams have a frequency difference of $2v/a$.

Fig. 28 Frequency changers: (a) a moving transmission grating; (b) a rotating $\lambda/2$ plate.

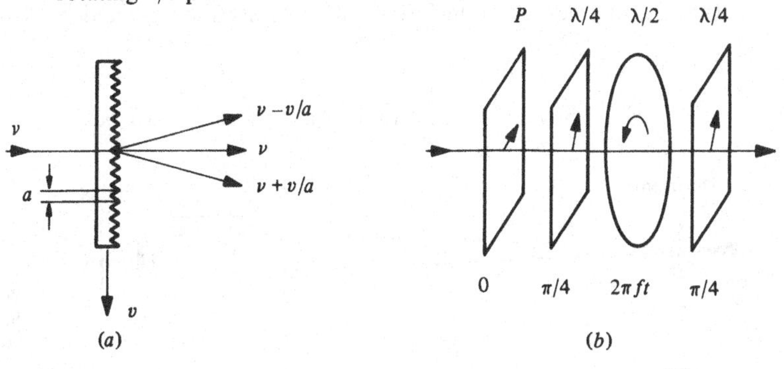

A continuously moving grating is provided by a rotating radial grating [Stevenson, 1970], or by an ultrasonic wave in a suitable medium, such as a plate of vitreous silica [Korpel, 1981]. This is known as an elasto-optical modulator, or a Bragg cell, when it is used at the Bragg angle to concentrate the diffracted light into one order.

The polarization equivalent of the rotating radial grating is a rotating $\lambda/2$ retarder, shown in fig. 28(*b*). If this is used between fixed $\lambda/4$ plates and is rotated at a frequency f, it produces a frequency difference $4f$ between the two orthogonal linear polarizations.

4.7.3. Spatial filters

The amplitude of the wave from a laser is not uniform across the wavefront but drops off from the centre as a gaussian function. In practice the wave amplitude has further irregularities, diffraction patterns caused by surface defects or dust on the various components. These can be removed by passing the light through a *spatial filter*, shown in fig. 29.

The beam from the laser is spread by a divergent lens to fill the aperture of a microscope objective. This then focuses it to a point image, of gaussian form rather than an Airy disk. A pinhole at this image allows the direct beam to pass but stops any light diffracted by defects. The pinhole should have a diameter several times that of the Airy disk produced by an objective of the same numerical aperture and should be in thin foil so that it does not obstruct the beam because of its depth. It is mounted so that it can be adjusted both laterally and in focus.

Spatial filters are often used without the divergent lens, but then a higher-power objective is needed to give the same spread to the beam. They are employed routinely with laser sources whenever uniform beams are wanted.

4.7.4. Optical isolators

When a laser is used with a reflex interferometer such as the Michelson, one of the two complementary interferograms produced is reflected back into

Fig. 29 A spatial filter, used to filter and spread a laser beam.

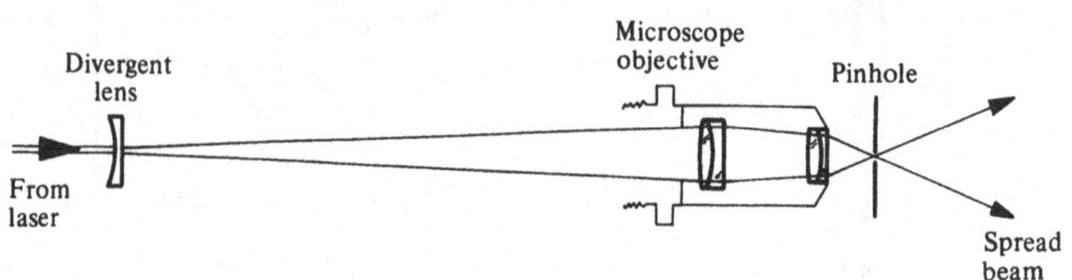

the laser and can alter its operation. This can be serious when the laser is being stabilized by a method involving the measurement of powers, or when a very stable laser frequency is required. To prevent this reflected light reaching the laser, an optical isolator is used.

The simplest consists of a polarizer followed by a $\lambda/4$ plate with its axis at $\pi/4$. Polarized light passing through this plate twice has its direction of polarization changed by $\frac{1}{2}\pi$ and is absorbed by the polarizer. But the light reaching the interferometer is circularly polarized, and should remain so if the isolation is to be effective. This is not a suitable state of polarization for many interferometers.

Other isolators use the Faraday effect [Aplet & Carson, 1964]. This effect, the rotation of the direction of polarization within a medium in a magnetic field, is direction sensitive, unlike the rotation in optically active media, such as quartz crystals and sugar solutions. If the laser is followed by a polarizer and then a Faraday cell that gives a $\pi/4$ rotation, returning light is rotated a further $\pi/4$ and does not pass the polarizer. The light to the interferometer is linearly polarized at $\pi/4$, but can be converted to horizontal or vertical polarization by either a $\lambda/2$ plate or an optically active rotator made of quartz.

5

Holograms

An interference pattern can be recorded on a photographic plate. It then becomes a hologram. When it is illuminated by a wave of the same form as one of the two that produced it, some of the light is diffracted to re-form the other wave. A hologram is usually made from the irregular *object wave* reflected by an object and a uniform spherical or plane *reference wave* coming from the same source. The recorded interference fringes (4.21) have variations of intensity that follow the amplitude variations of the object wave and variations from straightness that follow its phase. When this irregular pattern diffracts light, the amplitude and phase of the first-order diffracted wave reproduce both variations, so that this wave is a complete *reconstruction* of the original object wave. Observers receiving it see the object as if it were still there. If the object was three dimensional, they see in the reconstruction all the properties that show three dimensions: they can move their heads to look around near parts to see what lies behind, and they can change the focus and convergence of their eyes to focus on parts at different distances. Other attempts at providing three-dimensional images have relied on some of these clues only, and a stereoscopic picture has different views for each eye, but the focus and general direction of viewing are fixed. Only a hologram gives a complete reconstruction.

Holography is now an important branch of optics. It is treated in more detail by Collier, Burckhardt & Lin [1971], and Hariharan [1983].

5.1. Gabor holograms

Considered as an interferometer, the layout for making holograms has two paths that are far from matched, and a laser source is needed. But the first holograms were made by Gabor [1949] without a laser, and interferograms recorded even earlier have been reconstructed as holograms [Bryngdahl & Lohmann, 1968]. One of Gabor's aims was to improve the resolving power of electron microscopes by recording holograms with electrons and reconstructing them with visible light. While many spectacular advances have occurred in other fields, this application has yet to be achieved.

Gabor holograms, now known as *in-line holograms*, are still used to study droplets in clouds [Tyler & Thompson, 1976], and test optical components

[Mercier & Lowenthal, 1980]. They have the disadvantage that the reference wave is the light passing around the object and it travels in the same direction as the object wave. On reconstruction, the direct and diffracted rays all leave the hologram in the same general direction and the reconstruction has these other waves superimposed on it.

Some important advances were made during the succeeding years, but it was only when lasers became available and Leith & Upatnieks [1962, 1963, 1964] made holograms with the reference wave at an angle to the object wave that good enough reconstructions were obtained to show the value of the technique and capture popular imagination.

5.2. Holograms with an inclined reference beam

Modern holograms follow those of Leith and Upatnieks and have an inclined reference beam and so are records of interferograms with a large tilt. Their production is shown in fig. 30(*a*). The light from the laser is separated into two beams, both of which are then spread through spatial filters, one to illuminate the object, the other, the reference beam, to illuminate the photographic plate directly. It is simplest to use a divergent spherical reference wave, as shown, but a convergent or plane wave can equally well be used.

The exposed and developed plate is then re-illuminated by the reference beam, as shown in fig. 30(*b*). Some of the light passes directly through the plate; the rest is diffracted. The first-order diffracted wave has the same form as the original wave from the object. The minus-one order, at an equal angle on the other side of the direct wave, produces a conjugate image.

The photographic plate needs to resolve the fine fringes produced by the large tilt. Special fine-grain emulsions are used. The plate can be developed normally to produce variations in transmittance, or bleached to give variations in optical path, caused by variations of refractive index. They are not the only recording medium used. Dichromated gelatin produces index variations with negligible grain, while photoresists also give phase holograms, now produced by variations of thickness.

These are the main media used for making permanent holograms. For hologram interferometry, where the record may be needed only for a short time, erasable media are used, such as photo-thermoplastics [Saito *et al.*, 1980].

5.3. Theory

A hologram formed by the wave reflected from an irregular object and a plane reference wave is a record of the intensity distribution given by (4.20).

This can be rewritten as

$$I(\mathbf{u})=a_1^2\{1+a_2^2(\mathbf{u})/a_1^2+[a_2(\mathbf{u})/a_1]\exp i[\kappa(\mathbf{r}_2-\mathbf{r}_1)\cdot\mathbf{u}-\psi_2(\mathbf{u})] \\ +[a_2(\mathbf{u})/a_1]\exp -i[\kappa(\mathbf{r}_2-\mathbf{r}_1)\cdot\mathbf{u}-\psi_2(\mathbf{u})]. \quad (5.1)$$

The photographic plate, or other recording medium, converts this varying intensity into a varying transmittance, $\mathscr{T}(\mathbf{u})$. The process is usually not linear, and one representation of it is

$$\mathscr{T}(\mathbf{u})=I^{-\gamma}(\mathbf{u}), \quad (5.2)$$

where $-\gamma$ is the slope of the characteristic, or H and D curve of the photographic plate, which relates the logarithm of the transmittance to the logarithm of the exposure, that is, intensity multiplied by time. The slope is negative for a photographic negative.

Fig. 30 Transmission holograms: (*a*) making the hologram; (*b*) reconstruction.

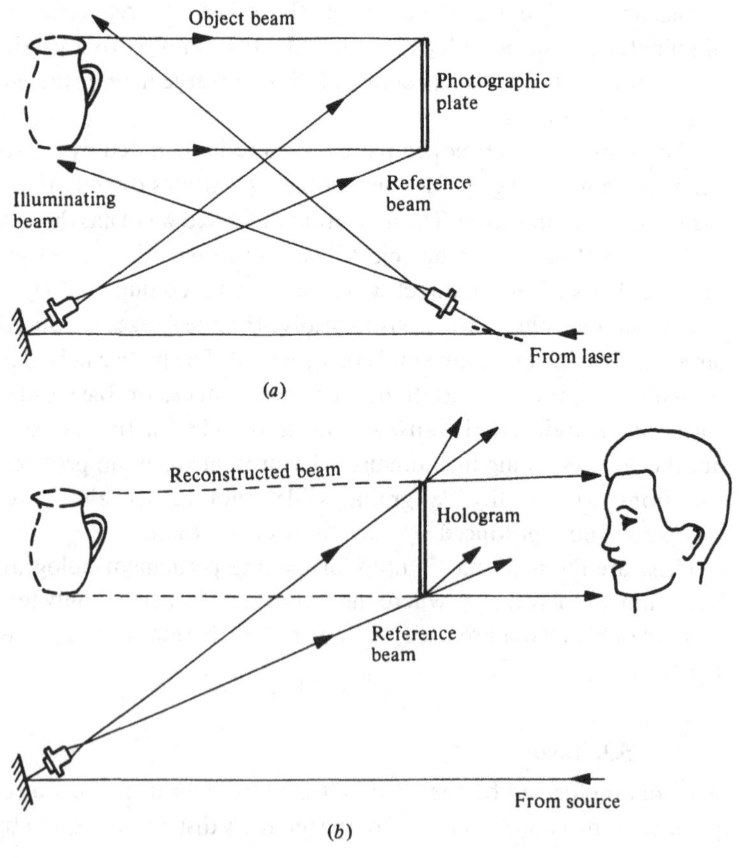

The transmission factor for amplitudes is $\mathscr{T}^{1/2}$. But associated with the variations of transmittance of a developed plate there are usually variations of thickness or refractive index that introduce phase variations into the transmission factor. These are included if γ, and hence $\mathscr{T}^{1/2}$, are allowed to have complex values.

When the developed hologram is illuminated by the reference wave $U_1 = a_1\,e^{i\kappa\mathbf{r}_1\cdot\mathbf{u}}$, the complex amplitude of the transmitted radiation is

$$U''(\mathbf{u}) = U_1 \mathscr{T}^{-(1/2)\gamma}(\mathbf{u}).$$

To avoid phase reversals of the interferogram, which would occur wherever $a_2(\mathbf{u}) > a_1$, the hologram is usually made with the intensity of the reference wave at least three times the average intensity of the object wave. Then the transmitted amplitude can be expanded as the power series

$$\begin{aligned} U''(\mathbf{u}) = a_1^{-1-\gamma}\{&[a_1^2 - \tfrac{1}{2}\gamma a_2^2(\mathbf{u})]\exp i\kappa\mathbf{r}_1\cdot\mathbf{u} \\ &-\tfrac{1}{2}\gamma a_1 a_2(\mathbf{u})\exp i[\kappa\mathbf{r}_2\cdot\mathbf{u} - \psi_2(\mathbf{u})] \\ &-\tfrac{1}{2}\gamma a_1 a_2(\mathbf{u})\exp i[\kappa(2\mathbf{r}_1 - \mathbf{r}_2)\cdot\mathbf{u} + \psi_2(\mathbf{u})] \\ &+\text{higher orders}\}. \end{aligned} \tag{5.3}$$

5.3.1. Interpretation

The first term of the expansion represents the direct beam, the portion of the reconstructing wave that continues in the direction $\mathbf{r}_1$. The next two terms are the two first-order diffracted waves that produce the reconstruction and the conjugate image in the directions $\mathbf{r}_2$ and $2\mathbf{r}_1 - \mathbf{r}_2$. With a plane reference wave, the phase of the conjugate wave is the reverse of that of the reconstructed wave. This means that the conjugate image is pseudoscopic and looks like a hollow replica of the object. Since the first holograms were made with the object behind the plate, their reconstructed image was virtual, being also behind the plate, and the conjugate image was in front. It became known, therefore, as 'the real image' but it need not be real if other recording arrangements are used: a spherical reference wave can give both as virtual images.

In equation (5.3) the γ of the plate appears as a constant multiplier, and a similar result is obtained if other approximations than (5.2) are used to represent the photographic process. Holograms are linear, and the same reconstruction is obtained whether the plate is developed as a negative (γ negative), a positive (γ positive), or even bleached (γ complex). Bleaching is most common since the *phase holograms* so produced are much more efficient at diffracting light into the reconstruction than holograms with varying transmittance.

5.3.2. Thick holograms

A further increase in efficiency is obtained by the use of a thick layer of recording material. The thickness of the photographic emulsion on plates is from 15 to 40 times the wavelength of light in the emulsion, and the standing waves of the interference pattern are not lines but surfaces, as shown in fig. 31. The simple theory (5.3) indicates that the two first-order waves have equal amplitude. But for thick holograms, Bragg theory of the diffraction from planes applies, and the light is concentrated into the reconstructed wave at the expense of the conjugate wave and other higher-order diffracted waves.

These higher-order terms in (5.3) represent waves leaving at larger angles to the reconstructing wave. They are, therefore, well separated from the reconstruction, but intermodulation effects between them can lead to scattered light around the reconstructed wave. Their reduction by the use of a thick recording medium is, therefore, desirable.

5.3.3. Other reconstructions

As well as reconstructing the object wave when it is illuminated by the reference wave, a hologram will reconstruct the reference wave when illuminated by the object wave. Thus a hologram may be regarded as an

Fig. 31 Reconstruction of a thick hologram by Bragg reflexion from the fringe planes.

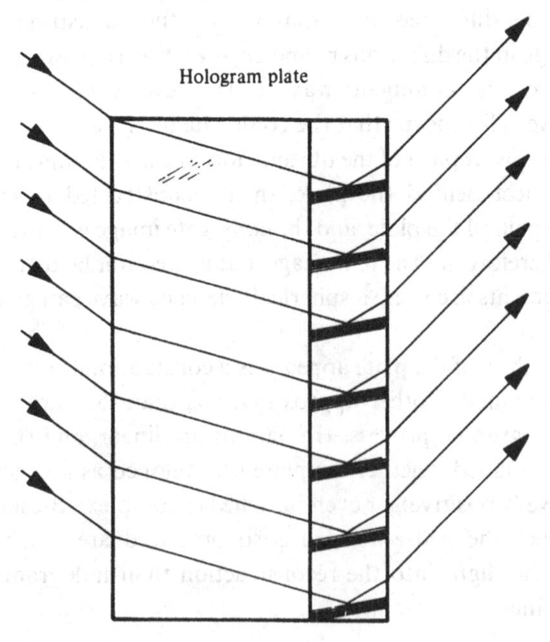

optical component that converts a wave of one shape into a wave of another shape and whose action is reversible.

A hologram can also be reconstructed with the waves reversed. If it is illuminated from the back by a wave that is the conjugate of the reference wave, that is, convergent if the reference wave was divergent, it reconstructs the conjugate of the object wave, as shown in fig. 32. This process is equivalent to phase conjugation, described in §3.1.1, but with the time delay needed to develop the hologram.

Conjugate images can be used as objects to make a second hologram, which, when reconstructed again in reverse, produces a true orthoscopic image. Since the object for the second hologram is an aerial image, the final reconstruction can be arranged to be behind, at, or in front of the hologram.

5.3.4. Classification

Holograms are described as either *Fresnel* or *Fraunhofer* according to whether they are recorded in the near or far field of the object. A third class of *image holograms* has been added later; its reconstruction is at the hologram.

A further name used is *Fourier hologram*, for those in which the basic fringe pattern, on which irregularities are superimposed, consists of straight lines, a single spatial Fourier component. These are obtained when the object and the source point for the reference wave are at the same distance from the hologram.

5.4. Reconstruction

While a laser is needed to make a hologram in nearly all practical circumstances, it is not necessary for its reconstruction. Light from a thermal source can be used, provided it has the same form as the reference wave, and the source is small and has a small spectral bandwidth: a super-pressure mercury lamp, for example, with a filter to isolate the green line.

Fig. 32 The conjugate reconstruction: a pseudoscopic image.

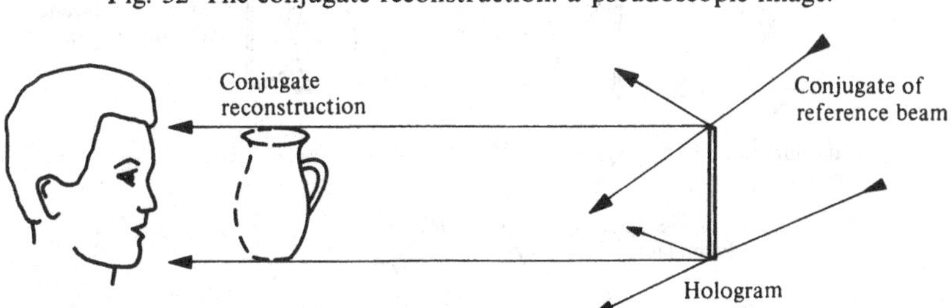

The eye does not detect the small amount of blurring caused by such a source, although a camera with a larger aperture may do so. The closer the reconstruction is to the plane of the hologram, the smaller is this blurring.

In fact, such a reconstruction often looks better and more natural than one produced by a laser, because the blurring reduces the speckle. This effect, a consequence of using a laser as light source to make the hologram, is discussed in §5.7.

5.5. Other holograms

The holograms described so far are *transmission holograms*. These have been modified to produce *rainbow holograms*, and there is also another basic type, the *reflexion hologram.*

5.5.1. Reflexion holograms

These were developed by Denisyuk [1962, 1963, 1980]. They differ from the transmission holograms of Leith and Upatnieks in having the object and reference waves arriving on opposite sides of the plate. Two methods are shown in fig. 33. The reference beam can be separate from the beam illuminating the object, or it may be the same beam, the fraction of it that passes through the plate being used to illuminate the object. The latter method has been used by Denisyuk with dichromated gelatin, which scatters very little light, to make very spectacular holograms. In either method the standing waves are approximately planes parallel to the surface of the plate, and they form a series of layers in the recording medium. After processing these reflect the reconstructing wave to produce again the object wave. If reconstructed by white light, the reflexions from these layers

Fig. 33 Making reflexion holograms: (*a*) with a separate beam to the object; (*b*) with the same beam serving as illuminating and reference beams.

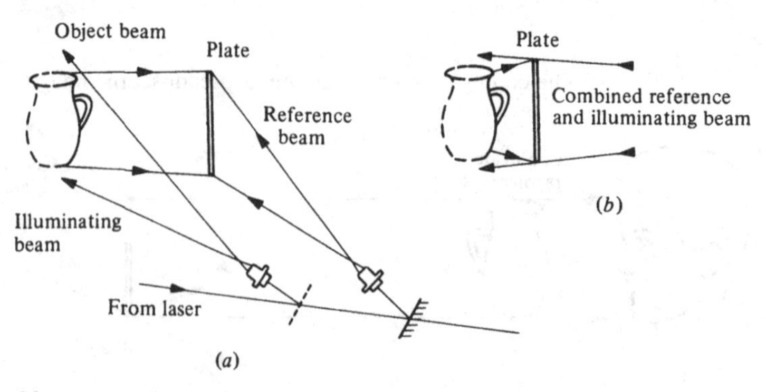

reinforce only for the colour for which the spacing of the layers is a half wavelength. The hologram acts as a reflecting interference filter to produce a reconstruction in one colour. Since processing may cause the emulsion to shrink, this may not be the original colour, but it is possible to re-swell the emulsion to return to this [Hariharan, 1982*b*].

As suggested by Leith & Upatnieks [1964], three colours can be recorded on one emulsion, or better, on superimposed plates to produce a reconstruction in natural colour. Methods of making coloured holograms are reviewed by Hariharan [1982*b*].

5.5.2. Rainbow holograms

Benton [1977] has introduced a modified transmission hologram that also reconstructs in white light. It is essentially a hologram of an object that is sending all the light it reflects to a horizontal slit in front of it, which acts as an exit pupil. When reconstructed, the image can be seen only when the viewer's eye is placed at the reconstruction of this slit. Sideways eye movements are possible, but not vertical, so vertical parallax has been sacrificed, but a brighter reconstruction is obtained. When reconstructed with white light, the slit is spread out vertically into a spectrum and observers see the reconstruction in the colour that corresponds to the height at which they place their eyes.

Since a real object does not occur associated with an exit pupil, rainbow holograms must be made from an image. This can be produced by a conventional optical system, as shown in fig. 34, or by another hologram. In the latter case the first hologram is reconstructed in reverse, through a slit, to produce a real but pseudoscopic image, the conjugate image, that is the object for the second hologram. This hologram is again reconstructed in reverse to give an orthoscopic image with a slit image out in front of it. To given an undistorted image of a three-dimensional object the magnification of the optical system must be ± 1 and should not vary noticeably over the depth of the object. A hologram achieves this automatically, but it is difficult to achieve with the single mirror shown in fig. 34, unless the mirror is very large compared with the object. But it applies when the optical system used is a telescope, since this has a magnification that does not change with object distance.

Three rainbow holograms can be made in each of three primary colours and cemented together with a transparent cement to give a hologram that gives a reconstruction in natural colour when viewed from the correct height [Hariharan, Hegedus & Steel, 1979; Hariharan, 1982*b*].

5.5.3. Computed holograms

The interference pattern used to make a hologram need not be produced optically. It is possible to compute and plot it, then reduce it photographically to the correct scale [Lohmann & Paris, 1967; Lee, 1978]. It is reconstructed in the usual way. By this means it is possible to produce a hologram of a shape that is being manufactured for the first time, and to compare the reconstruction with the object as it is being made. Since a plotter cannot reproduce anything like the amount of information that a fine-grain photographic plate can store, computed holograms are limited to objects of a simple form.

5.6. Holographic optical components

Since the general description of a hologram is that it is a component that changes a wave of one shape into a wave of another shape, a hologram can perform the function of other optical components. A hologram can be made that changes the curvature of a wave in the same way as does a lens; it differs, however, from a lens in that it also deviates the light. A holographic lens has a large lateral chromatic aberration but this, and other aberrations, can be corrected [Weingärtner & Rosenbruch, 1982].

A hologram can also be made to act as a beam splitter and divide a wave into two or more waves of a prescribed form. Thus an interferometer can be built entirely of specially made holograms [Matsuda, 1980].

Another holographic optical component is the *Vander Lugt filter*

Fig. 34 Making a rainbow hologram from a real image formed through a slit.

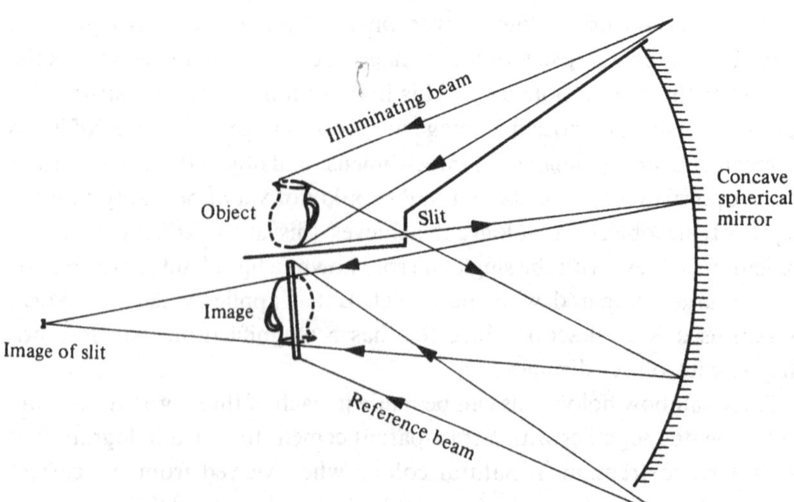

[Goodman, 1968]. This is used in the coherent optical system of §3.5.1. A filter for coherent waves should be capable of changing both the amplitude and the phase. While it can be made as an absorbing filter combined with one with thickness or index variations that change the phase, their registration is difficult. But both amplitude and phase changes can be produced simply by a hologram.

5.6.1. Holographic gratings

The first optical component made as a hologram was the *holographic grating*, produced by the interference of two plane waves. It is used in spectroscopy in the same way as a ruled grating. It predates lasers, having been made by Burch [1960] as an *interference grating.*

Holographic gratings are now produced routinely [Hutley, 1976; Schmahl & Rudolph, 1976], being recorded on photoresist and replicated in the same way as ruled gratings. They have less scattered light than ruled gratings and no periodic errors. They can be blazed to a certain extent by the choice of the angle at which the standing-wave planes penetrate the photoresist.

5.6.2. Holograms in interferometry

As well as being a spectacular application of interference in their own right, holograms have had a profound effect on the techniques of interferometry. As well as their use as optical components to make up an interferometer, they can store details of an object for later measurement. When the measurement is by interferometry, this is then the field of *hologram interferometry*, described in Chapter 12.

5.7. Speckle

When reconstructions from a hologram were first seen, they did not appear true to nature. The surface of a rough object is covered with a high-contrast pattern of light and dark spots, known as *speckle.* A photograph of this is shown in fig. 35. It is not a defect, since it appears on the object itself and has been faithfully recorded by the hologram. It appears unnatural only because we are not used to viewing objects in coherent illumination.

Speckle occurs when a rough surface is illuminated by a laser or any other very small source. It is not necessary for the light to be monochromatic; white light produces coloured speckle. Speckle is caused by the large and rapid variations of phase produced by the rough surface on transmission or reflexion. Waves leave the surface with uniform amplitudes but these rapidly become non-uniform as the waves propagate. An observer focuses on these amplitude variations, rather than on the low-contrast surface itself.

The turbulence in the atmosphere can also produce a speckle pattern when a short exposure is made on a star; long exposures give a large blurred image.

Speckle is not just a defect that prevents holograms giving an aesthetically pleasing reproduction or astronomical telescopes giving sharp images. Speckle patterns contain information on the position and surface structure of a rough object or on a star and the atmosphere through which it is viewed. From these patterns information can be derived, either on displacements of surfaces or on the form of stars, by the technique of *speckle interferometry*; the same name is used for the two different fields.

5.7.1. Properties of speckle patterns

Although random, speckle has well defined statistical properties. The mean size of the granulations is the same as that of the central part of the Airy disk that would be produced by the same aperture. For a circular aperture it has a mean diameter given by (3.34) of

$$\langle d \rangle = 1.22\lambda/\sin\alpha' \tag{5.4}$$

This equation applies to the two cases, where an illuminated rough disk of given size produces a speckle pattern on a screen some distance away, and where an image is formed of the rough surface by an optical system such

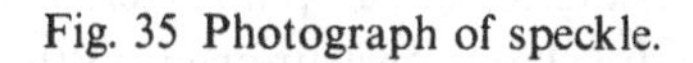

Fig. 35 Photograph of speckle.

Fig. 36 Probability distribution $p(I)$ of the speckle intensity: (1) a single speckle pattern; (2) a speckle pattern with a coherent uniform wave added of the same average intensity; (3) the incoherent sum of two similar speckle patterns.

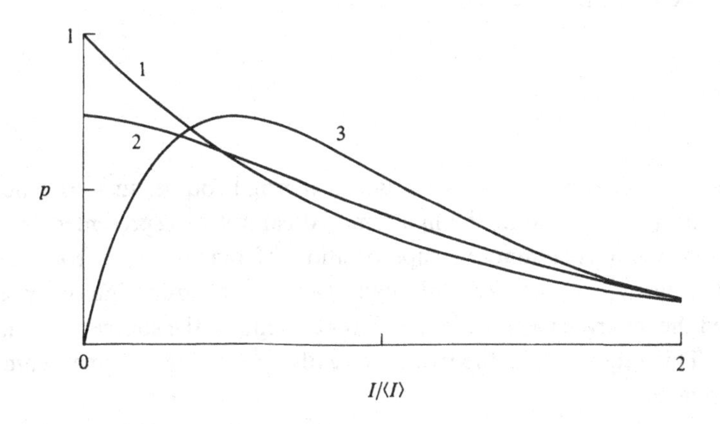

as a camera or the eye. These are sometimes distinguished as objective and subjective speckle. For the first, α' is the angular radius that the illuminated surface subtends at the speckle pattern. For the second, α' is the angular semi-aperture of the imaging system. The size of the speckles at an image can be controlled by an iris diaphragm at the imaging lens.

The probability distribution of intensity across a speckle pattern is given by

$$p(I)=[I/\langle I\rangle]\exp[I/\langle I\rangle], \tag{5.5}$$

where $\langle I\rangle$ is the mean intensity. This distribution is shown in curve 1 of fig. 36. The most probable intensity is zero and a speckle pattern has a large number of dark regions, which give it a high contrast.

If the complex amplitudes of two waves from similar rough surfaces illuminated by the same source are added together, the resultant speckle has the same distribution. But if a uniform wave of intensity $\langle I\rangle$ is added to a speckle pattern produced by the same source, the probability function is that of curve 2. The contrast has been slightly reduced but is still high.

Curve 3 shows the probability distribution when the intensities rather than the amplitudes of two speckle patterns are added. This curve applies to two speckle patterns with the same average size but orthogonal polarizations. There are now no completely dark regions and the contrast is lower.

6

Coherence theory

When the source of radiation is a classical thermal source, such as a hot filament or a gas discharge, the light emitted cannot be represented by a simple wave but is a random superposition of many waves from the different emitting atoms. Yet this light can still produce interference, provided the separate patterns from different points on the source are all in register. This requires that the two apparently random interfering beams are correlated.

6.1. General concepts

Coherence theory is a statistical description of radiation, expressed in terms of correlation functions. As interference is the oldest and simplest example of correlations between light beams, coherence theory has been developed first as a description of interference. Only recently have other correlation effects become of interest, and one of these will be discussed later under higher-order correlations.

The outline of coherence theory given in this chapter is intended to meet the needs of a study of interferometry. Fuller treatments are given by Beran & Parrent [1964], Mandel & Wolf [1965], and Peřina [1971], and an extension to radio astronomy is given by Cole [1977, 1979].

Although coherence is a measure of the correlation between two radiation fields, the term has been extended to apply to a source, which is said to be coherent if all beams from it are highly correlated.

6.1.1. Time coherence

For the Michelson interferometer described in § 1.5, the interference appears as a cosinusoidal variation of intensity when expressed as a function of the path difference p or the delay $\tau = p/c$ for the two arms. For radiation of a single frequency ν, this variation has the form $1 + \cos 2\pi\nu\tau$. When the radiation has a range of frequencies, the total intensity is proportional to the integral of this over all frequencies. Thus if the source is a spectral line of profile $g(\nu - \nu_0)$, with a bandwidth $\Delta\nu$ and mean frequency ν_0, the intensity can be expressed in terms of the cosine transform G_c,

$$I(\tau)=G_c(0)+G_c(\tau)\cos 2\pi\nu_0\tau. \tag{6.1}$$

As for a single frequency the modulation is still cosinusoidal, but the amplitude now decreases with increasing τ. The delay for which fringes may be observed is limited: it is the width of the function $G_c(\tau)$ and this is related to the bandwidth of the line by the uncertainty relation (2.34),

$$\Delta\tau\,\Delta\nu \gtrsim 1. \tag{6.2}$$

This time $\Delta\tau$ is called the *coherence time* of the radiation. A *coherence length* can also be defined as

$$\Delta p = c\,\Delta\tau,$$

and this can be related to bandwidth of the radiation, now expressed as a wavenumber, by

$$\Delta p\,\Delta\sigma \gtrsim 1. \tag{6.3}$$

These results are said to show the *temporal coherence* of the radiation, which increases as the bandwidth is decreased.

6.1.2. Spatial coherence

The Michelson interferometer also shows the effects of *spatial* coherence, for the contrast of the fringes depends on the size of the source. This effect is illustrated better by Young's interference experiment, shown again in fig. 37. A screen $\mathscr{P}$ is illuminated by a source that subtends a solid angle $\Delta\Omega$ at it. Two pinholes P_1 and P_2 in $\mathscr{P}$ let light through to a second screen $\mathscr{O}$.

Fig. 37 Young's experiment with an extended source.

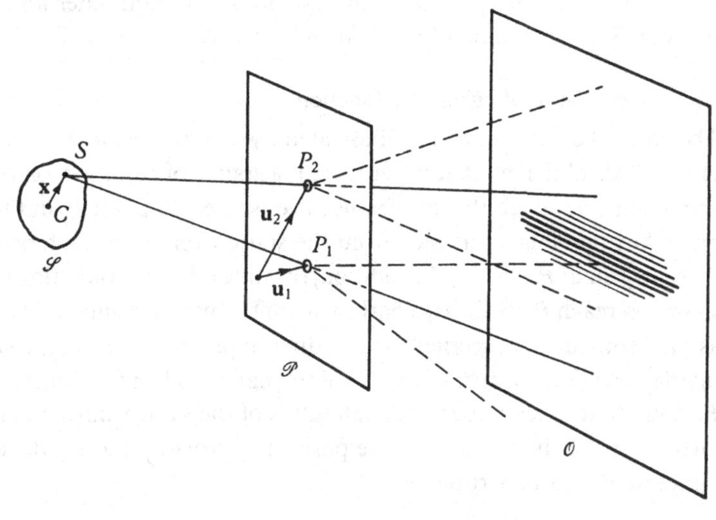

Provided these two holes are sufficiently close together, interference fringes are observed near the centre of $\mathcal{O}$. If Δu is the separation of the pinholes and λ the mean wavelength of the light, the fringes are observed provided Δu is limited by the relation

$$(\Delta u)^2\,\Delta\Omega \sim \lambda^2. \tag{6.4}$$

Then $(\Delta u)^2$ is called the *area of coherence* of the source $\mathcal{S}$.

When the spatial and temporal effects are combined, the product of this area and the coherence length is the *volume of coherence*. The number of photons in the same state of polarization found in this volume plays an important role in the description of radiation. It is called the degeneracy δ and, for a source of spectral radiance $I(\nu)$, expressed in photons per unit area, per unit solid angle, per unit frequency interval, per unit time, it is given by

$$\delta \sim \tfrac{1}{2}(c^2/\nu^2)I(\nu), \tag{6.5}$$

where the factor $\frac{1}{2}$ has been added to describe natural (unpolarized) light, which can be regarded as a mixture of equal amounts of two orthogonal polarizations. The narrowest emission lines have a bandwidth $\Delta\nu \approx 10^8$ Hz and so their coherence time $\Delta\tau$ is 10^{-8} s and their coherence length about 3 m. This radiation has a degeneracy $\delta \approx 10^{-3}$. But a laser running in a single TEM_{00} longitudinal mode can have a power of several watts and a bandwidth of 100 kHz. For it, $\Delta p \approx 3$ km and $\delta \approx 10^{14}$.

The revolutionary effect that lasers have had on interferometry is due, not so much to their high spatial and temporal coherence, but to the large power that they can provide with this coherence. In theory at least, the light from a thermal source can be made as coherent, spatially and temporally, as that from a laser, by passing it through a narrow-band filter and a small pinhole. But the amount of power then left would be too small to be useful.

6.2. Mutual coherence function

The basis of coherence can be illustrated by any two-beam interferometer. On one side of the interferometer there is a source of radiation, on the other some plane at which the interference is observed. A point P' in this latter plane has two images as seen from the source, one from each beam, and these appear at P_1 and P_2. In addition, the time taken for radiation from the source to reach P' by the two paths may differ by an amount τ. Then, as far as the source is concerned, the radiation arriving at P' through the interferometer is the sum of the amounts that would, in free space, arrive at P_1 and P_2 at times τ apart through filters of the same transmission as the interferometer. If P_1 and P_2 have position vectors $\mathbf{u}_1$ and $\mathbf{u}_2$, the analytic signal at P' can be written as

$$V(P',t)=k_1V(\mathbf{u}_1,t)+k_2V(\mathbf{u}_2,t+\tau), \tag{6.6}$$

where the factors k_1 and k_2 are the amplitude transmission factors for the two beams of the interferometer.

The quantity measured in optics is the intensity, the time average of the square of the real signal. To a factor of 2,

$$I(\mathbf{u})=\langle V^*(\mathbf{u},t)V(\mathbf{u},t)\rangle, \tag{6.7}$$

so that

$$I(P')=|k_1|^2I(\mathbf{u}_1)+|k_2|^2I(\mathbf{u}_2)+2\mathscr{R}[k_1^*k_2\Gamma(\mathbf{u}_1,\mathbf{u}_2,\tau)], \tag{6.8}$$

where $\mathscr{R}$ denotes the real part, and Γ is defined by

$$\Gamma(\mathbf{u}_1,\mathbf{u}_2,\tau)=\langle V^*(\mathbf{u}_1,t)V(\mathbf{u}_2,t+\tau)\rangle. \tag{6.9}$$

In this definition, P_1 is regarded as the reference point to which P_2 is referred.

The function $\Gamma(\mathbf{u}_1,\mathbf{u}_2,\tau)$ is the *mutual coherence function.* It is the correlation function for the essentially random signals $V(\mathbf{u},t)$ at two points $\mathbf{u}_1$ and $\mathbf{u}_2$ at times τ apart and is written for short as $\Gamma_{12}(\tau)$ and its real part as $G_{12}(\tau)$. It is an observable quantity of optics, being measured by simple interference experiments such as Young's experiment of fig. 37, with pinholes placed at $\mathbf{u}_1$ and $\mathbf{u}_2$ and the interference observed at a position where the time difference is τ. At radio frequencies, where the real signal $V^{(r)}(t)$ can be detected directly, the random nature of the radiation still requires the use of statistical methods and the same correlation function is employed.

The mutual coherence function has the dimensions of intensity. From it can be defined the dimensionless *degree of coherence*

$$\gamma(\mathbf{u}_1,\mathbf{u}_2,\tau)=\{I(\mathbf{u}_1)I(\mathbf{u}_2)\}^{-1/2}\Gamma(\mathbf{u}_1,\mathbf{u}_2,\tau). \tag{6.10}$$

The transmission factors in (6.8) can be eliminated by using the intensities $I_1(P')$ and $I_2(P')$ that are produced at P' by each beam separately; $I_1(P')=|k_1|^2I(\mathbf{u}_1)$, etc. Then

$$I(P')=I_1(P')+I_2(P')+2[I_1(P')I_2(P')]^{1/2}\mathscr{R}[\gamma(\mathbf{u}_1,\mathbf{u}_2,\tau)]. \tag{6.11}$$

The expression has the same form as (4.2), a uniform field on which an interferogram is superimposed. The expression (6.11) represents an *addition interferometer* in which the signals are added in phase. With a phase difference of π between the signals, the interferogram is subtracted from the uniform background. As shown by Ryle [1952], and Mills & Little [1953], this phase difference can be switched rapidly in and out of a radio interferometer to give, as an alternating signal, the interferogram separated from the background. This is now a *multiplication interferometer*.

The degree of coherence satisfies the relation

$$0 \leqslant m \leqslant 1,$$

where

$$m = |\gamma|. \tag{6.12}$$

Radiation is *fully coherent* if $m=1$, *incoherent* if $m=0$, and *partially coherent* in other cases. This modulus is related to the visibility of the interference fringes, being equal to it when the interfering beams are of equal intensity. In other cases the visibility is given by

$$\mathscr{V}(P') = \frac{2I_1^{1/2}(P')I_2^{1/2}(P')}{I_1(P')+I_2(P')} m(\mathbf{u}_1, \mathbf{u}_2, \tau), \tag{6.13}$$

and m is the optimum visibility, obtained when $I_2 = I_1$.

6.3. Temporal coherence and quasi-monochromatic radiation

When P_1 and P_2 coincide, the mutual coherence function reduces to the autocorrelation function $\Gamma(\mathbf{u}_1, \mathbf{u}_1, \tau)$ or $\Gamma_{11}(\tau)$, which contains only temporal effects. Its real part represents an interferogram with the fringes spread out, as in (4.9). As it is an autocorrelation function, the Wiener–Khinchin theorem (2.47) states that its Fourier transform is the power spectrum $g(\mathbf{u}_1, \mathbf{u}_1, \nu) \equiv g_{11}(\nu)$ of the radiation. Since Γ as well as V is an analytic signal and has no components of negative frequency,

$$g(\nu) = \int_{-\infty}^{+\infty} \Gamma_{11}(\tau) \, \mathrm{e}^{2\pi i \nu \tau} \, \mathrm{d}\tau \quad (\nu \geqslant 0) \tag{6.14}$$

and

$$\Gamma_{11}(\tau) = \int_0^{\infty} g_{11}(\nu) \, \mathrm{e}^{-2\pi i \nu \tau} \, \mathrm{d}\nu. \tag{6.15}$$

Comparison with (2.38) and (2.39) shows that $g_{11}(\nu)$ is the transform of $G_{11}(\tau)$, the real part of $\Gamma_{11}(\tau)$ that describes the intensity modulation, that is, the interferogram. The interferogram and the power spectrum are thus Fourier transforms of each other, and bandwidth and coherence time are related by the uncertainty relation (2.10).

The same result, but from a different viewpoint, is given by (6.1). When radiation is considered in terms of the field $V^{(r)}(t)$, it is the power spectrum of this quantity that is measured by the interferometer. But if we confine ourselves to observables at optical frequencies and think of the radiation in radiometric units, it is the direct transform of this that is measured.

In addition to the power spectrum $g_{11}(\nu)$, the *cross-power spectrum* can be defined as the transform of Γ_{12},

$$g_{12}(\nu) \equiv g(\mathbf{u}_1, \mathbf{u}_2, \nu) = \int_{-\infty}^{+\infty} \Gamma(\mathbf{u}_1, \mathbf{u}_2, \tau)\, e^{2\pi i \nu \tau}\, d\tau, \tag{6.16}$$

so that

$$\Gamma_{12}(\tau) = \int_0^{\infty} g_{12}(\nu)\, e^{-2\pi i \nu \tau}\, d\nu. \tag{6.17}$$

For quasi-monochromatic radiation ($\Delta\nu \ll \nu_0$), the cross-power spectrum has an appreciable magnitude only for frequencies close to the mean frequency ν_0. With a change of origin to ν_0, we can write

$$\Gamma_{12}(\tau) = e^{-2\pi i \nu_0 \tau} \int_{-\nu_0}^{\infty} \tilde{J}_{12}(v)\, e^{2\pi i v \tau}\, dv, \tag{6.18}$$

where $v = \nu - \nu_0$ and $\tilde{J}_{12}(v) = g_{12}(\nu)$. Then $\tilde{J}_{12}$ will have appreciable values only around $v = 0$, and its transform $J_{12}(\tau)$, the *mutual intensity*, will have only low-frequency Fourier components, and so is a slowly varying function of τ, changing little for

$$|\tau| \ll 1/\Delta\nu. \tag{6.19}$$

When the interferometer contains dispersive media, the time difference is no longer constant, but is a function of frequency, $\tau(\nu)$, and equation (6.17) is not a simple Fourier transform. However, $\nu\tau$ can be expanded as

$$\nu\tau(\nu) \simeq \nu_0 \tau(\nu_0) + (\nu - \nu_0)\tau_g, \tag{6.20}$$

where

$$\tau_g = \left[\frac{d}{d\nu}(\nu\tau)\right]_{\nu=\nu_0} = \left[\tau + \nu \frac{d\tau}{d\nu}\right]_{\nu=\nu_0} \tag{6.21}$$

Hence τ_g is a *group delay*, τ being the *phase delay*. Equation (6.18) can be rewritten for dispersive media with τ_g in the exponent inside the integral, and J_{12} becomes a function of τ_g. The centre of the interferogram, where the visibility is maximum, is thus given by $\tau_g = 0$ rather than $\tau = 0$, and is the position where the fringes of equal chromatic order, described in §1.5.2, are parallel to the direction of dispersion.

The mutual coherence function for quasi-monochromatic radiation can then be expressed as

$$\Gamma_{12}(\tau)=J_{12}(\tau_g)e^{-2\pi i\nu_0\tau}, \tag{6.22}$$

and the degree of coherence as

$$\gamma_{12}(\tau)=\mu_{12}(\tau_g)\,e^{-2\pi i\nu_0\tau}, \tag{6.23}$$

where

$$\mu_{12}(\tau)=(I_1I_2)^{-1/2}J_{12}(\tau_g). \tag{6.24}$$

Both J_{12} and μ_{12} are functions that vary only slowly with τ in comparison with the rapid fluctuations of $\exp(-2\pi i\nu_0\tau)$. The modulus of μ_{12} is the same as that of γ_{12}, namely $m_{12}(\tau_g)$. If its phase is written as $\beta_{12}(\tau_g)$ the coherence can be expressed as

$$\gamma_{12}(\tau)=m_{12}(\tau_g)\exp i[\beta_{12}(\tau_g)-2\pi\nu_0\tau]. \tag{6.25}$$

This expression shows the significance of β_{12} as an effective phase difference between the radiation at P_1 and P_2 over and above the phase due to the time-difference τ, namely $-2\pi\nu_0\tau$.

If the real part of equation (6.25), which describes the interferogram, is compared with that for monochromatic waves, (4.9), it is seen that the change to an extended quasi-monochromatic source has introduced this phase term β_{12} and has also reduced the visibility by the factor m_{12}. But the cosinusoidal variation, $\cos 2\pi\nu_0\tau$, is unaltered.

Equation (6.19), read as a condition on bandwidth, is more specific than the one (3.15) given earlier for quasi-monochromatic radiation. It shows when the approximations valid for quasi-monochromatic light can be applied to an interferometer of delay τ.

6.4. Spatial coherence

The coherence between two points illuminated by a source of incoherent light is given by the *van Cittert–Zernike theorem.* For a more rigorous treatment than that given here, see Beran & Parrent [1964]. The notation is shown in fig. 38.

Two points P_1 and P_2 are illuminated by an extended source $\mathscr{S}$ of quasi-monochromatic radiation of mean frequency ν. Imagine to be divided up into small elements centred at points $\mathbf{x}_1, \mathbf{x}_2, \ldots$. The signals at P_1 and P_2 are then the sum of the components $V_n(\mathbf{u}, t)$ due to each element of the source,

$$V(\mathbf{u}, t)=\sum_n V_n(\mathbf{u}, t). \tag{6.26}$$

The functuations of the radiation from different elements may be assumed to be statistically independent so that

$$\langle V_m^*(\mathbf{u}_1, t_2) V_n(\mathbf{u}_2, t_1) \rangle = 0 \quad \text{for all } m \neq n. \tag{6.27}$$

If each element of source has a strength A_n and R_{n1}, R_{n2} are the distances from $\mathbf{x}_n$ to P_1 and P_2, the analytic signals at these points are given by (3.18) so that

$$V_n^*(\mathbf{u}_1, t) V_n(\mathbf{u}_2, t+\tau) = \frac{A_n^*\left(t - \frac{R_{n1}}{c}\right) A_n\left(t + \tau - \frac{R_{n2}}{c}\right)}{R_{n1} R_{n2}} \cdot \exp\left[-2\pi i\nu\left(\tau - \frac{R_{n2} - R_{n1}}{c}\right)\right]. \tag{6.28}$$

If both $|(R_{n2} - R_{n1})/c|$ and $|\tau|$ are small compared with the coherence time $1/\Delta\nu$, the differences in the arguments of A_n^* and A_n may be neglected and the intensity $\langle A_n^* A_n \rangle$ written as the radiance $L(\mathbf{x})$ multiplied by the area of the element $d\mathbf{x}$. The coherence due to the whole source is then obtained by summing over all elements and applying (6.27),

$$\Gamma(\mathbf{u}_1, \mathbf{u}_2, \tau) = J(\mathbf{u}_1, \mathbf{u}_2)\, e^{-2\pi i\nu\tau}, \tag{6.29}$$

where

$$J(\mathbf{u}_1, \mathbf{u}_2) = \int_{\mathscr{S}} L(\mathbf{x}) \frac{e^{ik(R_2 - R_1)}}{R_1 R_2}\, d\mathbf{x}. \tag{6.30}$$

Fig. 38 Notation for the van Cittert–Zernike theorem.

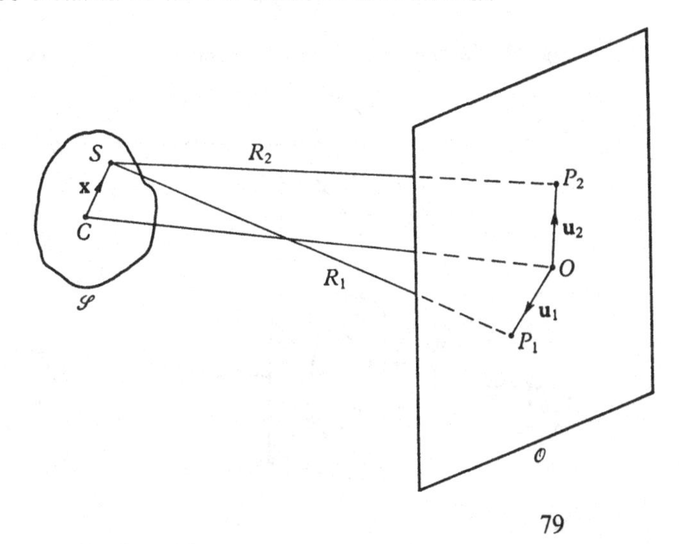

This is the van Cittert–Zernike theorem. If (6.30) is compared with (3.21) it is seen that, apart from constant factors, the calculation of the mutual coherence between P_1 and P_2 is equivalent to the calculation of the distribution of complex amplitude in the neighbourhood of a focus located at P_1. The theorem is illustrated in fig. 39. The source is replaced by a diffracting aperture of the same shape over which there is a spherical wave centred on P_1 with an amplitude distribution across it equal to the intensity distribution across the source. The complex amplitude in the analogue at a point displaced from the focus by the vector distance of P_2 from P_1 is, to a constant factor, the mutual intensity $J(P_1, P_2)$. An important point to note is that, to the approximations used, the coherence depends on the displacement of P_2 from P_1 only and not on the actual positions of P_1 and P_2.

The importance of this theorem in interferometry is that the results of a large number of calculations of amplitude distributions for various diffracting apertures can be taken over directly to give the fringe modulation in interferometers with these source shapes. A Fourier transform relation is also obtained as an approximation for the coherence, when the path difference $R_2 - R_1$ is expanded by (3.24), just as it is an approximation for complex amplitudes. The area of coherence due to a source is related to the area covered by the central maximum of the transform of the source and, by the uncertainty relation, the larger the source, the smaller the area of coherence. The size of the source is expressed as the solid angle it subtends at the two points and, if this is sufficiently large, the coherence function is sufficiently narrow to be treated as a δ-function: the illumination is effectively incoherent.

Fig. 39 The analogue given by the van Cittert–Zernike theorem.

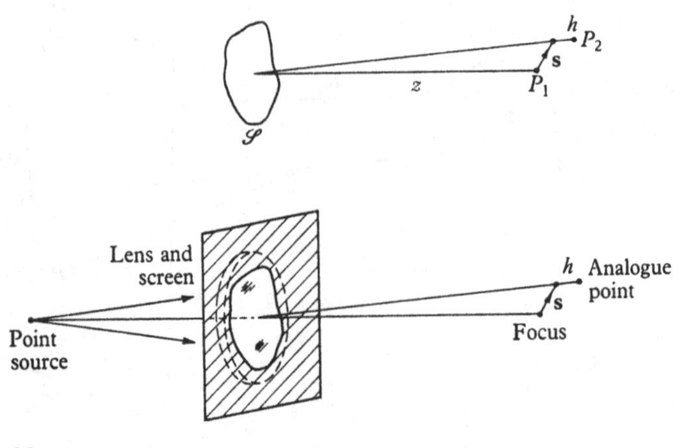

6.4. *Spatial coherence*

6.4.1. Imaging systems

A generalization of the van Cittert–Zernike theorem has been given by Hopkins [1951] for the case where there is not a uniform medium between the source and the two points. The simple propagation factor of (3.18) is replaced by a more general transmission factor $K(\mathbf{x}, P)$ that gives the amplitude at P due to a point source of unit strength at $\mathbf{x}$. Then

$$J(P_1, P_2) = \int L(\mathbf{x}) K^*(\mathbf{x}, P_1) K(\mathbf{x}, P_2)\, \mathrm{d}\mathbf{x}. \tag{6.31}$$

When P at $\mathbf{x}'$ is the image of the source formed by an optical system, the transmission factor K is the amplitude spread function $F(\mathbf{x}' - \mathbf{x})$ for this system. Thus, if the optical components possess aberrations (including defects of manufacture), the coherence is equal to the amplitude given by an analogue system having aberrations equal to the difference between those in the two paths from the source.

From this can be derived an important comparison between interferometers and image-forming optical systems. In both instruments defects will misdirect a certain fraction of the wave amplitude. In an interferometer the intensity in the interferogram will be in error by the same fraction, but for the intensity of a simple image, obtained from the square of the amplitude, the fractional error will be squared and thus reduced greatly. An interferometer will therefore require higher precision in its components and their adjustment than an image-forming instrument. Thus the driving mechanism that moves the mirror of an interferometer used for spectroscopy must be more precise than that used to rule a diffraction grating applied to the same purpose. Again, the light reflected from the surfaces of optical components may add negligible scattered-light intensity to an optical image, but in an interferometer it will interfere with the main beams to produce considerable modification to the interferogram.

6.4.2. Effective source

An interferometer, other than one used to study the spatial distribution of radiation, seldom has the actual emitter of radiation as its source. Usually there is some aperture illuminated by the actual source and this is the *effective source* for the interferometer. Since such effective sources are not incoherent, the van Cittert–Zernike theory does not, strictly, apply.

The two common methods for illuminating the effective source are shown in fig. 40. In one case the aperture defining it is illuminated directly by the actual source while, in the second case, an image of the actual source is formed there by a condenser. As seen earlier, for a source subtending a sufficiently large solid angle, the illumination in the first case is approximately incoherent. In the second case, the coherence at the effective source is given by (6.31) with K replaced by the amplitude spread function,

$$J(\mathbf{x}'_1, \mathbf{x}'_2, \tau) = \int_{\mathscr{S}} L(\mathbf{x}) F^*(\mathbf{x}'_1 - \mathbf{x}) F(\mathbf{x}'_2 - \mathbf{x})\, d\mathbf{x}. \tag{6.32}$$

If the solid angle subtended by the condenser is sufficiently large, this spread function is approximately a δ-function and the illumination is again incoherent. Thus, in both cases, it is the angle at which the light arrives that determines its coherence, whether this angle is subtended directly by the source or indirectly by a condenser. The intensity distributions, however, may differ in the two cases. The proof of this equivalence by Hopkins [1951] was of considerable importance in microscopy, where previously the two methods of illumination, known as Köhler and critical illumination, had been considered to be basically different.

There remains only the question of how much larger the angle subtended by the illumination should be than the aperture of the interferometer, for the approximation of incoherence to be valid for the effective source. While the work of Som [1963] suggests that a ratio of about four is desirable to give a good approximation, this does not seem to be necessary in practice. In microscopes, where the condensers and objective have the same numerical aperture, although the illumination is certainly not incoherent, the image does not depart greatly from that given by incoherent illumination. In interferometry the images observed are usually rather diffuse, being those of sine-wave fringes, and departures from incoherence should have even less importance. Hence, if the effective source is illuminated over an angle not less than the angular aperture of the interferometer, as is usual in practice, the approximation of an incoherent source seems satisfactory.

Fig. 40 Alternative methods of illuminating the effective sources $\mathscr{S}$ by the actual source $\mathscr{L}$.

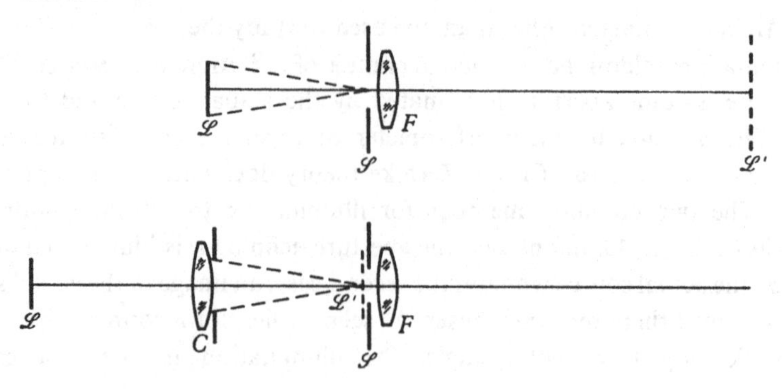

6.5. Combined effects

When the light is not quasi-monochromatic, temporal effects are combined with spatial effects. An argument similar to that used for the derivation of the van Cittert–Zernike theorem is applied to each spectral component and the result integrated over the source, of spectral radiance $I(\mathbf{x}, \nu)$. The mutual coherence function is given by

$$\Gamma(\mathbf{u}_1, \mathbf{u}_2, \tau) = \int_0^\infty e^{-2\pi i \nu \tau} \, d\nu \int_{\mathscr{S}} \frac{I(\mathbf{x}, \nu) \, e^{2\pi i \nu (R_2 - R_1)/c} \, d\mathbf{x}}{R_1 R_2}. \tag{6.33}$$

With the usual approximation (3.24) for path difference, this becomes a three-dimensional Fourier transform. The coherence volume, of base the coherence area and height $c\tau$, is connected by an uncertainty relation with the volume in reciprocal space, defined by the product of the solid angle subtended by the source and its bandwidth.

6.6. Diffuse reflexion

Although the source used in hologram interferometry is a laser, this is another case where a simple wave representation cannot be used. When objects with rough surfaces are studied, the reflected waves have random variations of amplitude and phase produced by the random irregularities of the surfaces. A full wave treatment would give the very fine interference pattern that showed these irregularities. But the essence of hologram interferometry is that the interfering waves come from the same object and what is wanted is the changes to it. There is obviously an upper limit to the changes that can be studied. If they are too large, the two reflected waves no longer match each other.

A theory of hologram interferometry of rough surfaces should, therefore, be independent of the detailed structure of the surface. It has a close resemblance to coherence theory, which is independent of the detailed fluctuations of the radiation field from a thermal source. The theory, given by Walles [1969], is based on average intensities and cross correlations of the reflected waves, but these are no longer time averages but ensemble averages taken over all possible surfaces. The form of the theory follows closely the coherence theory given earlier.

6.6.1. Homologous rays

The analogy between classical interferometers with extended sources and hologram interferometry would be exact if both the object and the laser source received the same displacement between the formation of the two waves. But this is most unlikely. Usually the source remains fixed, so that, as well as a displacement of the object with respect to the plane of interference, there is also a displacement of the source relative to the object. The two waves that leave the object, before and after the displacement, do not necessarily match in detail when viewed in the same direction. They match for the pairs of directions that take account of the relative movement of the source. The rays in these two directions are known as *homologous rays* [Walles, 1970].

Fig. 41 is a representation of a rough object illuminated at an angle α to the normal $\mathbf{n}$ to the mean plane of its surface. We consider a ray leaving it at an angle β. It is seen that the path variation across a length a of the surface is

$$a(\sin\alpha+\sin\beta)$$

β being negative as drawn.

If the direction of illumination is changed to $\alpha+\Delta\alpha$, the same path difference now occurs for the scattered wave at $\beta+\Delta\beta$, where

$$\Delta\beta=(\cos\alpha/\cos\beta)\,\Delta\alpha. \tag{6.34}$$

This equation is, in differential form, the grating equation (3.40).

An alternative derivation considers a Fourier analysis of the spatial distribution of the surface roughness. There is one Fourier component which, acting as a grating, diffracts a wave incident at an angle α in the direction β. When the direction of the incident wave is changed to $\alpha+\Delta\alpha$, the change $\Delta\beta$ to the diffracted direction is given as (6.34) by the grating equation.

When $\Delta\alpha$ is no longer coplanar with α and β, the two-dimensional form (3.40) of the grating equation is required.

Fig. 41 The change of direction $\Delta\beta$ to the wave scattered from a rough surface that does not change the path difference across the wavefront, when the direction of illumination is changed by $\Delta\alpha$.

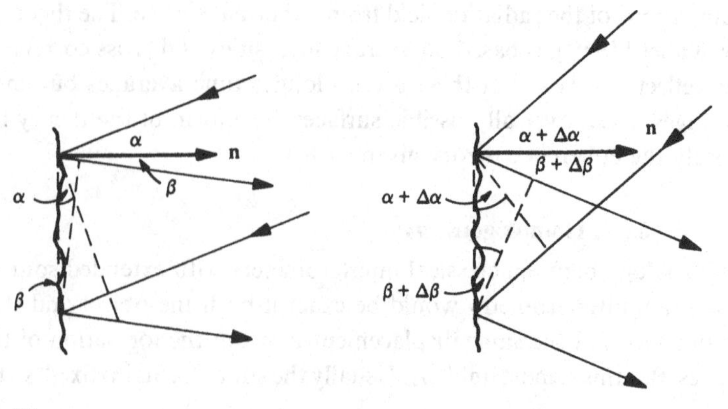

6.7. Polarization

So far in the discussion on coherence, the vector nature of radiation has been ignored and with it the polarization of the light. When two beams of light interfere, the visibility of the fringes, as well as depending on the spatial and spectral properties of the source, depends on the differences in their states of polarization and interferometers will also measure polarization. A unified tensor theory of coherence and polarization is discussed by Mandel & Wolf [1965].

When radiation is quasi-monochromatic, but not necessarily fully polarized, its coherence and polarization properties can be represented by the *coherence matrix*

$$\mathbf{J}_{12}=\begin{pmatrix} J_{xx} & J_{xy} \\ J_{yx} & J_{yy} \end{pmatrix}, \tag{6.35}$$

where the elements are the four correlations between the x- and y-components of the radiation at P_1 and P_2. Hence

$$J_{xx}=\langle E_{x1}E^*_{x2}\rangle, \quad J_{xy}=\langle E_{x1}E^*_{y2}\rangle, \text{ etc.} \tag{6.36}$$

The polarization properties of interferometers can be derived in terms of these matrices [Fymat, 1972]. But most of the problems of polarization effects in interferometers can be treated in terms of the simpler polarization vectors and the Jones calculus of §3.6.

In visual optics, interferometry and polarimetry are usually considered to be separate fields, although closely related. But in radio astronomy, one of the methods of measuring polarization is by a direct application of interferometry.

6.8. Higher-order correlations

The mutual coherence function may be described as a second-order correlation function, since it involves two analytic signals. Glauber [1963] has generalized this concept to define higher-order correlations.

The most important practical example is the correlation of intensity developed by Hanbury Brown & Twiss [1957]. In their experiments, the instantaneous intensity is detected at $\mathbf{u}_1$ and $\mathbf{u}_2$ at times t and $t+\tau$ to give

$$V^*(\mathbf{u}_1,t)V(\mathbf{u}_1,t) \quad \text{and} \quad V^*(\mathbf{u}_2,t+\tau)V(\mathbf{u}_2,t+\tau).$$

These are detected separately, multiplied together, and the result averaged in time to give the correlation

$$\Gamma^{(2,2)}_{12}(\tau)=\langle V^*_1(t)V_1(t)V^*_2(t+\tau)V_2(t+\tau)\rangle. \tag{6.37}$$

If the radiation is from a thermal source and $V(t)$ is a gaussian random process, this can be simplified to give [Mandel, 1963*b*]

$$\Gamma_{12}^{(2,2)}(\tau)=I_1 I_2(1+|\gamma_{12}^{(1,1)}(\tau)|^2), \tag{6.38}$$

where I_1 and I_2 are the average intensities at each detector and $\gamma_{12}^{(1,1)}$ is the second-order degree of coherence defined by (6.10). Intensity-correlation experiments can thus give the modulus of $\gamma_{12}(\tau)$ and this can be related to both the spatial and spectral distributions of the radiation. The solution, however, is not complete since the phase of γ is not given. Experiments of this type are known as *intensity interferometry* or *correlation interferometry*.

In practice, the efficiency is not as high as (6.38) suggests unless the detectors can resolve times less than the coherence time. Detectors of visible intensity, photomultipliers, have a spread in transit time for electrons of the order of 10^{-8} s, long compared with the coherence time of the radiation. The fluctuations of the instantaneous intensity are smoothed considerably and the final correlation is considerably reduced, being given, for unpolarized light, by Mandel [1963*b*] as

$$\rho=\frac{\frac{1}{2}\delta}{1+\frac{1}{2}\delta}|\gamma_{12}(0)|^2, \tag{6.39}$$

where δ is a degeneracy parameter giving the average number of photoelectrons emitted by the photocathode in a time equal to the coherence time, if it were coherently illuminated. For thermal light sources δ is small and long integration times are necessary to make a measurement. But for radio waves, δ is large.

A variant of this experiment is that of Twiss & Little [1959] where, instead of measuring correlations between photocurrents, they recorded coincidences between photon counts by the two detectors. Again Mandel [1963*b*] has given the theory for this case; it closely resembles that for intensity correlations.

Because of its low signal-to-noise ratio, interferometry based on higher-order correlations has only a few applications, the chief of which is the measurement of stellar diameters. Its potentialities have been reviewed by Twiss [1969].

6.9. Phase retrieval

Whether correlation interferometry is used to study spectral or spatial distributions of radiation, it yields no information on the phase of the coherence but gives only its squared modulus. Other methods of interferometry, which in principle can measure phases, sometimes cannot do so in practice, since it is more difficult to measure fringe position, which gives the phase, than visibility, which gives the modulus. Examples are the

Michelson stellar interferometer and Michelson's measurements of the profiles of spectral lines. These give $|\gamma_{12}(0)|$ and $|\gamma_{11}(\tau)|$ respectively. The same problem, known as *phase retrieval*, occurs in other branches of physics. In crystallography, the modulus of an X-ray or an electron diffraction pattern can be measured, but not the phase. More extensive treatments of the phase problem in optics are given by Kohler & Mandel [1973], and Burge *et al.* [1976].

The result desired from the interferometer measurement is a spatial or spectral distribution that is the Fourier transform of γ. Without the phase, only its autocorrelation can be obtained as the transform of $|\gamma|^2$. This difficulty disappears when the distribution is known to be symmetrical, or is assumed to be so. Then the phase of γ is zero. A solution is also obtained when a reference point can be added near the object, such as an unresolved star near the star being imaged, or a very narrow spectral line near the spectral region studied. Then the Fourier transform of $|\gamma|^2$ shows three distinct parts. At the centre is the autocorrelation of the distribution, and on either side are the cross correlations of the distribution with the point [Liu & Lohmann, 1973]. One of these is the distribution itself, the other is the inverse of the distribution.

Related to this method are those of Gamo [1964]. He has shown that, if triple correlations are taken at three points or at three times, phase information can be obtained. In another, less promising method he has shown that, if a background beam is added to each of the two beams to be correlated, phase information is restored if the two background beams are coherent. This method involves the techniques of both standard and correlation interferometry.

Another method uses the analytic properties of coherence. It has commonly been assumed that the determination of the phase of γ requires an independent measurement from that made to determine the modulus. But work, reviewed by Mandel & Wolf [1965], has shown that this is not necessarily so and, at least in some special circumstances, the phase can be derived from the modulus. The degree of coherence $\gamma(\tau)$ is an analytic signal, analytic and regular in the lower half-plane of complex τ. If it has no zeros there, the function

$$\ln \gamma(\tau) = \ln |\gamma| + \mathrm{i} \arg \gamma \tag{6.40}$$

is also analytic so that the phase of γ is the Hilbert transform of the logarithm of its modulus. If, however, there are zeros of γ, their location must be known to derive the phase [Burge *et al.*, 1976; Bates, 1982]. It is known, however, that the image intensity must be real and not negative.

With such additional information, computing methods have been developed that are giving promising results, for example, those of Knox [1976], Fienup [1978], and Brames & Dainty [1981].

7

Theory of two-beam interferometers

The simple theory of two-beam interferometers, given in Chapter 4, can now be extended to include sources of finite size. This general theory provides not only a description of these instruments but also a tabulation of the fields of application that is valid for multiple-beam interferometers as well.

7.1. Reference planes

Just as an image-forming system has two independent sets of reference planes, the object plane and the pupil plane, so has an interferometer. Different names are used however.

An example of a two-beam interferometer is shown in fig. 42. The plane of observation, where the interference fringes are observed, has images back through the two arms of the interferometer. The example illustrated is the Linnik interference microscope [1933*a*] for examining reflecting surfaces, and the plane of observation $\mathcal{O}'$ is conjugate with the surface studied.

Fig. 42 A two-beam interferometer with a source at $\mathcal{S}$ and the interference observed at $\mathcal{O}'$.

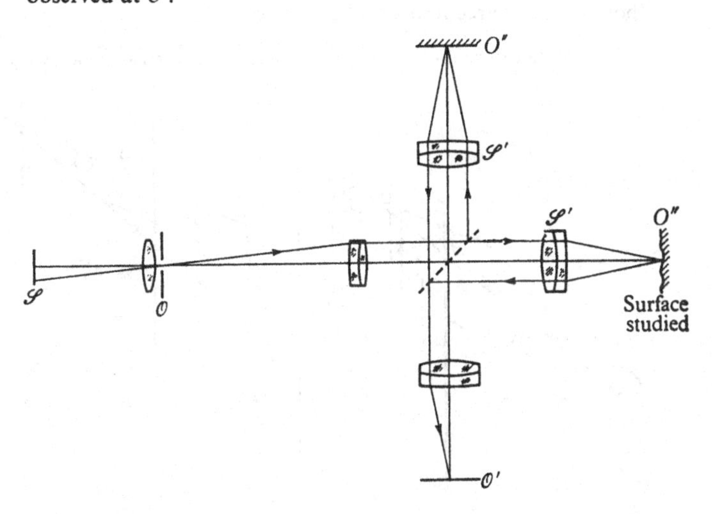

The amount of radiation reaching each point in the plane of observation is determined by some aperture away from this plane. It may be the boundary of the actual light source, an aperture illuminated by the source, or the pupil of the system that observes the interference, for example, the eye or a camera. Whatever form it takes, it has an image in the space in front of the interferometer, which I shall call the *source*, and which lies in the *source plane*. The distinction between the two sets of planes, source and observation, while analogous to that between field and pupil planes in an imaging system, is more arbitrary. If the surface being studied by the interferometer in fig. 42 is replaced by a plane mirror, the instrument can be used to compare the aberrations of the two lenses. The planes are now interchanged and fringes are observed at what was previously an image of the source.

The important fact remains, however, that in any interferometer there are two separate and independent sets of planes containing limiting apertures that together define the étendue of the interferometer, the product of the area of one aperture and the solid angle subtended at it by the other.

7.2. Geometry of the general interferometer

A general representation of a two-beam interferometer is shown in fig. 43. Again the interference is observed in a plane of observation $\mathcal{O}'$ and there is a source in a plane $\mathcal{S}$. Each of these planes has two images when viewed from the other, one image from each beam of the interferometer. Fig. 43 shows

Fig. 43 A general representation of a two-beam interferometer with the two pairs of images, those of the plane of observation in source space, and those of the source in observation space.

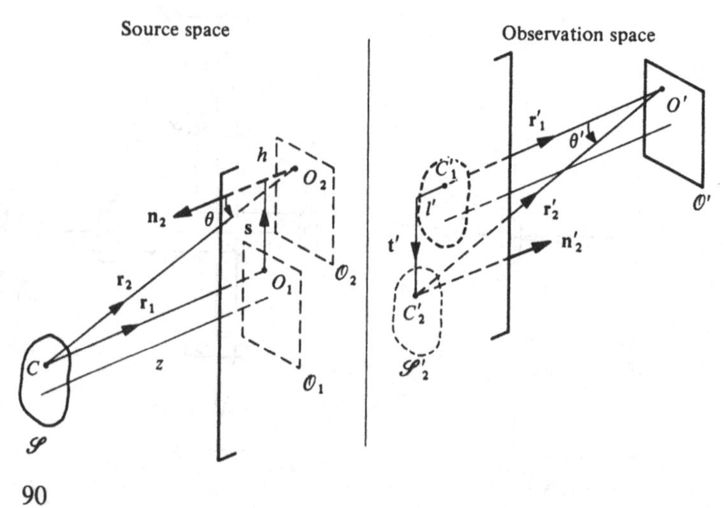

both sets of images, those of the plane of observation in source space and those of the source in observation space.

The four parameters, shear, shift, tilt, and lead, introduced in Chapter 4, are defined in terms of these images. In source space the shear $\mathbf{s}$ is the lateral separation of the two images O_1 and O_2 of a point O' on the plane of observation, and the shift h is their longitudinal separation. In observation space the tilt $\mathbf{t}'$, is the lateral separation of the two images C_1' and C_2' of a point C on the source, and the lead l' is their longitudinal separation. The angular shear and tilt are $\boldsymbol{\sigma} = \mathbf{s}/z$ and $\theta' = \mathbf{t}'/z'$. They can each be identified in the other space. The two images of the plane of observation in source space, O_1 and O_2, are inclined to each other at the angle of tilt $\boldsymbol{\theta}$, the image of $\boldsymbol{\theta}'$. As drawn, the ray $\mathbf{r}_1$ is normal to both the source and plane of observation, so $\boldsymbol{\theta}$ is the angle between $\mathbf{r}_2$ and $\mathbf{n}_2$, the normal to $\mathcal{O}_2$. Similarly, in observation space, $\boldsymbol{\sigma}'$ is the angle between $\mathbf{r}_2'$ and $\mathbf{n}_2'$, the normal to $\mathcal{S}_2'$.

The fifth parameter that completes the description of the interferometer is the delay. It is derived from the *optical path difference p*, the difference of the two sums, for each arm of the interferometer, of the products of the path length in each medium and the refractive index:

$$p = \sum_2 nd - \sum_1 nd \equiv \sum_{2-1} nd. \tag{7.1}$$

From this are derived the two related quantities, the order of interference

$$N = p/\lambda_v, \tag{7.2}$$

and the delay

$$\tau = p/c. \tag{7.3}$$

In interferometers made up of plane mirrors and beam splitters only, as studied by Lambert & Richards [1978], the three longitudinal parameters, h, l', and p, are all equal. But, when dispersive media are present, p will differ from h and l'.

All three will differ when there are focusing lenses or mirrors inside the interferometer. In addition, the shear and tilt may not be constant but may vary across their respective planes, $\mathbf{s}$ as a function of $\mathbf{u}$, $\mathbf{t}'$ as a function of $\mathbf{x}'$, to give the varying shears, radial, rotational, and folding, introduced in Chapter 4, and corresponding varying tilts. The last two shears, rotational and folding, can occur also in an interferometer made up of plane surfaces only.

These shears are illustrated again in fig. 44, which shows the two images of the plane of observation as seen from the source. The total shear is given by

$$\mathbf{s} = \mathbf{s}_0 + \mathbf{u}_2 - \mathbf{u}_1, \tag{7.4}$$

the sum of any uniform lateral shear $\mathbf{s}_0$ and the varying shears, which are given by the difference between $\mathbf{u}_2$ and $\mathbf{u}_1$, the two images of the position vector $\mathbf{u}'$. In fig. 44 the rotational shear is shown by the angle θ between the two images of a reference axis $\mathbf{i}$, the folding is about this axis, and the ratio of the lengths of the two vectors is the radial shear.

If the lengths are related by

$$|\mathbf{u}_2| = |\mathbf{u}_1|/r,$$

r is the *shear ratio*. In terms of §4.4, the wave through path 1 of the interferometer is magnified r times relative to that through path 2, so that, in the view back at the source, the inverse of this ratio applies.

When the two beams through the interferometer are limited by the same apertures and so have the same étendue, any differences between one set of images will be accompanied by the inverse differences at the other set. Thus a radial shear of ratio r will be associated with a radial tilt of ratio $1/r$.

The representation of fig. 43 cannot give the optical path difference, p. This must be found by the summations (7.1) along the actual paths. But it does give the variations of this path difference across the two planes. The departures from the path difference p_0 for the central points C and O' are

$$\Delta p_x = \Delta R_x(\mathbf{u}_2, z+h) - \Delta R_x(\mathbf{u}_1, z),$$

$$\approx -\mathbf{s}\cdot\mathbf{x}/z - h|\mathbf{x}|^2/2z^2 + \ldots, \tag{7.5}$$

Fig. 44 Shears that vary with position: rotational, folding, and radial shears.

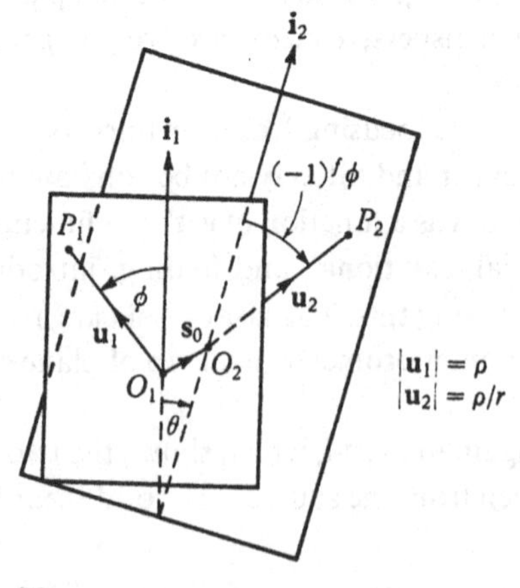

and

$$\Delta p_u \approx -\mathbf{t}'\cdot\mathbf{u}'/z' - l'|\mathbf{u}'|^2/2z'^2 + \ldots \tag{7.6}$$

A two-beam interferometer is thus described by five parameters: the shear, shift, tilt, lead, and delay. These may be constants or they may vary across the source or the plane of observation. For any particular interferometer, they are found by forming the geometrical pictures that correspond to fig. 42. Images are found through the two beams by the methods given in Chapter 3, and some examples will be given in Chapter 8.

When an interferometer is used with a source that is effectively a point, a laser or a source of microwaves, the shear and shift have no significance, when measured from the source. (Measured from the specimen, they determine how the interferogram represents the phase variations across the object.) When the interferogram is studied at only one point in the plane of observation, the tilt and lead can be ignored.

7.3. General theory

When the five parameters describing an interferometer are known, the intensity distribution in the plane of observation follows immediately from coherence theory. Equation (6.11) gives immediately

$$I(\mathbf{u}') = I_1 + I_2 + 2(I_1 I_2)^{1/2}\mathscr{R}\{\gamma(\mathbf{u}_1, \mathbf{u}_2, \tau)\}\cos c, \tag{7.7}$$

where the factor $\cos c$, from (3.47), has been added to include the possibility that the two beams, initially in the same state of polarization, may leave the interferometer in different states. Equation (7.7) holds for any two-beam interferometer with radiation of any bandwidth. Only the effects of diffraction have been ignored in the derivation of the geometrically equivalent system. In a practical instrument, there would be apertures between $\mathscr{S}$ and $\mathscr{O}'$ and diffraction at these would alter the coherence at $\mathscr{O}'$, only slightly for visual radiation but considerably for microwave interferometers.

The discussion in §6.4.2 has shown that diffraction in the condenser system may also be neglected if radiation arrives at the effective source over a reasonably large angle. Then the effective source can be taken to be incoherent, and the degree of coherence γ_{12} depends only on the separation of P_1 and P_2 and not on their individual positions. In fact, as (6.30) shows, the spatial variations of γ_{12} for a particular point P' at $\mathbf{u}'$ depend only on the intensity distribution across the effective source and the path variation Δp_x, itself a function of shear and shift:

$$\gamma(\mathbf{u}_1, \mathbf{u}_2, \tau) \equiv \gamma[\Delta p_x, \tau] \equiv \gamma(\mathbf{x}, \mathbf{s}, h, \tau). \tag{7.8}$$

This applies strictly only for quasi-monochromatic radiation and, for this, further expansion of (7.7) is possible.

7.4. Quasi-monochromatic radiation

When an interferometer is used to measure the phase variations across a sample placed inside it, quasi-monochromatic radiation is used. To include such measurements in a general theory, the disturbances of the optical path due to the sample must be added to the variations, Δp_x and Δp_u, which are functions of the geometry of the interferometer. The samples causing disturbances may be present in both beams and located at any plane. It is necessary, however, if a simple interpretation of the result is to be obtained, to lump together the disturbances into those at planes conjugate with the source, represented by Δw_x, and those in planes conjugate with the plane of observation, represented by Δw_u. These should then be added to Δp_x and Δp_u respectively. In either case, Δw represents the difference in optical path for the two beams:

$$\Delta w = w_2 - w_1. \tag{7.9}$$

Often the reference beam has no disturbance, so that $w_1 = 0$. In other cases, the same disturbance w appears in both beams, but at points with a relative shear, so that

$$\Delta w_u = w(\mathbf{u}' - \mathbf{s}_i) - w(\mathbf{u}'). \tag{7.10}$$

The shear $\mathbf{s}_i$ is that measured from the actual intermediate plane at which the sample is located, not from the source. A shearing interferometer used in double passage has a shear at the intermediate image and the disturbances in the optical path have the form (7.10). But this shear cancels on the return passage and there is no shear at the plane of observation.

The radiation is quasi-monochromatic when its bandwidth $\Delta \nu$ is small compared with its mean frequency ν, and also small compared with $1/\tau$ for all delays τ that occur in the interferometer. Then the discussion leading to equation (6.23) has shown that the degree of coherence can be separated into the product of two terms, one of which is a rapidly varying phase term and the other a slowly varying function, $\mu = m\,e^{i\beta}$.

The rapidly varying phase depends directly on the phase delay to the actual point $\mathbf{u}'$ that is being considered on the interferogram. This is the delay between the centres of the source and the plane of observation plus the variations of the delay with $\mathbf{u}'$, and the corresponding phase can be written as

$$\psi = \psi_0 + \Delta\psi_u - \beta, \tag{7.11}$$

where

$$\psi_0 = 2\pi\nu\tau = \kappa p, \quad \kappa = 2\pi/\lambda_v,$$

$$\Delta\psi_u = \kappa[\Delta p_u + \Delta w_u]. \tag{7.12}$$

The remaining term μ that varies only slowly with the delay is a function of the geometry of the interferometer, as expressed by the path variations Δp_x and any disturbance Δw_x. It can therefore be written as a function of the phase

$$\Delta\psi_x = \kappa[\Delta p_x + \Delta w_x]. \tag{7.13}$$

It also has some dependence on the delay, but it is the group delay τ_g (6.21) that is involved [Pancharatnam, 1963]. Thus the fringe visibility is maximum when this group delay is zero, and any measurement that involves setting an interferometer to maximum visibility is a setting to $\tau_g = 0$ at the mean frequency ν. The group delay can be derived from a group optical path,

$$\tau_g = \sum_{2-1} n_g d/c, \tag{7.14}$$

where

$$n_g = n + \nu\frac{dn}{d\nu} \approx n - \lambda\frac{dn}{d\lambda} \tag{7.15}$$

is the *group refractive index* of the media inside the interferometer.

The degree of coherence can now be written as

$$\gamma(u_1, u_2, \tau) = m(\Delta\psi_x, \tau_g)\,e^{-i\psi}$$

and the general expression for the intensity in the interferogram as

$$I(\mathbf{u}') = I_1 + I_2 + 2(I_1 I_2)^{1/2} m(\Delta\psi_x, \tau_g)\cos c \cos\psi, \tag{7.16}$$

where, as above,

$$\psi = \psi_0 + \Delta\psi_u - \beta(\Delta\psi_x, \tau_g),$$

$$m = |\mu|,$$

$$\beta = \arg\mu. \tag{7.17}$$

This formula is the basis for the listing of the fields of interferometry in Table 1.1. It shows first that three classes of measurement are possible: those that locate the positions of the maxima or minima of intensity and hence give the phase term, those in which the visibility of the fringes is measured by measurements of intensities, and those that give the complete interferogram and hence both visibility and phase. The last measurements give the real part of the coherence γ_{12} and hence, by the theory of Chapter 6, the spatial and spectral distributions of the radiation at the source; they will also include polarization effects within the interferometer, if these are present.

It is apparent that an interferometer measurement involves several physical quantities. When an attempt is made to measure one of the quantities that enter into (7.16), the result will be affected by all the other terms in this equation. The technique of interferometry is to design experiments, and the equipment to make them, in which unwanted effects are eliminated as far as possible or are of a known magnitude. Thus an ideal interference spectrometer would be an interferometer in which the optical path difference could be varied without introducing either shears or shifts that would make the results dependent also on source size.

The design of an interferometer to be insensitive to unwanted effects is *interferometer compensation.* We can speak of an interference spectrometer that is compensated for source size. Interferometers for measuring phase should have sufficient compensation for both spectral and spatial effects, so that the fringes have sufficient visibility to be seen clearly.

7.5. Spatial relations

The van Cittert–Zernike theorem relating the mutual coherence to the intensity distribution across the source, (6.30), may be rewritten in terms of the parameters of the general two-beam interferometer as

$$\mu(\Delta\psi_x, \tau_g) = \int_{-\infty}^{+\infty} L(\mathbf{x})\, e^{i\Delta\psi_x}\, d\mathbf{x} \Big/ \int_{-\infty}^{+\infty} L(\mathbf{x})\, d\mathbf{x}, \tag{7.18}$$

where

$$\Delta\psi_x = \kappa[-\mathbf{s}\cdot\mathbf{x}/z - h|\mathbf{x}|^2/2z^2 + \Delta w_x], \tag{7.19}$$

and $L(\mathbf{x})$ is equal to the radiance across the effective source and is zero outside the source boundary. The introduction of the disturbance Δw_x is equivalent to the addition of aberration to the analogue system; the diffracting aperture of fig. 39 should now have a phase distribution $\kappa\,\Delta w_x$ across it. The coherence is then given by the complex amplitude near the image point of the analogue, with the shear $\mathbf{s}$ corresponding to a lateral displacement in the image plane and the shift h to a displacement normal to this plane, a defocusing. The fringe visibility is maximum, $m = |\gamma| = 1$, when $|\mathbf{s}| = h = 0$, and falls off if either a shear or a shift is introduced.

Equation (7.18) is another example of a physical process that can be represented by a Fourier transform. As a function of the shear, the coherence and hence the modulation of an interferogram is the Fourier transform of the intensity distribution of the source. Simple diffraction gives Fourier transforms in amplitude and requires coherent illumination. Interferometers, however, give, in intensity, the transform of an intensity distribution of incoherent light, and are used to study these distributions. This gain in the ability to use incoherent light and to have a result that is linear in intensity is obtained at the expense of a uniform background being added and the greater complexity of an interferometer as compared with a simple diffracting system.

To give a complete two-dimensional picture of the transform, the interferometer should have a total shear, $\mathbf{s}=\mathbf{s}_0+\mathbf{u}_2-\mathbf{u}_1$, that is linearly related to the position vector $\mathbf{u}'$ of the interferogram. As Mertz [1965] has indicated, two types of shear give the desired result. A radial shear of shear ratio r has, directly,

$$\mathbf{s}=(1/r-1)\mathbf{u}_1$$

while for a small rotational shear of angle θ,

$$|\mathbf{s}|\approx\theta|\mathbf{u}_1|,$$

the direction of $\mathbf{s}$ being normal to $\mathbf{u}_1$. A one-dimensional transform is given by an interferometer with a folding shear.

A simple lateral-shearing interferometer, with a constant shear, cannot give such a direct picture but it is the most convenient instrument for giving single values of the Fourier transform. Since the shear is independent of $\mathbf{u}'$, the interferogram can be integrated over a finite area of the plane of observation.

7.5.1. Fringe localization

Fringes have good visibility only where both the shear and shift are small. If the position of the plane of observation is moved from such a position, the shear and shift change and the visibility decreases. Fringes are said to be *localized* at the position of maximum visibility. Localization requires an extended source; with a laser, fringes are unlocalized and can be seen almost anywhere.

When both the shear and shift are zero, all parts of the source produce fringes at the same place. Then, as the direction of viewing is changed so that different parts of the source are used, the fringes do not move: no parallax is seen. Welford [1969*b*, 1970] has used this definition of localization, which corresponds to the usual meaning of the term. But an interferometer may only have a position of the plane of observation at which the shear and shift have minimum, but not zero, values. The fringe visibility is maximum there, although there is still parallax. I shall use the more common definition of the position of localization as that of maximum visibility.

This position of localization is found by a well-known construction. Since a shear causes a larger reduction of visibility than a shift of the same magnitude, the construction locates the position of minimum shear. A ray from the source, split into two at the beam splitter, is traced through the two arms of the interferometer to observation space. Here the two rays have a position of closest approach, which is the position of least shear. If they intersect there, the two images of this point of intersection are seen from the source along the same ray, that is, with no shear. But there may be a shift present, and this determines whether the localized fringes have good visibility.

The factors that determine whether there is a point of intersection are shown in fig. 45, which repeats the essential features of fig. 43. The ray drawn from the source is the normal $\mathbf{n}$ and it appears in observation space as $\mathbf{n}'_1$ and $\mathbf{n}'_2$. These appear to leave points on the two images of the source that have a separation $\mathbf{t}'$, the tilt, and are inclined to each other at $\boldsymbol{\sigma}'$, the angular shear. They will intersect if the tilt and shear are coplanar. If the shear is changed by some adjustment to the interferometer, the position of localization changes [Steel, 1980]. It can also be changed by a change of tilt, but this is a design to be avoided, since it is often wanted to change the number of fringes without affecting their visibility.

When there is a uniform lateral shear, rays from the source to other points on the plane of observation will usually intersect at the same distance, to give a plane of localization. But when there are varying shears,

Fig. 45 The general representation of a two-beam interferometer: localization of the fringes.

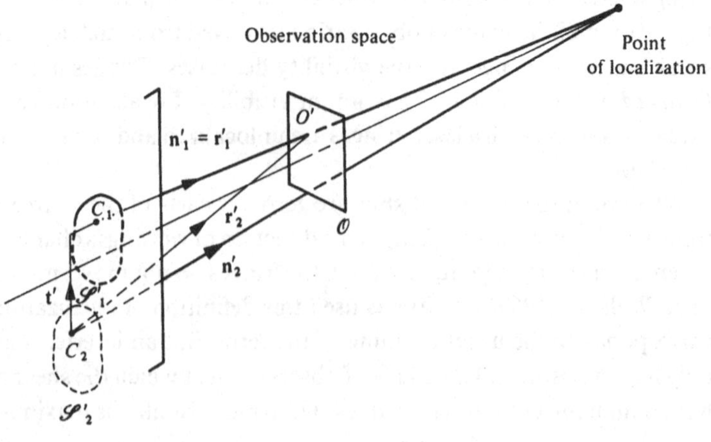

the localization may be limited to a line or a point. A radial or a rotational shear always leaves one point of the plane of observation with no shear, and a folding shear leaves a line. Away from this point or line, the shear increases linearly and the visibility drops. This is true for any position of the plane of observation. A position of maximum visibility can occur, however, if the shift varies with the position of the plane.

If the plane of observation is moved from a region of localization, or if the adjustment of the interferometer is altered to introduce a shear or shift, the visibility will drop. The fringes, however, will have some *range of visibility*, a function either of the position of observation or of the adjustment of the interferometer. For a shift, this range of visibility is analogous to the depth of focus of an imaging system and, by the van Cittert–Zernike theorem, the range of visibility of a source is equal to the depth of focus of an aperture of the same size; Hansen [1942] has pointed this out. For a shear, the range of visibility is analogous to the distance over which the amplitude in a point image drops off laterally; it is proportional to the smallest distance resolved by an imaging system. This is normally much smaller than the depth of focus and, as stated earlier, a shear is more important in fixing the localization of fringes. If an interferometer has a certain shift, the source size should be chosen so that, considered as an aperture, it has sufficient depth of focus to cover this shift. If it has a shear, the source size should be such that the equivalent aperture has a sufficiently poor resolving power so that it cannot resolve two points separated by the shear.

7.5.2. Circular source

Results for the diffraction analogue of a circular source, the circular aperture, have been given by equation (3.37). The interference is represented by the coherence function

$$\mu_{12} = C(\zeta, \rho)$$

where

$$C(\zeta, \rho) = \frac{e^{-\pi i \zeta}}{\pi \zeta}[U_1(2\pi\zeta, 2\pi\rho) + iU_2(2\pi\zeta, 2\pi\rho)], \tag{7.20}$$

where U_1 and U_2 are the Lommel functions of two variables discussed by Watson [1944]. The reduced coordinates ρ and ζ are defined for an interferometer in terms of the parameters shear and shift and the size of the source. These should all be measured in the same space. Since ρ and ζ involve the invariants (3.9) and (3.10), this space can be chosen to be at whatever part of the interferometer is most convenient. If the circular source has an angular radius α and subtends a solid angle $\Omega = \pi\alpha^2$ at O_1, then

$$\rho = \alpha|\mathbf{s}|/\lambda,$$

and

$$\zeta = \alpha^2 h/\lambda. \tag{7.21}$$

The modulus and phase of (7.20) are plotted against these coordinates in fig. 46; results over a larger range are given by Born & Wolf [1959], and Minkwitz & Schulz [1964]. This figure shows that m, the maximum visibility for $I_1 = I_2$, decreases at about the same rate for equal departures of ρ or ζ from zero. But, since α is usually small, equation (7.21) shows that equal amounts of shear and shift give much larger values of ρ than of ζ.

For a rectangular source, the two principal semi-angles α_i and α_j may differ. Two sets of reduced coordinates are then required,

$$\left.\begin{aligned} \xi_n &= \alpha_n s_n/\lambda \\ \zeta_n &= \alpha_n^2 h/\lambda \end{aligned}\right\} \quad (n = i, j), \tag{7.22}$$

where s_i and s_j are the components of the shear **s**. For each direction, we can define

Fig. 46 The (maximum) visibility m_{12} (full curves) and the phase β_{12} (broken curves) for an interferometer with a uniform circular source. The phase curves are taken from Minkwitz and Schulz [1964] and are labelled with the values of β_{12}/π.

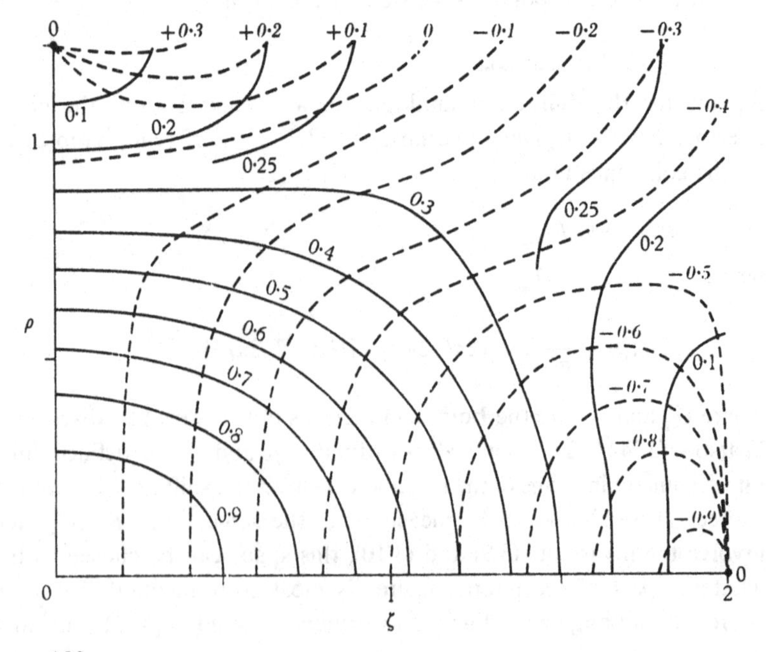

$$F(\zeta,\zeta)=\frac{1}{2}\int_{-1}^{1} e^{-2\pi i(\xi t+(1/2)\zeta t^2)}\,dt \tag{7.23}$$

and express the result in terms of Fresnel integrals. The coherence is then given by

$$\mu_{12}=F(\xi_i,\zeta_i)F(\xi_j,\zeta_j). \tag{7.24}$$

For a long narrow slit, one of these functions is approximately unity.

7.5.3. Integrated values

The intensity distribution given by (7.16) applies in all cases when there is a point-by-point analysis of the interferogram—by eye, by photography, or by scanning or sampling with power detectors of small area. When a detector that collects radiation from an appreciable area is used, integrals of (7.16) are required. Such integrals have been evaluated only for very simple cases; these are given in Chapter 8. It is no longer possible to use the analogy with diffraction and take over earlier results, for integrals in diffraction theory have been of intensity distributions while, for interferometry, integrals of amplitude distributions are the equivalent that is required.

If the detector has a sensitivity $D(\mathbf{u}')$ at a point $\mathbf{u}'$ in the plane of observation, the total signal received is

$$S=\int m(\Delta\psi_x,\tau_g)D(\mathbf{u}')\cos\left[\beta(\Delta\psi_x,\tau_g)-\psi_0-\Delta\psi_u\right]d\mathbf{u}'. \tag{7.25}$$

With (7.19) this gives

$$S=e^{-2\pi i\nu_0\tau}\iint_{-\infty}^{+\infty} L(\mathbf{x})D(\mathbf{u}')\exp i[\Delta\psi_x-\Delta\psi_{u'}]\,d\mathbf{x}\,d\mathbf{u}'. \tag{7.26}$$

These expressions are based on the assumption of incoherence, both at the source and at the detector. Although this is not physically possible, these equations are still useful first approximations.

7.6. Spectral effects

The general theory given in (6.14) and (6.15), relating the spectral distribution of the radiation to the mutual coherence function, describes the intensity distribution in the interferogram as a function of the delay τ. An interferometer can therefore be used to measure spectral distributions. With two-beam interference, the Fourier transform is obtained but, when a large number of beams interfere, sufficient Fourier components are obtained to give a result that approaches a direct representation of the spectrum.

When an interferometer is used with a source of large bandwidth at delays greater than allowed by the condition for quasi-monochromatic radiation, no fringes are seen in the interferogram. But if the radiation is focused on the slit of a spectroscope, the spectrum is seen to be crossed by bands resembling interference fringes. These are *Edser–Butler bands* or *fringes of equal chromatic order* and the resulting spectrum is a *channelled spectrum.* Their intensity represents the cosine of the phase difference at each frequency of the radiation, and hence their angle to the direction of dispersion in the spectroscope gives the change of phase with frequency. When the phase is stationary at some frequency, the fringes there being parallel to the dispersion, there is zero group delay at that frequency.

The fringes are straight only when the dispersions of the materials in the two beams are equal, and the delay can be made zero at all frequencies. Strongly curved fringes would denote that one beam has gone through more glass but less air than the other. A less pronounced curvature would be caused by glasses of different dispersion in the two beams. Thus, to correct a given interferometer so that the delay is zero over a wide range of frequencies, compensators of more than one material may be required [Johnson & Scholes, 1948].

Fringes of equal chromatic order provide the most sensitive method of adjusting the interferometer delay to obtain maximum visibility with a source of finite spectral bandwidth. It is an old technique, implied by Walker [1904], used by Michelson & Pease [1921], and frequently rediscovered since. It contains its own criterion: for good visibility the deviation of the fringes across the bandwidth of the source, normal to the direction of dispersion, should be less than the fringe spacing. This is the spectral analogue of Hansen's condition for source size, discussed in §7.8.

Other methods of setting to maximum visibility involve directly searching for it while changing the delay. It is made easier if sources of increasing bandwidth can be used in turn: first a low-pressure mercury lamp, then a medium-pressure lamp, the bandwidth of which decreases as it warms up. But, provided a spectroscope of sufficient resolving power is used, fringes of equal chromatic order can be seen when the delay is too large for the direct fringes to be visible.

7.6.1. White-light fringes

When the two paths of an interferometer match well enough to give fringes of equal chromatic order that are straight lines across the visible spectrum, fringes are obtained with white light. There is a central white or black fringe, with coloured fringes on both sides. The white-light fringe, because it can be readily identified, is very useful for measuring phase steps over sharp discontinuities.

If the mean delay, but not its dispersion, is corrected, there is no clear white or black fringe, but a larger number of coloured fringes is seen.

7.7. Combined effects

The combined effects of the size of the source and its spectral distribution on the interferogram are given by (6.33). From this we can list the conditions that must be satisfied to obtain fringes of high visibility. These are:

(i) $\tau \Delta\nu \lesssim 1$, (7.27)

either the bandwidth or the delay should be small;

(ii) $\alpha_s|\mathbf{s}| \lesssim \lambda$, (7.28)

either the shear should be small or the source should subtend a small angle in the direction parallel to the shear; and

(iii) $\alpha^2 h \lesssim \lambda$, (7.29)

either the shift or the angle subtended by the source at any azimuth should be small. In addition, there are the further conditions given by (7.16):

(iv) $I_1 = I_2$, (7.30)

the two arms of the interferometer should transmit equal intensities; and

(v) $\cos c = 1$, (7.31)

the two beams should not undergo any relative changes in polarization.

The first three conditions can be satisfied in two ways: either the interferometer is designed to have small shear, shift, and delay, or the size and bandwidth of the source are limited. A laser meets these limitations and has plenty of energy, particularly when it is oscillating in a single longitudinal mode. Where it used to be necessary to design a specially compensated interferometer, it is often possible now to use a much simpler instrument with a laser source. But, because lasers give unlocalized fringes, it is sometimes difficult to separate the desired interferogram from the fringes produced by scattered light or reflexions from the surfaces of components. To separate these by their localization, an extended source is needed. A laser may still be used, made spatially incoherent by a moving diffuser, such as a rotating glass disk or a colloid in brownian motion. The source then has a finite size but retains its narrow spectral bandwidth, so that the delay need not be zero, although the shear and shift should.

With thermal sources, very little energy is available when the size and bandwidth are small, and there are then considerable advantages in designing the interferometer so that a large source of greater bandwidth can be used. This design, often involving methods of changing one parameter without affecting another, is discussed in §8.3.

7.8. Alternative interferogram

It has been mentioned earlier that the choice of the plane of observation is often arbitrary and the integrations of the interferogram (7.26) can equally well be obtained as integrals over the source of an interferogram formed there. If a stop is placed at some plane conjugate to the plane of observation, this becomes the effective source for the alternative fringes discussed in § 1.5.1. Although these alternative fringes are observed in some plane, conjugate with the source, that follows the interferometer, they can be described in terms of their image at the source. The intensity distribution there is then

$$I(\mathbf{x}) = I_1 + I_2 + 2(I_1 I_2)^{1/2} m(\Delta\psi_u, \tau_g) \cos c \cos \psi, \tag{7.32}$$

where, now

$$\psi = 2\pi\nu\tau + \Delta\psi_x - \beta(\Delta\psi_u, \tau_g), \tag{7.33}$$

and the intensities I_1 and I_2 now refer to the source plane.

A comparison of this expression with (7.16) shows that the roles of shift and shear alternate with those of lead and tilt, between the two sets of fringes. Where the visibility of the test fringes (7.16) is a function of $\Delta\psi_x$, and hence of the shear and shift, and their phase contains $\Delta\psi_u$, a function of the tilt and the lead, these roles are interchanged for the source fringes. Thus, as Hansen [1942] has pointed out, the visibility of one set of fringes depends on the spacing of the alternative fringes, this spacing being measured relative to the aperture that will be used as the effective source. If a diaphragm conjugate with the plane of observation is closed down, so that it is small compared with the spacing of the fringes in that plane, high-visibility source fringes will be seen. This result is of considerable practical importance, for it is simpler to adjust an interferometer in terms of the fringe spacing than the fringe visibility.

Hansen's criterion, that the source should be limited within the first source fringe, will be studied further in Chapter 8. It can, however, be justified by qualitative argument. If the source covers a bright and a dark source fringe, it contributes equal amounts of in-phase and out-of-phase radiation to the test fringes. The out-of-phase regions of the source will form fringes with a phase shift of π with respect to those formed by the in-phase regions, the maxima of one set falling on the minima of the other. The resultant interference has zero visibility.

7.9. Polarization effects

So far it has been assumed that the light is fully polarized when it enters the interferometer. Even when the interferometer affects the polarization so that the two beams leave in different states, as defined by (3.47), the entering and emergent beams are taken to be both fully polarized.

Although interferometers are not commonly used to measure polarization, [but see Fymat, 1981], fringe visibility may be affected by it and an interferometer should be compensated for these effects. As stated in §3.6 oblique reflexion at metal mirrors is characterized by two complex reflexion coefficients $r_{\parallel}$ and $r_{\perp}$ for radiation that is linearly polarized either parallel or perpendicular to the plane of incidence. These coefficients normally differ slightly in modulus but considerably in phase: the mirror acts as a retarder. Radiation entering the interferometer linearly polarized may not remain so, and the states of polarization of the two emergent beams may differ. There will be a reduction in the visibility of the interferogram by the factor $\cos c$ of (7.16) or (7.32), and no interference when the states are orthogonal. (But the interference can be obtained by special recording techniques, such as that of Jonathan & May [1980].)

When the light entering is unpolarized, it can be treated in terms of the coherence matrix $\mathbf{J}_{12}$ [Steel, 1965*b*]. Alternatively, it can be regarded as made up of two separate orthogonally polarized components. Each of these should give rise to outgoing beams in the same state of polarization if the final interferogram is to have full visibility. The incoming and outgoing states of polarization in each arm of the interferometer are related by a matrix equation

$$\boldsymbol{\varepsilon}' = \mathbf{A}\boldsymbol{\varepsilon}, \tag{7.34}$$

where $\mathbf{A}$ represents the combined effects of the mirrors in that arm. To obtain the two outgoing polarizations in the same state, $\boldsymbol{\varepsilon}_2' = \boldsymbol{\varepsilon}_1'$ for any incident polarization $\boldsymbol{\varepsilon}$, and hence for unpolarized light also, we require

$$\mathbf{A}_2\boldsymbol{\varepsilon} = \mathbf{A}_1\boldsymbol{\varepsilon}$$

for all $\boldsymbol{\varepsilon}$. This requires that the same matrix shall represent each arm, or

$$\mathbf{A}_2 = \mathbf{A}_1. \tag{7.35}$$

Each matrix **A** is the product of diagonal matrices of the form (3.45), representing reflexion at a mirror or reflexion or transmission at a beam splitter, and rotation matrices (3.43), which are not diagonal. Even if the two arms of the interferometer contain identical components, their matrix products are not the same, since one starts with transmission through the beam splitter and ends with reflexion there, while the other starts with reflexion and ends with transmission. (This is true for a Michelson interferometer; for other systems the arrangement of beam splitters may differ, but it remains unsymmetrical.)

Matrix products do not commute unless all the matrices are diagonal. If all the normals to the mirrors and beam splitters are coplanar, or in perpendicular planes, there are no non-diagonal rotation matrices, and the products commute. As is well known, plane interferometers with identical arms have no polarization effects.

Mirrors with oblique planes of incidence give rise to diagonal matrices only when they occur in pairs, the angles of incidence being the same at each but the planes of incidence perpendicular to each other. The pair can then be reduced to a diagonal matrix. In the notation of §3.6,

$$\mathbf{A}=\mathbf{S}(-\tfrac{1}{2}\pi-\phi)\mathbf{M}\ \mathbf{S}(\tfrac{1}{2}\pi)\mathbf{M}\ \mathbf{S}(\phi),$$

$$=r_{\|}r_{\perp}\begin{pmatrix}1 & 0\\ 0 & 1\end{pmatrix}. \qquad (7.36)$$

The general interferometer in three dimensions is not, however, compensated, even when the arms are identical. The most common examples are Michelson interferometers with the plane mirrors replaced by trihedral mirrors or cube corners [Lenhardt & Burckhardt, 1977; Korotaev & Pankov, 1981] or by inclined roof mirrors [Roddier, Roddier & Demarq, 1978].

Even though an interferometer is not compensated for use with unpolarized incident light, there may be particular polarizations of the incident light for which the two emergent beams are in the same state of polarization. If these states are linear polarizations, the interferometer may be used with a single polarizer, either in front of it or following it, to give fringes of unit visibility. These particular polarizations are solutions of (7.34), rewritten to include the possibility of a phase difference between the two emergent beams:

$$\mathbf{A}_1\boldsymbol{\varepsilon}=\mathbf{A}_2\boldsymbol{\varepsilon}\,\mathrm{e}^{\mathrm{i}\psi}. \qquad (7.37)$$

In general this equation has two orthogonal solutions $\boldsymbol{\varepsilon}_1$ and $\boldsymbol{\varepsilon}_2$, or two eigenvectors, and for each there is a different phase change, or eigenvalue, ψ.

The two states give interference patterns in different positions, corresponding to this difference of phase, and any addition of these patterns gives a reduction in visibility.

We have seen that some interferometers may be used with unpolarized light while others require one polarizer. There is a third class: polarization interferometers require two polarizers, one before and one after the interferometer. In this way polarization compensation is ensured: the incident light is fully polarized and the two emergent beams are fully polarized in the same state. If the other conditions of §7.7 are satisfied, the fringes have unit visibility. This brute-force method of obtaining compensation is not often necessary for interferometers with partially reflecting beam splitters.

7.10. Division of wavefront

An interferometer with division of wavefront can be represented by an optical system similar to that shown in fig. 43 for interferometers with division of amplitude, and the general theory applies to both interferometers. There are, however, two important differences. An interferometer with division of wavefront has no beam splitters, and hence none of the losses that these cause, and it always has a lateral shear. As the two apertures that separate the light into two beams cannot overlap, the width of the beams is limited by the amount of shear. The presence of this shear makes it necessary to use a source that is narrow in the direction of shear, a slit, otherwise good visibility cannot be obtained. This requirement can be avoided by illuminating the instrument through an interferometer with division of amplitude that introduces a compensating shear.

A special case of division of wavefront has been given in fig. 37. The theory of Young's experiment is normally expressed [Born & Wolf, 1959] in terms of the coherence between points in the plane containing the two pinholes, while our general theory has given the result in terms of the coherence between two beams to the images of the plane of observation, as seen from the source. However, Young's experiment can be included in the general theory if it is noted that, as the radiation across the first plane is sampled by two small pinholes, the coherence between the two beams at the plane of observation is the same, to a phase factor, as that between them at the pinholes.

8

Practical two-beam interferometers

The general theory developed in Chapter 7 can now be illustrated by the properties of the more important two-beam interferometers with division of amplitude. In addition, some practical aspects of their construction are given.

8.1. The Michelson interferometer

Although Michelson used the interferometer named after him without any focusing system, this interferometer is most useful when the radiation is collimated so that the stop defining the effective source is at infinity. Twyman & Green [Candler, 1951] used this form for testing optical components, and it is commonly used also for length measurement and for spectroscopy. From the viewpoint of compensation, it is most suitable for measuring either phase differences in double passage or spectral distributions, since its delay can be varied simply. The treatment applies also to other interferometers, such as the Fizeau, in which the light is both collimated and reflected back on itself.

The interferometer is shown in fig. 47 with the two equivalent optical systems, as seen from either the source or the plane of observation. The collimating systems are represented by lenses of focal lengths f_1 and f_2, both at a distance d from the reference position of the two mirrors at the ends of the arms. The effective source is defined by a stop at the focus of the first of these lenses, while the plane of observation is at a distance g from the focus of the second. The mirror 2 is assumed to have a longitudinal displacement e from its reference position and a rotation θ about an axis normal to the plane of the figure.

The plane of observation is seen from the source at an apparent distance

$$z=f_1^2/(f_1+f_2-2d+f_2^2/g), \tag{8.1}$$

with a shear

$$s=2\theta(z-f_1-zd/f_1), \tag{8.2}$$

and a shift

$$h=2ez^2/f_1^2. \tag{8.3}$$

The shift is always present, but the shear vanishes when

$$g = f_2^2/(d - f_2), \quad z = f_1^2/(f_2 - d). \tag{8.4}$$

This is just the condition that the plane of observation should be conjugate with the mirrors, and the test fringes or Fizeau fringes are localized there. But, when one mirror is displaced from the image of the other, it is not possible to decide at which mirror they are localized. For the fringes to have sufficient visibility to be seen with the shift (8.3) present, the source should be small enough to have a range of visibility equal to this shift. By the van Cittert–Zernike theorem, it has then the size of an aperture whose depth of focus would equal the shift. The fringes are imaged at the plane of observation by an optical system with an aperture the size of the effective source, and this would then have a depth of focus that extends from one mirror to the other. Fringes can be seen over the whole range of focus.

For a uniform, circular source, the visibility is given by the modulus of the Lommel functions of fig. 46, with $\rho = 0$. It is most convenient to work in the

Fig. 47 The Michelson interferometer with collimated radiation and its two equivalent optical systems.

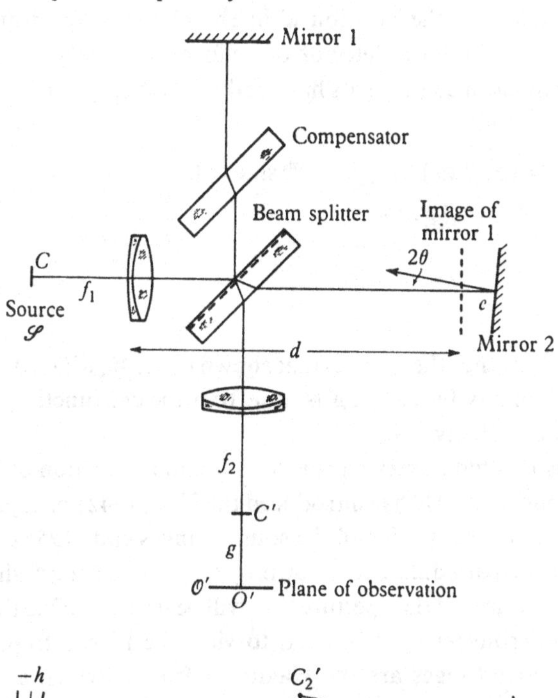

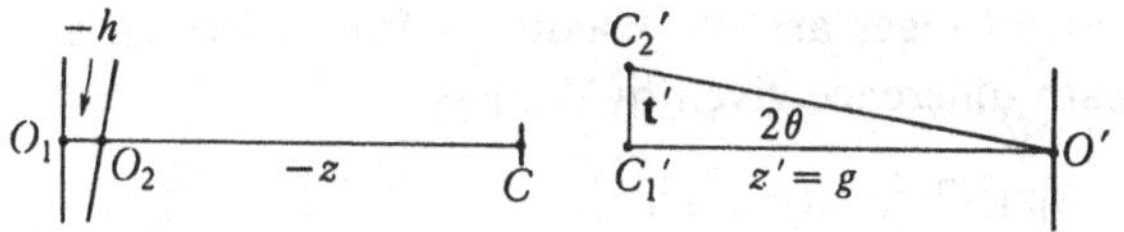

space containing the mirrors at the ends of the two arms, in which the shift is $2e$, the angular tilt is 2θ, and the source, which appears at infinity, subtends a solid angle Ω. Then

$$\zeta = 2\Omega e/\pi\lambda \tag{8.5}$$

and

$$m = \operatorname{sinc}\tfrac{1}{2}\zeta. \tag{8.6}$$

The delay,

$$\tau = 2ne/c,$$

is a direct multiple of the shift and (8.5) can be written as

$$\zeta = \Omega\nu\tau/\pi.$$

Since the two images of the source are at infinity, there is no lead.

The coherence function for this interferometer is then

$$\mu_{12} = \mathrm{e}^{-(1/2)\pi i\zeta}\operatorname{sinc}\tfrac{1}{2}\zeta. \tag{8.7}$$

This is independent of the position $\mathbf{u}'$ in the plane of observation. The integrated signal (7.25) for a detector of uniform sensitivity, the image of which at the interferometer mirrors has a radius a, is then, in the notation of (7.20).

$$\begin{aligned} S(\rho', \zeta) &= [\mathrm{J}_1(2\pi\rho')/\pi\rho']\{\mathrm{e}^{-(1/2)\pi i\zeta}\operatorname{sinc}\tfrac{1}{2}\zeta], \\ &= \mathrm{C}(0, \rho')\mathrm{C}(\zeta, 0), \end{aligned} \tag{8.8}$$

where

$$\rho' = 2\theta a/\lambda. \tag{8.9}$$

This function (8.8) is not the same as that shown in fig. 46, which depends on shears rather than tilts, but, when ρ' is equal to ρ, the two functions have the same values along the two axes.

The Michelson interferometer provides a useful illustration of Hansen's criterion for source size. He has introduced the idea [1942] of adjusting the size of the source to the spacing of the source fringes and [1955] states the condition that the path difference shown by the source fringes should not exceed $\tfrac{1}{2}\lambda$ over the area of the aperture that will be used to define the source. When the interferometer is to be used to view the Fizeau fringes at the mirrors, the source fringes are the Haidinger fringes, which are centred circles with a path difference given by (7.5) as

$$\Delta p_x = -h|\mathbf{x}|^2/2z^2, \quad h = 2ez^2/f_1^2, \tag{8.10}$$

no shear being present. This has a range of $\frac{1}{2}\lambda$ when the source has a radius δx given by

$$\left(\frac{\delta x}{z}\right)^2 = \frac{\lambda}{h}, \quad \text{for a centred source,} \tag{8.11}$$

or a width δx where

$$\frac{x\,\delta x}{z^2} = \frac{\lambda}{2h}, \quad \text{for an annular source off centre.} \tag{8.12}$$

In either case, the source subtends a solid angle

$$\Omega = \pi\lambda/2e. \tag{8.13}$$

When the source is a centred circle and the delay is such that the source fringes have a bright centre, the first dark fringe has a radius given by (8.11). For a source of this size, $\zeta = 1$, and (8.6) gives the visibility of the Fizeau fringes as

$$m = \operatorname{sinc}\tfrac{1}{2} = 0.64.$$

For an interferometer with a tilt but no lead, the variations of optical path for the Fizeau fringes are given by (7.6) as

$$\Delta p_u = -\mathbf{t}' \cdot \mathbf{u}'/z', \quad |\mathbf{t}'| = 2\theta f_2; \tag{8.14}$$

these are straight fringes parallel to the axis of tilt. A source in a conjugate plane has a variation of path of $\frac{1}{2}\lambda$ when its width u' is given by

$$|\mathbf{u}'| = u' = \lambda z'/4\theta f_2,$$

its length being unlimited. For a circular source of diameter equal to this width u', $\rho' = \frac{1}{4}$ and the visibility of the Haidinger fringes is

$$m = 4\mathrm{J}_1(\tfrac{1}{2}\pi)/\pi = 0.72.$$

If the source is a slit of width u', set parallel to the Fizeau fringes, the visibility of the Haidinger fringes is again 0.64.

This example shows that Hansen's criterion will give reasonable visibility for both sets of fringes in a Michelson interferometer. As would be expected, the visibility obtained varies slightly with the shape of the source and also with the form of the fringes.

8.1.1. Interferometer without collimation

A Michelson interferometer with no collimating optical systems is equivalent to the representation shown in fig. 48. The source and plane of observation are at distances d and g from the mirrors and again a displacement e and a rotation θ of mirror 2 will be considered. The interferometer then has the properties:

apparent distances	$z=d+g,$		$z'=-d-g,$
shift	$h=2e,$	lead	$l'=2e,$
shear	$s=2\theta g,$	tilt	$t'=2\theta d.$ (8.15)

A lead equal to the shift has now appeared; it was absent in the first interferometer entirely because of the collimation, for a source that appears to be at infinity is not affected by a finite displacement e. A shear is again absent when the test fringes are localized at the mirrors and $g=0$. In general, both the source fringes and test fringes have circular and straight-line components and appear as ellipses or hyperbolas.

If the interferometer is adjusted so that there is no shear nor tilt, the coherence function is still represented by (8.7) for a point on axis. But it is no longer independent of the position $\mathbf{u}'$ in the plane of observation and, in general, simple expressions for the integrated signal are no longer found. Fig. 49 shows the modulus and phase of $S(\zeta)$ for a Michelson interferometer with collimation, and for one without in which both the source and detector are limited by circular apertures of the same radius. For this,

$$S(\zeta)=4\int_0^1\int_0^1 \mathrm{e}^{-\pi i\zeta(r^2+r'^2)}\mathbf{J}_0(2\pi\zeta rr')r\,\mathrm{d}r\,r'\,\mathrm{d}r',$$

which reduces to

$$=1/\pi\zeta\{\mathrm{e}^{-2\pi i\zeta}[i\mathbf{J}_0(2\pi\zeta)-\mathbf{J}_1(2\pi\zeta)]-i\}. \tag{8.16}$$

Again

$$\zeta=\Omega\nu\tau/\pi,$$

Fig. 48 The Michelson interferometer without collimation.

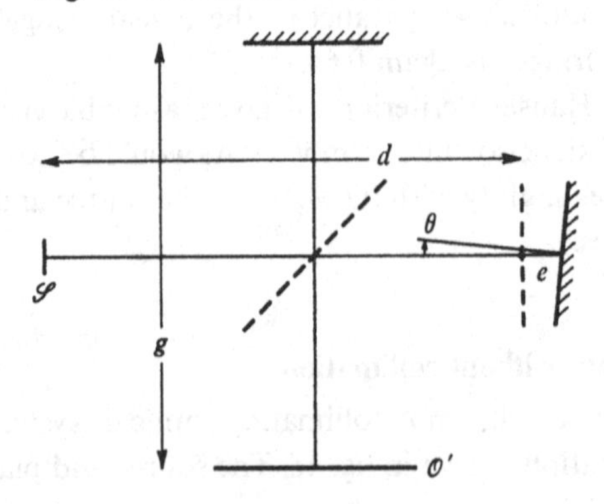

Ω being the angular size of the source. These two curves of fig. 49 represent the two extreme cases for circular apertures. Intermediate cases have been studied by Bottema [1972].

8.2. Interferometer with retroreflectors

If one mirror of a Michelson interferometer is replaced by a retroreflector, the two images of the plane of observation appear at relative orientations of π and, for every point except the centre, there is a shear that increases with distances from the centre. Visible fringes are obtained only with a very small source, and the smaller the source the further from centre the fringes can be seen. But when both mirrors are replaced by retroreflectors, there is no rotational shear and the interferometer resembles a simple Michelson interferometer. It has been studied in some detail by Peck [1957] and is shown in fig. 50 with the equivalent optical systems. In place of a tilt, which is not present, a lateral displacement d of one retroreflector is considered as well as the longitudinal displacement e. An alternative form is shown with the trihedral mirrors of the first form replaced by cat's eyes; the aberrations these introduce are given by Steel [1974].

With this interferometer, a lateral displacement of a retroreflector gives a shear for all positions of the test fringes, except infinity. Since it is customary

Fig. 49 Modulus (full curve) and phase (broken curve) of the integrated interferogram S for the Michelson interferometer: (*a*) with collimation; (*b*) with no collimation and two circular apertures of the same size [Steel, 1964*a*].

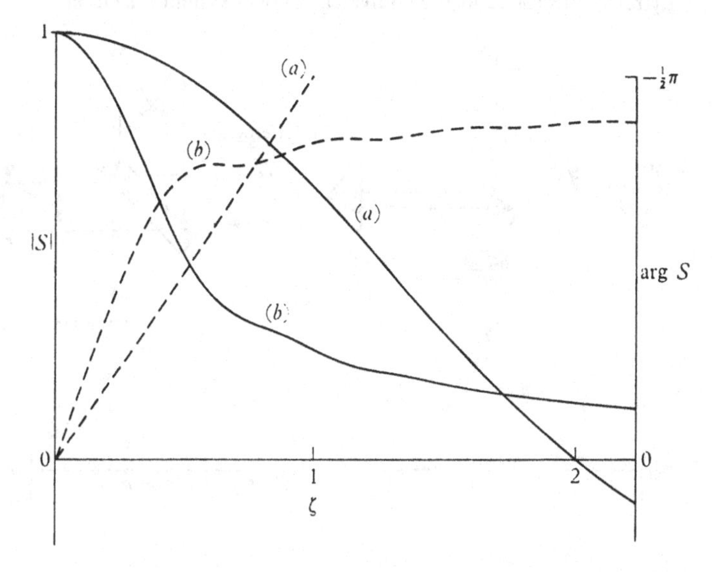

to have the source fringes at infinity, the test fringes will be taken to be at the apex of the trihedral mirrors, or at the beam splitter when cat's eyes are used (the smaller mirror of this system can have its radius chosen to image the beam splitter back on itself). As there are no tilts nor leads, these fringes are spread out and show no variation across the plane of observation. The intensity across the whole field varies uniformly as the displacement e is changed.

The interferometer in collimated light has a shear

$$s = 2dz/f_1 \tag{8.17}$$

and a shift equal to that of a Michelson interferometer (8.3). The source fringes have as their effective source the images of O', separated both laterally and longitudinally, and are off-centre ellipses, which look almost circular when the shear is small. The coherence function for a circular source is $C(\zeta, \rho)$ of (7.20), plotted in fig. 46. As stated earlier, integrated coherence is not the same as defined for the Michelson interferometer, for which it is given in (8.8). But along the two axes, and approximately elsewhere, the two have similar values if $\rho = \rho'$, that is, if the rotation θ of the Michelson mirror and the lateral displacement d of the retroreflector are related by

$$d\alpha' = \theta a, \tag{8.18}$$

where, as before, a is the radius of the mirrors and α' the angular radius of

Fig. 50 Two forms of the Michelson interferometer with mirrors replaced by retroreflectors and the equivalent optical system for either.

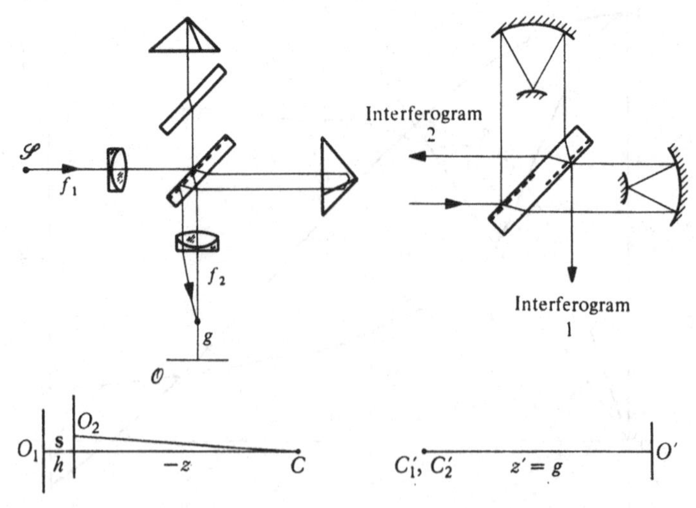

the source, as seen from either the mirrors or retroreflectors.

To change the delay of an interferometer, the mirror or retroreflector is moved longitudinally and any errors of this movement can cause rotation of the mirror or lateral displacement of the retroreflector. Equation (8.18) shows that the errors are less important for retroreflectors. To convert a lateral displacement, affecting a retroreflector, into an equivalent rotation of a mirror, this mirror would need to be mounted on a carriage of length a/α', too long to be used in practical instruments.

The interferometer with retroreflectors has another important difference from the Michelson interferometer. The latter is a reflex interferometer in which beams return from the mirrors along the same path as they entered. The same can be true with retroreflectors, as shown in the first example of fig. 50, where the beams return, inverted, along their original paths. But, as shown in the second example, the beams can enter the retroreflectors on one side of centre and leave on the other. They recombine at a separate part of the beam splitter and two emergent beams are obtained, each showing an interferogram. By conservation of energy considerations, these two patterns are complementary when the beam splitter is non-absorbing: maxima in one correspond to minima in the other. If both can be detected, their difference measures the interference to double the sensitivity given by a single interferogram, the background being eliminated. In the Michelson interferometer, one of these interferograms is sent back to the source and, to recover it, a second beam splitter would be required; about three quarters of the radiation is then lost. As well as giving two interferograms, the interferometer can be made symmetrical, for one of the interferograms, by coating the beam splitter as shown in the second example in fig. 50. In a Michelson interferometer, a beam splitter that consists of a simple coating on a transparent support introduces an asymmetry into the instrument; this is discussed later.

In another modification of the Michelson interferometer, the mirrors are replaced by pairs of roof mirrors rather than retroreflectors. This is a hybrid instrument. In one plane of the instrument, it has the properties of an interferometer with retroreflectors: it can have a shear and a shift and the outgoing beam can be separated. In the plane perpendicular to this it has the properties of a Michelson interferometer: it can have a tilt and the two beams are superimposed. In addition, a new type of adjustment is possible, for the roof mirrors can be rotated in azimuth about an axis parallel to the incident light. Such a rotation causes a rotational shear of one image of the plane of observation with respect to the other. Armitage & Lohmann [1965] have described an interferometer of this type.

8.3. Interferometer compensation

The theory has shown the advantages of designing interferometers in which one parameter can be varied at a time without affecting the others. Several examples can be given of cases where this is useful. It is often desirable to change the number of fringes across the field of observation or their curvature, by changing the tilt or lead, without at the same time introducing any shear, shift, or delay that would reduce the fringe visibility. In other applications, it is desirable to change either the delay or the shift without affecting the other. In Fourier spectroscopy the delay must be varied, preferably without introducing a shift or lead. When testing an optical component, its insertion into the interferometer introduces a shift, and it is desirable to remove this without introducing a delay.

Methods of uncoupling these parameters are known as interferometer compensation. That their complete separation is possible has been shown by Steel [1975*b*]. Some of the methods used have already been encountered in the study of the Michelson interferometer. When the source is at infinity as seen from the mirrors, a longitudinal movement of one mirror does not introduce a lead, although it does change the shift and the delay. Similarly, a longitudinal movement of a mirror that sees the plane of observation at infinity does not change the shift.

The Michelson interferometer shows also how the tilt and shear can be separated. When the fringes are observed in a plane conjugate to the mirrors, a change of the angle of a mirror changes the tilt without introducing a shear. With image-forming components inside the interferometer, it is possible to have separate regions where first the source then the plane of observation are imaged to infinity, and in each of these regions to have the other plane imaged on the surface of a mirror that can be tilted. Then all parameters except the delay are separately adjustable.

A second class of interferometer compensation is that of designing an instrument in which movement of a component to change one or more of the longitudinal parameters does not, through accidental tilts or lateral displacements, introduce either of the lateral parameters, shear or tilt.

8.3.1. Field-widened Michelson interferometer

In a simple Michelson interferometer, any longitudinal displacement of one of the mirrors changes both the shift and the optical path difference by the same amount, and the resulting interferogram depends on both the spatial and spectral distributions of the radiation. For purposes of compensation, it is desirable to separate these effects and vary the shift and optical path (or delay) independently. For the study of spectral distributions, a compensated interferometer would be one that gives a variable delay but no shift, and, as this would make it insensitive to the spatial distribution of the radiation, it could collect radiation over a large angle. Such an interferometer is said to be *field widened*. Another application of the same compensation is found in lens testing, where the image shift caused by introducing a focusing component into one arm of an interferometer can be corrected without changing the correction for delay.

For the Michelson interferometer with no tilt, the integrated coherence is given by (8.5) and (8.8) as a function of the delay:

$$S(\tau) = \operatorname{sinc} \Omega \nu \tau / 2\pi.$$

This vanishes when

$$\Omega = 2\pi / \nu \tau, \tag{8.19}$$

and this value Ω of the solid angle subtended by the source at the mirror is regarded conventionally as the maximum useful size of the source. For a field-widened interferometer, the corresponding limit can often be written as

$$\Omega = 2\pi G / \nu \tau, \tag{8.20}$$

where G is the gain factor due to the compensation employed.

An obvious method of obtaining a small gain is to increase the refractive index n of the interferometer arms. Then an increase d in the length of one arm increases the optical path there by nd, but increases the shift, on which the source size depends, by only d/n. The result is a gain

$$G = n^2. \tag{8.21}$$

The arms could contain a liquid, but this could introduce problems of non-uniform index. Alternatively, the effective thickness of a plane-parallel slab of a transparent solid could be varied by sliding two wedges past each other.

The different rates of variation of delay and shift obtainable when the thicknesses of two media are changed has been used by Mertz [1965] to give a larger amount of field widening. The method is shown in fig. 51. In one arm of a Michelson interferometer, the air path is altered by moving the mirror, while both it and a glass path are altered by moving a wedge. With the path lengths indicated on this figure, the shift and delay are given by

$$h = 2\,\delta e + 2(1/n - 1)\,\delta a, \tag{8.22}$$

$$c\tau = 2\,\delta e + 2(n - 1)\,\delta a. \tag{8.23}$$

If the mirror movement is linked to that of the wedge so that

$$\delta e = -(n_g - 1)\,\delta a, \tag{8.24}$$

the shift can be changed without changing the group delay. For spectroscopy, where the delay should be changed without introducing any shift, a different link is required that gives

$$\delta e = (1 - 1/n)\,\delta a, \tag{8.25}$$

and then

$$c\tau = 2(n - 1/n)\,\delta a. \tag{8.26}$$

This coupling need only involve moving a wedge and the mirror of the other arm on the same carriage, provided the two arms are at an appropriate angle (not a right angle, as shown in fig. 51).

When the interferometer has its maximum delay, one beam passes through an extra thickness $2\,\delta a$ of glass and it is the spherical and chromatic aberrations of this that now limit the angular size of the source. Bouchareine & Connes [1963] have shown that the limit due to spherical aberration is

$$\Omega^2 \leqslant 8\pi^2 n^2/\nu\tau; \tag{8.27}$$

it cannot be simply expressed in the form of (8.20). This limit applies when the condition (8.25) is used, taking into account axial distances only. When

Fig. 51 The use of a pair of wedges to change the shift and the delay independently.

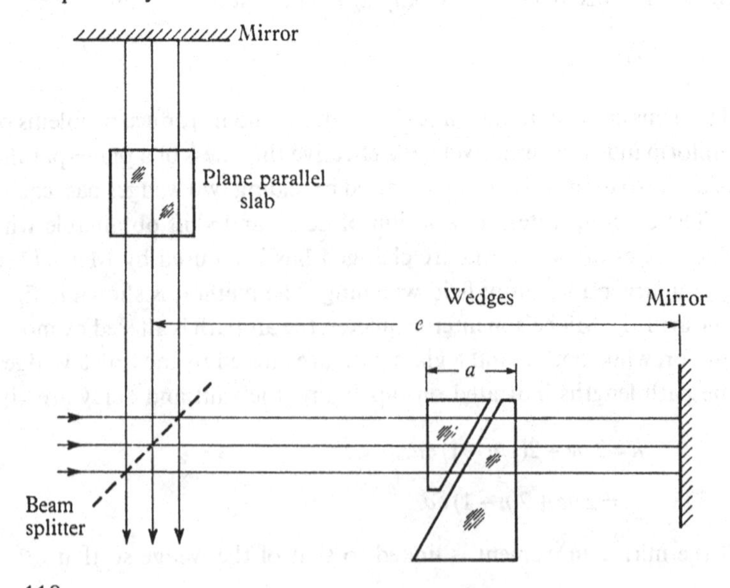

the actual solid angle is included in the condition, by increasing $\delta e/\delta a$ by $(n^2-1)\Omega/4\pi n^3$, twice the solid angle may be used.

The limit due to chromatic aberration applies when the interferometer covers a wide spectral range, for the condition (8.25) gives full compensation for only one frequency. If the dispersion of the glass across the range of frequencies studied is δn, and the interferometer is adjusted so that (8.25) applies for the mean refractive index, the gain is limited by chromatic aberration to

$$G=2n(n^2-1)/\delta n. \tag{8.28}$$

Since it is inconvenient to move both a wedge and a mirror, Bouchareine & Connes [1963] have modified this compensation, as shown in fig. 52(*a*), by placing one wedge in each arm. Each mirror is a reflecting coating on the back of the wedge and, seen from in front, appears at an inclination somewhere between those of the two surfaces of the wedge. If one wedge is moved in the direction of the apparent position of its reflecting surface, the apparent distance obviously remains fixed and no shift is introduced. The delay, however, is varied. The astigmatism of the wedges is not equal for the two arms, except for zero delay, and it is this that limits the usable solid angle. The source fringes are now hyperbolic instead of circular, provided the direction of movement corresponds to the best focus for the image of the reflecting surface, that is, the position midway between the two astigmatic foci, tangential and sagittal. The gain is given by

$$G=2\cot^2\theta, \tag{8.29}$$

where θ is the angle of the wedges.

Fig. 52 Field-widened interferometers with one single moving prism: (*a*) form of Bouchareine & Connes [1963]; (*b*) alternative form, of no advantage.

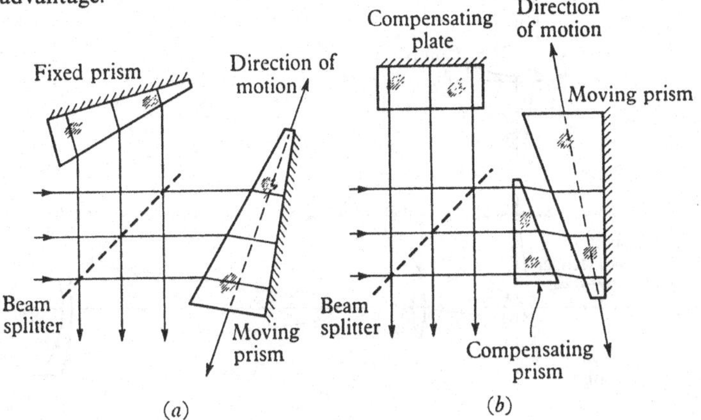

It might be thought that the addition of a second wedge to each arm, as shown in fig. 52(*b*), would correct the astigmatism. But now the two wedges do not keep in contact as one is moved in the direction of the apparent image. An inclined, plane-parallel region of air is introduced into one arm and the astigmatism due to this is equal to that given by the simpler system of fig. 52(*a*). A review of the various two-media systems is given by Ring & Schofield [1972].

Instead of plates or wedges of a dispersive medium, focusing systems can be used in an interferometer to give an independent control of shift; Golay [1973] has given the theory of these methods. The simplest one is shown in fig. 53, a lens-testing interferometer in which separate images of the source are formed in each arm. Different lenses could be placed at these images to control, separately for each arm, the apparent distance to the plane of observation. As the lenses are at planes conjugate with the source, any defects in their surfaces do not appear as phase errors in the test fringes, although they can reduce their visibility. In fact, a lens in this position can be regarded as a radially symmetrical phase disturbance Δw_x that can compensate the path variations Δp_x of the interferometer. For lens testing,

Fig. 53 A lens-testing interferometer in which the shift can be altered without changing the delay; a modified Williams interferometer.

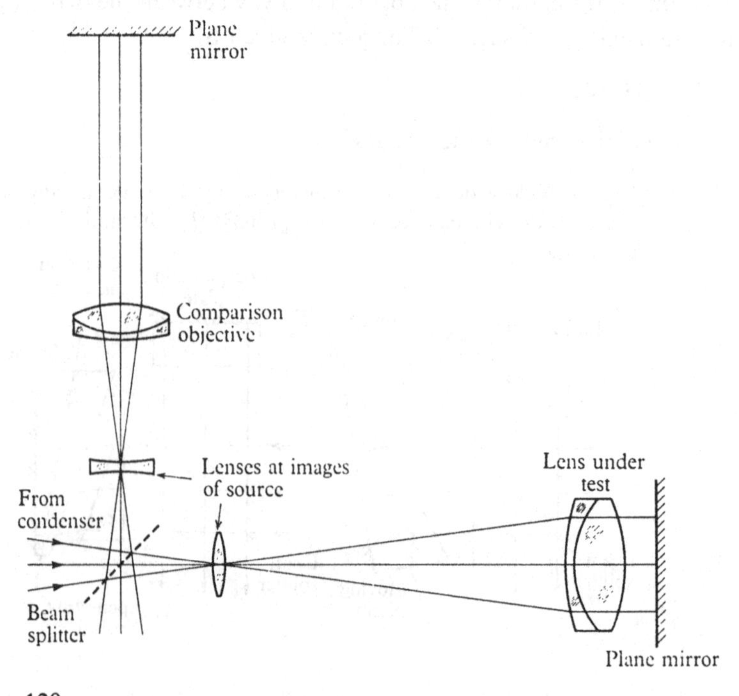

the compensation is obtained by selecting suitable lenses from an ophthalmic trial set, as a continuously variable adjustment is not warranted. But for spectroscopy, the compensation should be continuous. As it usually requires only a small change of power of the optical component at the source, Cuisenier & Pinard [1967] have proposed the use of spherical mirrors rather than lenses; these could be given the required change of power by mechanical deformation.

Another compensation based on focusing elements is shown in fig. 54. A telescope is placed in each arm of a Michelson interferometer. If one of these is displaced longitudinally, it does not alter the delay, but changes the shift. When this adjustment is combined with a movement of a mirror, there are again two variables available to change both the delay and shift. For lens testing, it is immediately applicable but, for spectroscopy, the two movements need to be coupled to keep the shift zero as the delay is varied. Connes [1956] has shown that, if the telescope has a magnification $2^{-1/2}$, it may be coupled directly to the mirror in the other arm, as shown in fig. 54. For use outside the regions where transparent materials for lenses are readily available, a mirror telescope may be used. But, as well as being limited by the aberrations of the telescope for objects not at infinity, the usable solid angle may be limited by the sheer length path in such a telescope.

8.3.2. Tilt compensation

A plane mirror is unaffected by lateral movements and a Michelson interferometer can have no shear, if the fringes are localized at the mirrors. A retroreflector is unaffected by rotations about its centre and, when it replaces the mirrors of a Michelson interferometer, the instrument can have

Fig. 54 A field-widened interferometer based on telescopes.

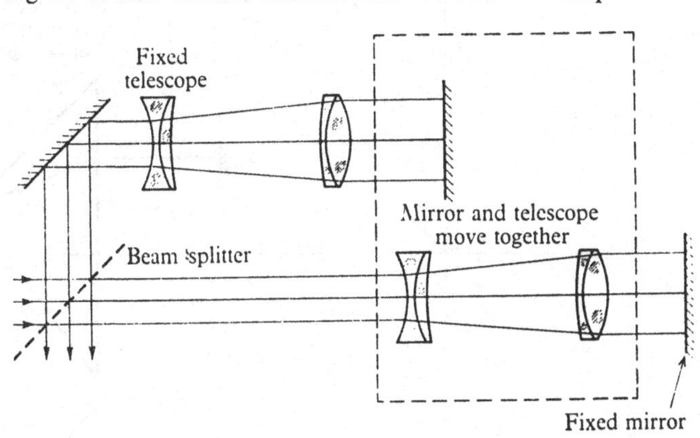

no tilt. To produce an interferometer for which errors in the motion of the component used to change the optical path can cause neither a shear nor a tilt, a fixed plane mirror and a movable retroreflector can be combined, as shown in fig. 55. This interferometer has been used for length measurements by Terrien [1959] and its properties have been discussed by Murty [1960].

Alternatively, an interferometer can be designed so that the tilts cancel for the two arms. The two mirrors are rigidly connected and are most conveniently set back to back. Such an interferometer is compensated against tilt when a beam through the two arms, represented by a flat strip, joins at this double mirror to form a Möbius band with π twist. An example is shown in fig. 56, the only form that lies in a plane, but, to do so, it requires a pair of roof mirrors. Practical experience has shown that these interferometers are liable to distortion with changes of temperature. There are advantages in sticking to a plane system. All Möbius-band interferometers have mirrors whose planes of incidence are not parallel to that of the beam splitter and, by the theory of §7.9, they will affect the polarization of the radiation. To be insensitive to the polarization of the incident radiation, more complex systems are necessary.

Tilt cancellation is simpler for interferometers with retroreflectors. No twist to the beam is required: the tilt or shear cancels when the retroreflectors at the end of each arm are joined back to back.

Another method of compensating for tilt is shown in fig. 57. Displacement of one or more mirrors is replaced by a rotation of part of one

Fig. 55 A Michelson interferometer immune to both shear and tilt.

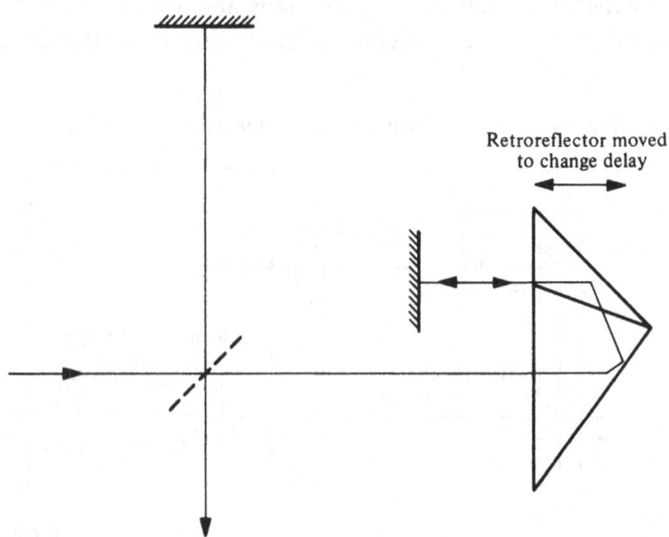

arm and the beam splitter about the centre of the latter. The delay is also changed by a rotation, without a tilt being introduced, in an interferometer due to Young and described by Mertz [1965].

8.4. The Mach–Zehnder interferometer

While the Michelson interferometer is a reflex system, the basic interferometer in which the two beams continue on to recombine at a second beam splitter is the Mach–Zehnder interferometer, shown in fig. 58. It has the property, discussed earlier, of separating the outgoing beams from the incident radiation and thus giving two accessible interferograms. Its geometry shows that both shears and tilts can be introduced independently, without introducing shifts or leads, but it is not an interferometer suited to giving a variable delay, except in the form shown in fig. 4(*a*).

Fig. 56 A Möbius-band interferometer, due to Masui [Steel, 1964*c*].

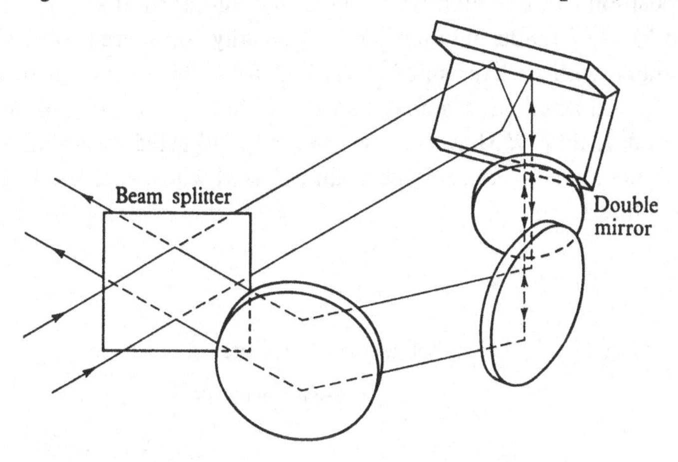

Fig. 57 Interferometer compensated for tilt [Sternberg & James, 1964].

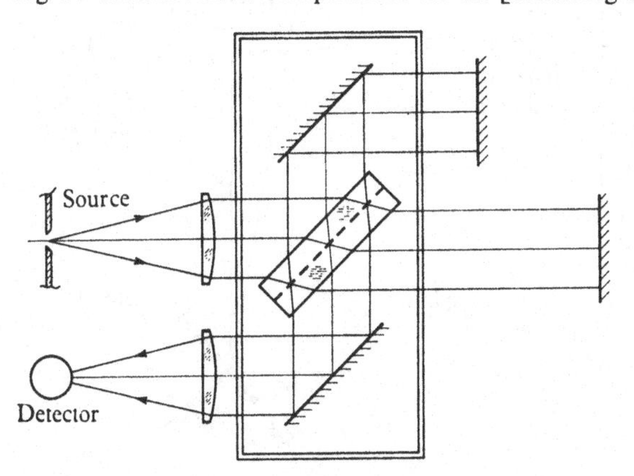

If the beam splitters or mirrors are displaced or rotated from their ideal position, the properties of the interferometer can be found from equation (3.4). Thus if the i-th component is given a rotation θ_i about an axis normal to the plane of the instrument, for the dimensions shown in fig. 58 there is a shear

$$s=2g(-\theta_1+\theta_2+\theta_3-\theta_4)+2a(\theta_2+\theta_3-2\theta_1),$$

and a tilt

$$t'=2f(-\theta_1+\theta_2+\theta_3-\theta_4)+2a(\theta_2+\theta_3-2\theta_4),$$

but, to the first order, no shift nor lead. The test fringes are localized where the shear is zero, that is, where

$$g=a(\theta_2+\theta_3-2\theta_1)/(\theta_1-\theta_2-\theta_3+\theta_4). \tag{8.30}$$

This position can be found by the construction given in §7.5.1.

The Mach–Zehnder interferometer is usually considered to be a plane instrument, with perhaps small deviations from this plane. But it can be made up as a three-dimensional instrument. Such a form is considered by Bennett & Kahl [1953], who show that, when adjusted, it should have its mirrors tangential to the surface of an ellipsoid whose foci are at the two beam splitters.

Fig. 58 The Mach–Zehnder interferometer.

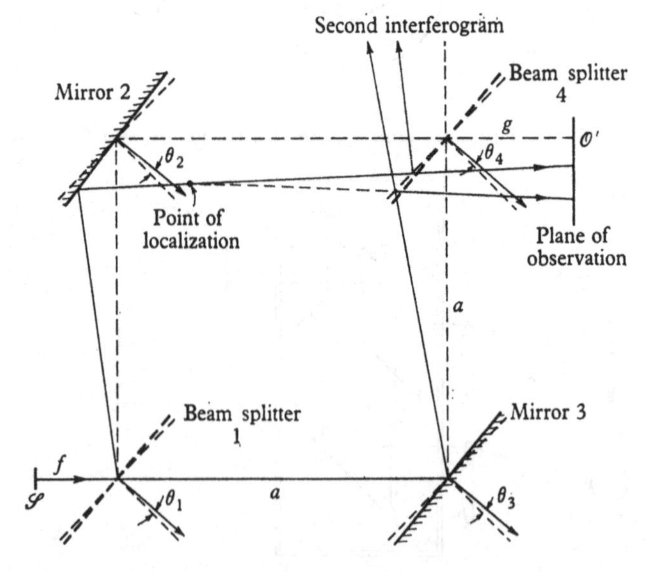

8.5. The cyclic interferometer

Two forms of the Sagnac or cyclic interferometer are shown in fig. 59. The two beams follow the same path round a closed circuit in opposite directions and their longitudinal properties are the same: there is no shift, lead, nor delay. Shears or tilts can be introduced, however, by rotations or displacements of the mirrors or beam splitter.

The properties of a cyclic interferometer differ according to whether there is an odd or even number of reflexions in the circuit. For the form shown in fig. 59(*a*), with three mirrors and a beam splitter, there is an odd number of reflexions for each beam. This form is insensitive to a displacement of any of the mirrors or of the beam splitter, but rotations of these components give a shear

$$s=4(2a+g)(\theta_1-\theta_2+\theta_3-\theta_4), \tag{8.31}$$

which vanishes for fringes localized at the mirror opposite the beam splitter, where

$$g=-2a.$$

When there is an even number of reflexions for each beam, as in the form with two mirrors, either displacements or rotations of the mirrors produce a

Fig. 59 Two forms of the cyclic interferometer: (*a*) with an odd number of reflexions in each beam; (*b*) with an even number; (*c*) notation for a displacement *e* and a rotation θ of the *i*-th component.

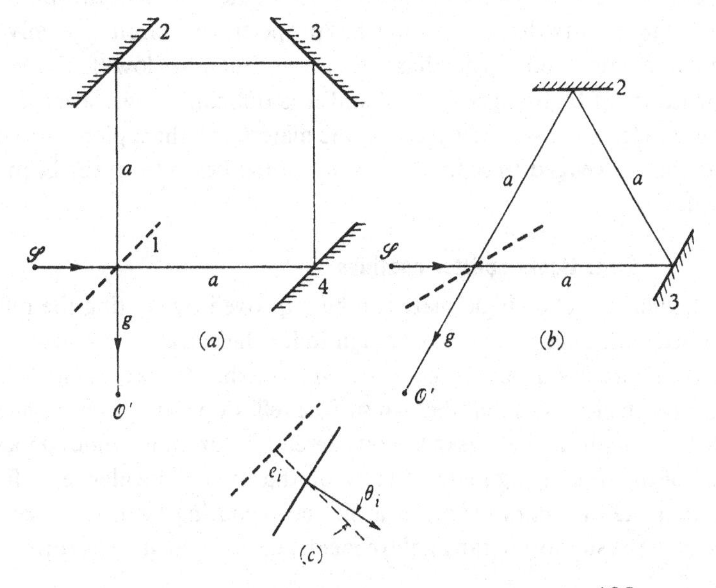

shear. If the i-th component is displaced a distance e_i in the direction of the normal and rotated θ_i about an axis perpendicular to the plane of the instrument, the shear is

$$s=2(e_2+e_3-\sqrt{3}e_1)+2a(3\theta_1-\theta_2-\theta_3); \quad (8.32)$$

it is a lateral separation of the beams that is independent of the distance g to the plane of observation. Finally, if a phase disturbance is placed within the interferometer, the image given by one beam is folded or reverted with respect to that from the other for the interferometer with three mirrors, but the two images are the same with two mirrors. A plane of observation following the interferometer, however, is seen by both instruments with no folding.

8.6. Beam splitters

Any interferometer with division of amplitude requires some form of beam splitter. The ideal partially reflecting beam splitter would transmit and reflect one half the incident radiation with a constant phase change, whatever its polarization. If used over a range of frequencies, it would also remain close to this ideal over this range and, in addition, it would introduce no phase changes that vary with frequency. In practice, all reflecting surfaces used at other than normal incidence have properties that vary with the polarization of the radiation, and beam splitters may depart far from this ideal.

The simplest partial reflector is a boundary between two transparent media. The surface of a glass plate in air would give a beam splitter that, although sensitive to polarization, has properties which change only slowly with frequency, and has no phase variations. But it has low efficiency, unless a material of high index is used, and it is difficult to get rid of the beam reflected from the second surface of the material. A thick plate is needed, or one that is wedged to separate this unwanted beam from the beam being used.

8.6.1. Beam splitter coatings

The efficiency of a single plate can be improved by coating the reflecting surface with either a metal or a high-index dielectric. For a beam splitter used at one frequency, a dielectric coating should have a quarter-wave effective thickness so that the two surface reflexions add in phase, and there are then no phase changes to be considered. The optimum index depends on that of the supporting plate and also on the angle of incidence. In fig. 60 is plotted the optimum index for a dielectric coating against the refractive index of the support for an angle of incidence of $\pi/4$ in air. The top curve has been calculated to satisfy the condition of maximum efficiency for both polarizations,

$$\mathscr{R}_{\parallel}(1-\mathscr{R}_{\parallel})+\mathscr{R}_{\perp}(1-\mathscr{R}_{\perp})=\text{max.},\tag{8.33}$$

where the reflectances are given by the standard theory for thin coatings, for example, in Born & Wolf [1959] or Heavens [1955]. These optimum indices may be compared with those of available materials, given by Heavens [1960], and it will be seen that the ideal index is higher than those of materials commonly available, except for the infrared, where silicon ($n\approx 3.5$) and germanium ($n\approx 4$) become transparent.

When a dielectric coating is used as a beam splitter for a range of frequencies, the thickness departs from a quarter wavelength and the efficiency drops for frequencies other than that for which it is designed. This drop may be reduced by an increase of the index of the coating. The curves shown in fig. 61 are for a germanium coating, which has an index greater than the optimum for a single frequency and, for unpolarized light, they show a high efficiency over a range of frequencies. A further effect introduced by the departure from a quarter wavelength of the thickness of coating is a phase change which varies with frequency. The difference in

Fig. 60 Optimum refractive index for a single dielectric coating, for use as a beam splitter at $\pi/4$.

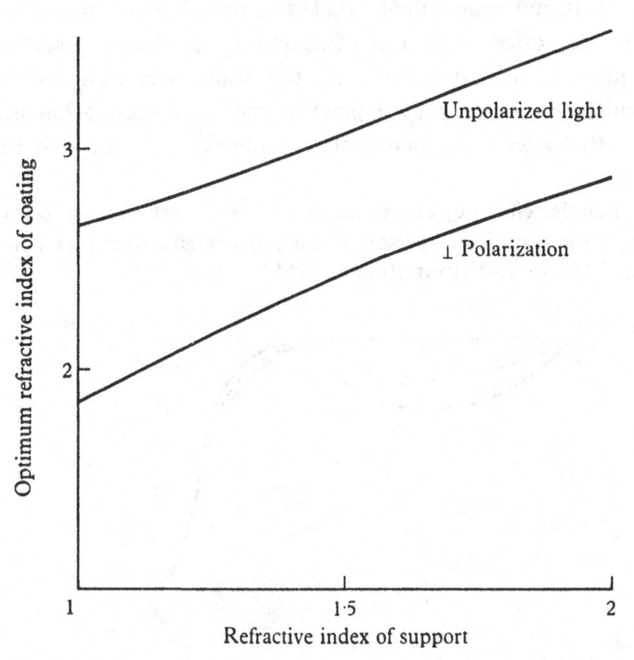

phase for the two polarizations causes a loss of visibility when unpolarized light is used, while the departures of the mean phase from a straight line are important in interference spectroscopy.

If a metal coating is used, it has a phase difference even for a single frequency: the two polarized components have different phases and the fringe visibility in unpolarized light is reduced. In addition, the two beams have unequal intensities and this further reduces the visibility. Although both have the same transmittance through the metal layer, their reflectances differ according to whether the reflexion is on the air side or the support side. This difference also varies with frequency.

This phase difference with absorbing beam splitters means that the two interferograms of §8.4 are not truly complementary, in that the phase difference of their modulations is not π. In fact, a beam splitter can be designed to give two interferograms in quadrature [Raine & Downs, 1978], which is useful for electronic fringe counting.

8.6.2. Beam splitters in convergent light

In the Michelson interferometer of fig. 47, the beam splitter is in collimated light and produces no aberration for the image of the source. But it introduces aberrations into the image of the plane of observation. This aberration matches in both arms and so does not affect the formation of the fringes, but it affects slightly the quality of their image.

An inclined plate in convergent light has aberrations, the chief of which is astigmatism. This can often be seen in a Fizeau interferometer. A second plate at the opposite inclination corrects the coma, but doubles the astigmatism, which is corrected by a plate at the same inclination but rotated in azimuth by $\frac{1}{2}\pi$. A sequence of four plates, each at $\frac{1}{2}\pi$ to the

Fig. 61 Relative efficiency of a beam splitter, $\mathscr{R}(1-\mathscr{R})$, over a range of frequencies (or coating thicknesses), for a coating of refractive index 4 on a substrate of index 1.45 [from Roland, 1965].

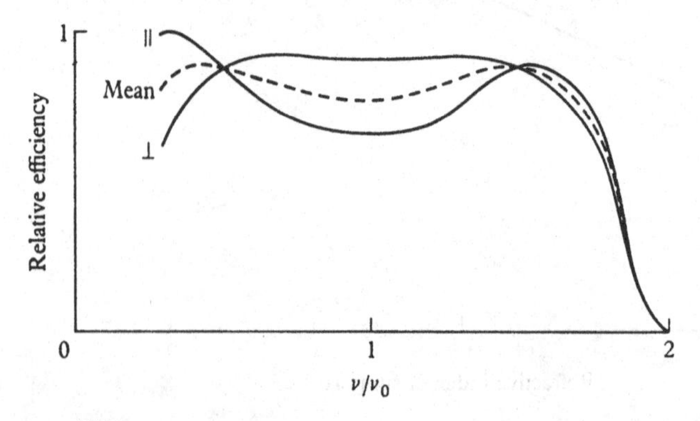

preceding one, corrects coma, astigmatism, and the lateral displacement of the beam. Some spherical aberration remains.

Another method of correcting the coma and astigmatism of a thick inclined plate has been given by Cuny [1955]. The plate is given an appropriate wedge angle, which effectively adds a prism to it. The aberrations of this correct those of the plate. Coma and astigmatism are both corrected only if the refractive index or the angle of tilt is used as a further variable, but when the astigmatism is corrected for a plate at $\pi/4$, the residual coma is small. The prism introduces a small amount of anamorphic magnification, so that circular fringes are seen as ellipses.

The notation is shown in fig. 62. For a beam splitter of centre thickness d, and refractive index n, inclined at an angle θ, the wedge angle required to correct the astigmatism is

$$\alpha \approx \frac{d}{2z} \frac{\sin\theta\cos^2\theta}{n^2 - \sin^2\theta},$$

which, for $\theta = \pi/4$, reduces to

$$\alpha = d/[2\sqrt{2}(2n^2 - 1)z], \tag{8.34}$$

where z is the distance to the source. Other methods of correction, involving spherical refracting surfaces, are given by Doherty & Shafer [1980].

8.6.3. Compensating plate

A Michelson interferometer with a beam splitter consisting of a coating on a supporting plate is unsymmetrical unless this is balanced by a *compensating plate* in the other arm, as shown in fig. 47. When this plate does not match the support of the beam splitter in both material and thickness, the two aberrations, astigmatism and chromatic aberration, are not compensated. If the plates differ only in thickness, either aberration, but not both, can be

Fig. 62 Notation for the theory of a wedged plate as support for a beam splitter.

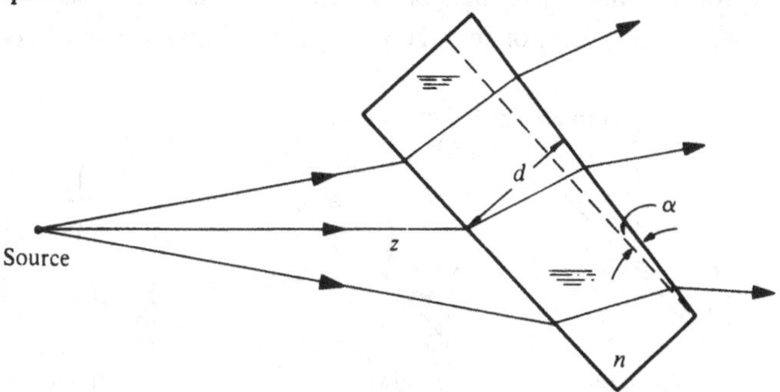

corrected by tilting the compensating plate to a different angle to that of the supporting plate. If the error in the thickness d is δd, and the chromatic aberration is corrected, the astigmatism remaining would limit the source size. For plates with a refractive index between 1.4 and 2, set at $\pi/4$, this limit has the approximate value

$$\Omega \leqslant 2\lambda/\delta d. \tag{8.35}$$

The error δd should be small enough for this solid angle to be large compared with the size of the source to be used.

If the plate is tilted instead to correct the astigmatism, the variation of optical path with frequency, due to the chromatic aberration, is given by

$$\Delta p \approx \Delta n\, \delta d. \tag{8.36}$$

where Δn is the dispersion of the plate over the range of frequencies to be covered. Obviously, in an interferometer used to measure phases in quasi-monochromatic light, the astigmatism is the aberration that should be corrected. This correction is tested by observing the source fringes, which are circular when there is no astigmatism. An interferometer used for spectroscopy to cover a large range of frequencies should have its chromatic aberration corrected since this represents a phase error dependent on frequency. The method of testing for this correction is to examine the channelled spectrum discussed in §7.6. When the fringes are parallel to the direction of dispersion of the spectroscope, there is no phase variation.

8.6.4. Symmetrical beam splitters

The compensating plate corrects the major asymmetry caused by the support of the beam splitter, but smaller asymmetries remain to cause the phase errors and mismatch of intensities discussed above. A completely symmetrical beam splitter having one of the forms shown in fig. 63 will eliminate these. For longer wavelengths, an unsupported dielectric film can be used, for example, Melinex or Mylar sheet used in the far infrared. Otherwise, the dielectric or metal coating can be sandwiched between two

Fig. 63 Symmetrical beam splitters.

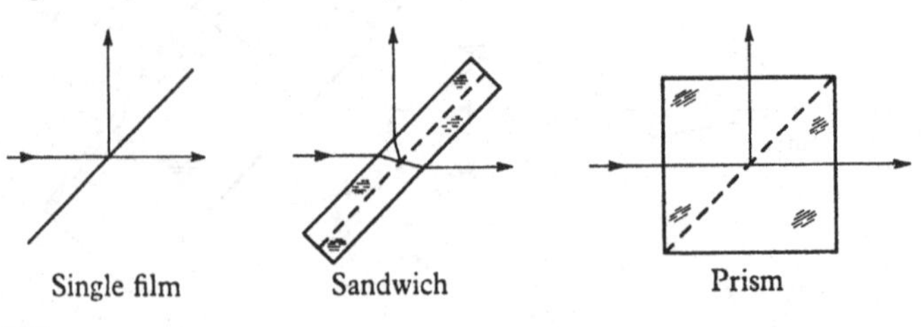

identical plates or two identical prisms. For complete symmetry, this sandwich should be made without a cement, the second plate being joined by optical contact, as described by Cuisenier & Pinard [1967] or Smartt & Ramsay [1964]. Since such contact is relatively permanent, care should be taken to set prisms correctly before contacting, for there is no separate compensator to adjust later. For the same reason, the two plates should be closely matched in thickness.

8.6.5. Prism beam splitters

There is no inclined plate in the third beam splitter shown in fig. 63 and the only aberrations are the spherical and chromatic aberrations of a parallel-sided block of glass. Other prisms with this same property are used to obtain different dispositions of the beams. The Kösters prism, described by Saunders [Strong, 1958], and shown in fig. 64(*a*), gives parallel output beams. It can be modified as shown to give all beams in line, or outputs in opposite directions. As drawn, all have an angle of incidence at the coating of $\pi/4$, but this can be modified, between at least $\pi/6$ and $\pi/3$.

The optimum refractive index for a single coating is higher for a sandwich, fig. 63, than for an unsymmetrical beam splitter. It is higher again for a prism, well above that of common materials used with visible light. The simplest solution is a metal coating, even with its absorption loss. Multilayer dielectric coatings can be designed, either to match both polarizations [Thelen, 1976], or to be achromatic [Clapham, 1971]. Both require many layers, and a design that combined both properties would be even more complicated.

8.6.6. Polarizing beam splitters

If a single surface or a symmetrical beam splitter is used at its Brewster angle, the reflected beam is linearly polarized. A Michelson interferometer with such a beam splitter used twice will transmit only this one component. The transmitted beam, however, is not fully polarized by one passage; to

Fig. 64 Prism beam splitters: (*a*) the Kösters prism; (*b*) modification for in-line beams; and (*c*) for opposed output beams.

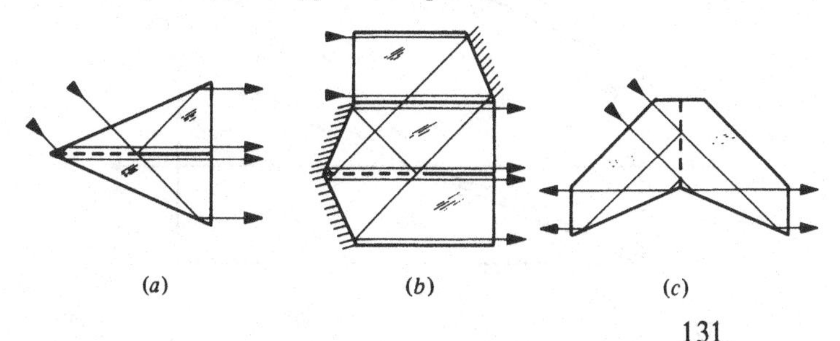

obtain both beams fully polarized, a beam splitter of the types shown in fig. 65 is required.

The first represents a Glan–Foucault prism, a pair of calcite prisms, with their optic axes normal to the incident light, separated by an air film at which the ordinary ray is totally reflected. The optimum angle for the prisms has been given by Thompson [1905]. It depends on the range of frequencies for which the prism is used. If a cement is used in place of the air film, prisms of much larger angle are required, similar to the second type shown. This can also be made in the form shown, consisting of two glass prisms with a thin plate of calcite between them [Steel, Smartt & Giovanelli, 1961].

The third polarizing beam splitter shown makes use of a series of layers at the Brewster angle. These are of alternately high and low index, n_h and n_l, and effectively $\frac{1}{4}\lambda$ thick. For the glass prisms between which they are mounted to have an angle of $\pi/4$, the glass should have a refractive index n_p

Fig. 65 Polarizing beam splitters.

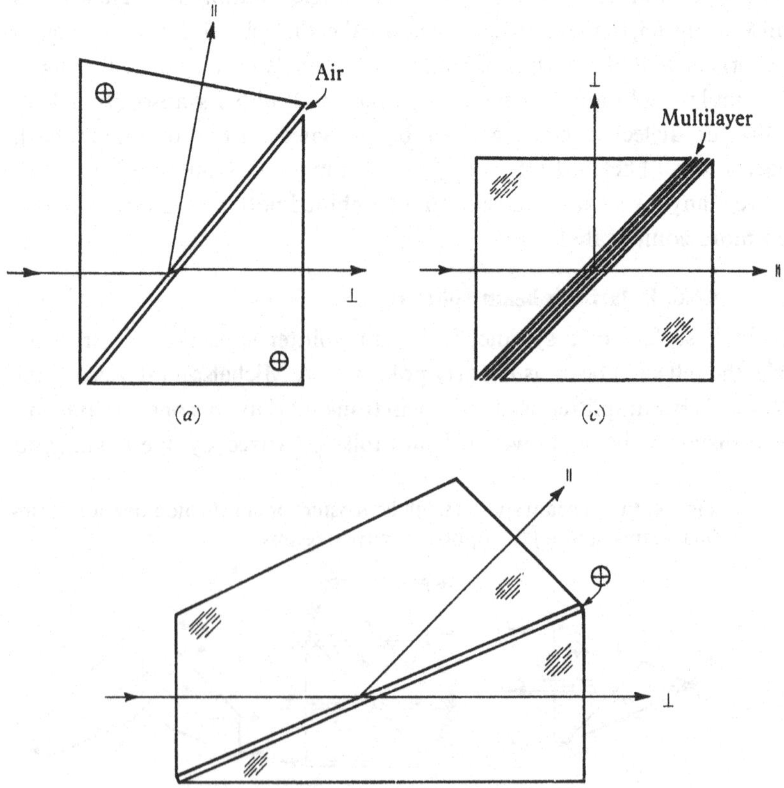

given by

$$2/n_p^2 = 1/n_h^2 + 1/n_l^2. \tag{8.37}$$

This beam splitter was introduced by McNeille, and has been described by Banning [1947], and Clapham, Downs & King [1969]. It requires a special glass with the index given by (8.37). Another form, designed by Netterfield [1977], does not have as wide a spectral range, but can be made of glass of any refractive index.

The second prism is usually cemented to the coated surface of the first. In the Netterfield type, the glass and cement can be chosen to have the same refractive index. It is also possible to join the prisms by optical contact [Smartt & Ramsay, 1964].

8.7. Polarization interferometers

As stated in §1.3.3, a polarization interferometer, in addition to having a polarizing beam splitter, gives the two beams different paths by making use of the birefringence of certain transparent crystals. Uniaxial crystals are most commonly used and the ordinary refractive index n_o and the birefringence $\mu = n_e - n_o$ are listed in Table 8.1 for the more useful materials with their useful transmission ranges, obtained mainly from Wolfe [1978]. The respective indices are for green light. In addition to these materials, mica is commonly used for quarter- and half-wave plates, but it is not suitable for thick components.

The birefringent form of a Michelson interferometer is a *retardation plate* or *linear retarder*, a plate of birefringent crystal with plane parallel faces. A uniaxial crystal is cut so that the optic axis is parallel to these faces. To

Table 8.1. *Refractive index and birefringence of some uniaxial crystals*

Crystal	Index n_o	Birefringence μ	Range of use ν (PHz)	λ (μm)
Ammonium dihydrogen phosphate	1.525	−0.045	0.25–1.5	0.2–1.2
Calcite	1.662	−0.17	0.14–1.0	0.3–2.2
Magnesium fluoride	1.378	+0.012	0.04–2.0	0.15–7.0
Potassium dihydrogen phosphate	1.510	−0.041	0.23–1.0	0.3–1.3
Quartz	1.546	+0.009	0.1–1.5	0.2–3.0
Rutile	2.649	+0.31	0.07–0.7	0.43–4.5
Sapphire	1.771	−0.008	0.12–1.5	0.2–2.5
Sodium nitrate	1.590	−0.25	0.1–0.9	0.33–3.0

produce interference fringes, this plate is placed between polarizers. Although it is a single-passage component, it is closely equivalent to a Michelson interferometer with a fixed delay and no tilt, and it may be used in collimated or uncollimated light. The astigmatism of the extraordinary ray gives an effect equivalent to shift: the source fringes limit the usable field, being now hyperbolic and not circular.

Normal to the faces, the path difference for a plate of thickness d and birefringence μ is

$$p_0 = \mu d. \tag{8.38}$$

The quantity $N = p_0/\lambda$, corresponding to the order of interference, is called the *retardation* for polarization interferometers. The variations of path-difference across the source, represented by $p(\mathbf{x}) = p_0 + \Delta p_x$, are given, for uniaxial crystals, by

$$p(\theta, \phi, d) \approx \mu d\left[1 - \frac{\theta^2}{2n_o}\left(\frac{\cos^2\phi}{n_o} - \frac{\sin^2\phi}{n_e}\right)\right], \quad \theta \text{ small}, \tag{8.39}$$

where the vector $\mathbf{x}$ at the source subtends an angle $\theta = |\mathbf{x}|/z$ and has an azimuth ϕ.

The coherence for a birefringent interferometer is given by the integral of $\exp[-2\pi i p(\mathbf{x})/\lambda]$, taken over the source. For a circular source of solid angle $\Omega = \pi\alpha^2$, where α is the maximum value of θ, the result is

$$\mu_{12} \approx \left[1 - \frac{\Omega^2 \nu^2 \tau^2}{12n^4} + \cdots\right] \exp\left[-2\pi i \nu\tau\left(1 - \frac{\mu\Omega}{8\pi n^3} + \cdots\right)\right], \tag{8.40}$$

where n_o and n_e are both replaced by the mean index n. To compare this with the expression (8.7) for a Michelson interferometer, the latter should also be given as a series, namely

$$\mu_{12} \approx \left[1 - \frac{\Omega^2 \nu^2 \tau^2}{24} + \cdots\right] \exp\left[-2\pi i \nu\tau\left(1 - \frac{\Omega}{4\pi}\right)\right]. \tag{8.41}$$

A comparison of the moduli of these two functions shows that the birefringent interferometer has a gain

$$G = n^2/\sqrt{2}; \tag{8.42}$$

this is, $1/\sqrt{2}$ times the gain that would be given by the use of a material of refractive index n in a Michelson interferometer.

Two forms of field-widened retarders have been given by Lyot [1944]. For the first he used composite plates, made of two different crystals with birefringence of opposite sign, for example, quartz and calcite. For his second form, each retarder is split into two halves which are set with their axes crossed with a half-wave plate between them at an azimuth of $\pi/4$. The source fringes then become almost circular, and are represented by the path difference

$$p(\theta, \phi, \tfrac{1}{2}d) + p(\theta, \phi + \tfrac{1}{2}\pi, \tfrac{1}{2}d) \approx \mu d\left[1 - \frac{\mu\theta^2}{4n_o^2 n_e} + \cdots\right]. \tag{8.43}$$

When this is used to derive the integral over the source, the gain is found to be

$$G = 2\sqrt{2n/\mu} \tag{8.44}$$

with respect to a single plate of the same material. For calcite, $G = 26$, for quartz $G = 480$; slightly different values have been given by Evans [1949], who used a criterion that did not involve integration, and gave the gain in angular radius ($G^{1/2}$ in our notation).

A retarder may be tuned over one order of interference by adding to it a rotatable half-wave plate, set between two quarter-wave plates with axes at $\pi/4$. This is the same as the frequency changer shown in fig. 28. If the azimuth of the half-wave plate is ϕ, the transmission of a birefringent interferometer of delay $\tau = \mu d/c$, fitted with this tuning, is

$$\tfrac{1}{2}[1 + \cos(2\pi\nu\tau + 4\phi)]. \tag{8.45}$$

Tuning over more than one order is obtained by making the retarder as two wedges which can be moved past each other to give a plate of variable thickness. If a variable delay that includes zero is required, this variable plate must be crossed with a fixed plate. If the fixed plate has a thickness D and the variable plate a thickness $D + d$, the path-difference is given in terms of (8.39) as

$$\begin{aligned} &p(\theta, \phi, D + d) - p(\theta, \phi + \tfrac{1}{2}\pi, D) \\ &\qquad \approx \mu d - \frac{\theta^2\mu}{2n^2}(2D + d)(\cos^2\phi - \sin^2\phi), \end{aligned} \tag{8.46}$$

where again n_o and n_e have been replaced by their mean n. The axial paths subtract but the off-axis variations add so that there is reduction in the usable angle by the factor

$$G = d/(2D + d) \tag{8.47}$$

over the gain given by (8.42). This loss can be overcome by the field-widening techniques given earlier.

Other interferometers have their birefringent form. A fixed shear, without shift or delay, is given by the Wollaston prism shown in fig. 66(*a*), and a tilt can be built in. Nomarski [1955] has shown how to modify the design to localize the fringes outside it, as shown in fig. 66(*b*). One of the component prims is now cut with its optic axis inclined. Françon & Sergent [1955] have introduced a field-widened form, fig. 66(*c*). For collimated light, a shear is given by a single plate of calcite with its axis inclined to the faces as shown in fig. 67(*a*). It has also a shift, due to the different apparent distances to the ordinary and extraordinary images, and a delay. These are corrected in the Savart polariscope, shown in fig. 67(*b*), which consists of two such plates at relative azimuths of $\frac{1}{2}\pi$. In the field-widened form, due to Françon [1957], the plates are at π azimuth with a half-wave plate between them. A variable shear can be obtained by two calcite plates or two polariscopes in series that are counterrotated. To prevent each beam from the first plate being split again as it enters the second, quarter-wave plates should be added to the adjacent faces of the polariscopes, so that the two beams are circularly polarized in the space between the two polariscopes. Two examples are shown in fig. 68. In each a rotating $\lambda/2$ plate has been added to the variable retarder to produce a frequency difference between the two polarizations.

Fig. 66 Forms of Wollaston prism, all giving an angular shear.

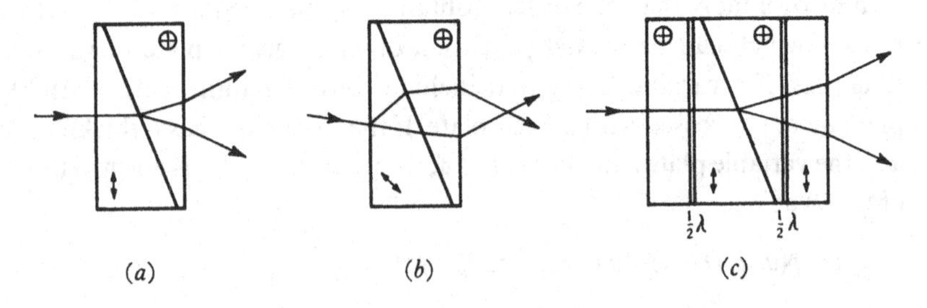

Fig. 67 Calcite plate and Savart polariscopes, giving a shear in collimated light.

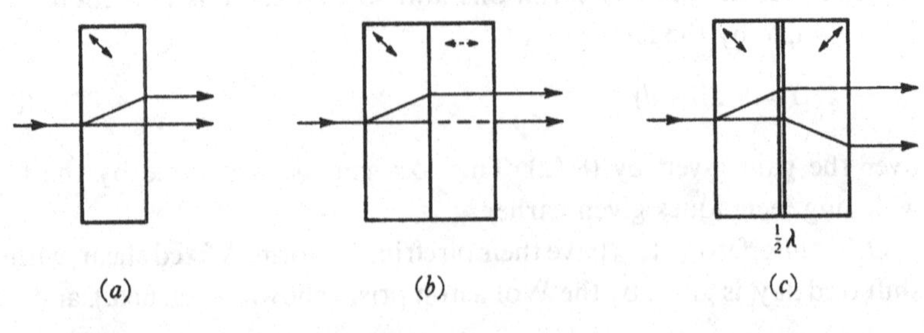

8.8. Diffraction forms

Interferometers can be made up of either transmitting or reflecting gratings. They can be used both as beam splitters and, instead of mirrors, as beam deflectors. Since a displacement of a transmission grating has much less effect on the optical path than a displacement of a mirror, this form of the interferometer is easier to adjust and less sensitive to disturbances. The gratings can be made holographically, tailored to suit the interferometer [Chang, Alferness & Leith, 1975]. Fig. 69 shows two diffraction forms of the Mach–Zehnder interferometer, due to Weinberg & Wood [1959].

8.9. Microwave interferometers

While many microwave interferometers are similar to the systems already described, microwave interferometry has provided several techniques, which have been adopted later for higher frequencies.

For millimetre and centimetre microwaves, lenses and mirrors can be made larger than the wavelength and microwave interferometers can have the same forms as their counterparts for visual radiation. Microwave interferometers have been reviewed by Tremblay & Boivin [1966]. The

Fig. 68 Method of obtaining a variable shear with calcite plates or with Savart polariscopes.

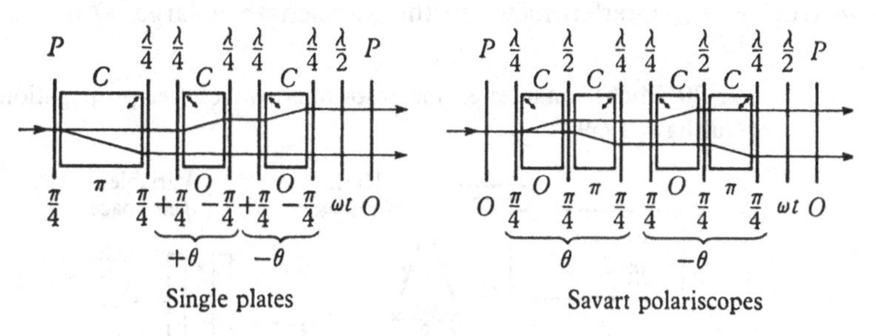

Fig. 69 Mach–Zehnder interferometers with gratings as beam splitters B and beam deviators D.

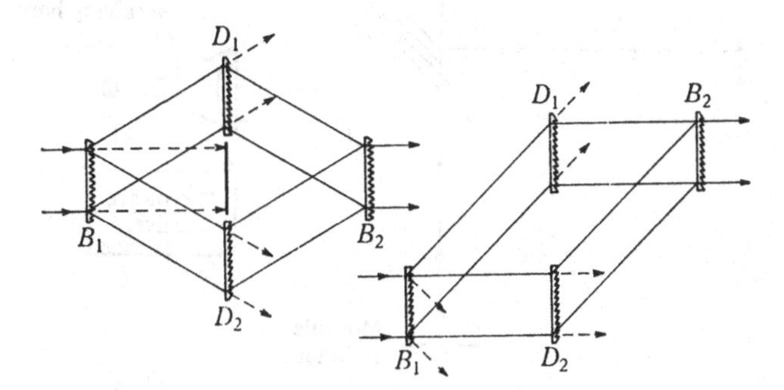

free-propagation form of the Michelson interferometer that follows closest to the optical model is that of Culshaw [1959], shown in fig. 70. The beam splitter consists of two dielectric plates, each a quarter-wavelength thick, at a variable separation that is adjusted to give equal transmitted and reflected intensities. A wire grid could also be used. An alternative beam splitter to that shown in fig. 70 consists of two dielectric prisms. While a single prism will reflect all the radiation by total internal reflexion, the second prism, when close to it, collects part of the evanescent wave. The amount of radiation collected can be controlled by adjustment of the spacing between the prisms, and it is possible to match the intensities of the transmitted and reflected beams. This beam splitter is one employing frustrated total internal reflexion.

Since microwave sources are coherent, the simple theory of Chapter 4 applies. Only the delay has significance, since the other parameters, needed to describe interferometers with extended sources and planes of observation, have no importance when the source and detector can be regarded as points. But the detector can be scanned across the plane of observation to build up an interferogram.

8.9.1. Diffraction corrections

The simple theory is a good approximation only when the ratio of the aperture of the interferometer to the wavelength is large. Otherwise the

Fig. 70 Michelson interferometer for microwaves: free propagation form [Culshaw, 1959].

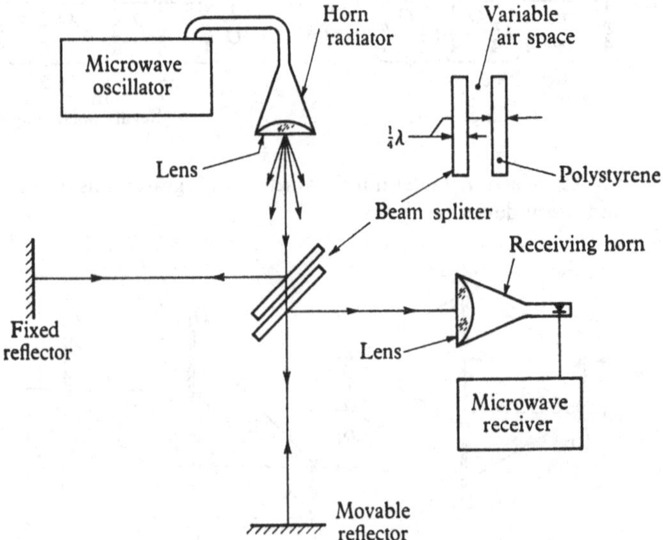

effects of diffraction become important and, for microwave interferometers, both diffraction and interference theory must be combined.

The effects of diffraction can be regarded as a correction to be applied to derive the effective delay from the value calculated from the axial optical path. As with all diffraction theory, the treatments can be given in either the Fraunhofer or the Fresnel approximation. In the former, the correction is smaller but more radiation is lost from the instrument. Froome [1954, 1958] has treated Fraunhofer diffraction, and Culshaw, Richardson & Kerns [1960] have treated Fresnel diffraction.

Initially ignored in visible-light interferometers, diffraction corrections are also needed for these when apertures are small and path differences large. Tango & Twiss [1974] have studied the diffraction effects in a Michelson stellar interferometer, in which path differences are large. Dorenwendt & Bönsch [1976], and Lyubimov, Shur & Etsin [1978] have considered diffraction corrections in interferometers designed for precise intercomparisons of wavelengths.

8.9.2. Guided waves

Microwaves can be transmitted by coaxial lines or waveguides and part of the interferometer can consist of guided propagation in these rather than following completely the optical model of free propagation in space. An example is shown in fig. 71. The beam splitter is now a hybrid junction, or a directional coupler. These and other microwave components are described by Harvey [1963].

Fig. 71 Mach–Zehnder form of microwave interferometer [Froome, 1954].

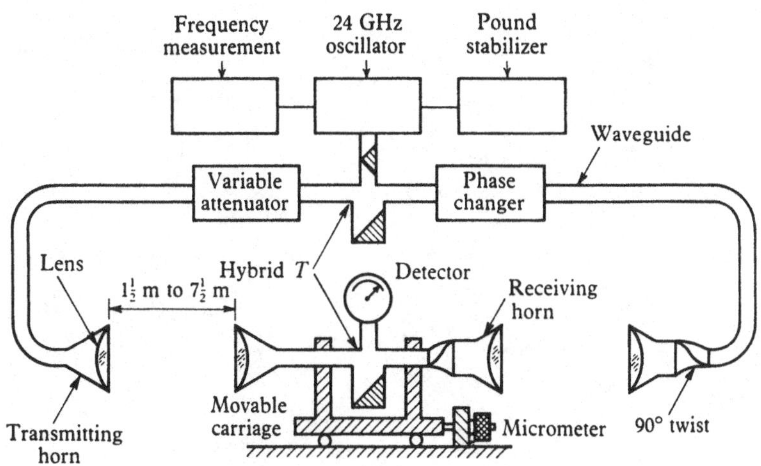

Guided-wave forms of traditional interferometers are now possible for visible and near-infrared radiation, for which optical fibres are available as dielectric waveguides. Most of these are multimode, and light that enters as a coherent beam leaves as a speckle pattern. For interferometers, a single-mode fibre is more commonly used, in which the higher-index core is so small that only the lowest-order mode can propagate.

Beam splitters can be the traditional types, with the light radiating to them, and then being refocused into the fibre. But they can be made within the fibre [Sheem & Giallorenzi, 1979], or by twisting two fibres together. This last method has been used by Imai, Ohashi & Ohtsuka [1981] in an all-fibre Michelson interferometer. Fibres have also been used in cyclic interferometers to produce a long light path [Vali & Shorthill, 1977].

Fibres, particularly single-mode fibres, can be birefringent because of strain, induced either during manufacture or in later use. They can thus be used as a sensor of the various causes of the strain [Giallorenzi, 1981].

9

Multiple-beam interferometers

The general conclusions of Chapter 7 remain true when more than two beams interfere: a multiple-beam interferometer measures either the phase variations across a sample inside it or Fourier components of the spectral or spatial distribution of the radiation, and it has the applications listed in Table 1.1. To obtain these distributions themselves a series of Fourier components are needed. These are measured one after another by a two-beam interferometer, but a multiple-beam instrument can measure them all simultaneously, an obvious advantage for the study of rapidly changing radiation.

Measurements of phase do not require such a series of readings and the effects of changing from a two-beam to a multiple-beam interferometer are less important. But the latter has a narrower fringe profile and allows the use of multiwavelength fringes, spaced at some fraction of a wavelength [Herriott, 1967; Schwider, 1968].

9.1. General characteristics

The equivalent optical system of a multiple-beam interferometer has as many images as there are interfering beams and a series of values is required for each parameter, instead of the single value that describes a two-beam interferometer. In principle, any series is possible, but, in practical interferometers, the values usually increase in an arithmetical progression. The Fourier components detected by the interferometer are then successive harmonics, and these are usually in phase and of roughly the same magnitude. Their sum, representing the instrumental function of the interferometer, approximates a Dirac comb. This means that, approximately, a multiple-beam interferometer gives direct readings of either spatial or spectral distributions, and not their Fourier transforms. These readings are repeated at the repetition interval of the Dirac comb so that the instrument has a series of *orders*.

The intensity distribution of multiple-beam fringes is also represented approximately by the same function. The fringes give sharper contours of phase than the sinusoidal two-beam fringes and, when their position is read visually or photographically, more precise results can be obtained. But this gain in precision is not fundamental. When photo-electric methods are used to set on the fringes, roughly the same limiting precision is obtained for either two-beam or multiple-beam fringes with the same amount of light [Hanes, 1959, 1963; Hill & Bruce, 1962, 1963]. The change to multiple-beam interference may best be regarded as the provision of one stage of amplification by the interferometer itself.

The Dirac comb represents an ideal interferometer. Practical instruments would be represented by a more general function

$$k(\psi)=\{\mathrm{III}(\psi/2\pi) * a(\psi)\}b(\psi), \tag{9.1}$$

where each spike is broadened by convolution with a function $a(\psi)$ and different orders have different intensities, denoted by $b(\psi)$. If the diffraction grating is taken as an example, it has a transmission (or reflexion) factor for complex amplitudes that is a periodic repetition of some function $P(u)$, limited to a width b;

$$F(u)=\{\mathrm{III}(u) * P(u)\} \operatorname{rect} u/b. \tag{9.2}$$

The complex amplitude in the far field (Fraunhofer diffraction) is then

$$f(x)=\mathrm{III}(x)p(x) * \operatorname{sinc} bx,$$

and the intensity is the squared modulus, which can be written in the form (9.1) with

$$a(\psi)=\operatorname{sinc}^2 bx, \quad b(\psi)=|p(x)|^2. \tag{9.3}$$

The width of each spike is proportional to $1/b$, the reciprocal of the width of the grating, while the relative magnitudes of the different orders, denoting the *blaze* of the grating, depend on the transform of P, the function describing the shape of the grooves.

A useful term, used chiefly for Fabry–Perot interferometers but applicable to others, is the *finesse*, which is the ratio of the spacing of these spikes to their width at half peak intensity. For a diffraction grating, it is found to be

$$\mathcal{F}=1.13 \text{ m}. \tag{9.4}$$

Thus the finesse is approximately equal to m, the number of rulings, that is, the number of interfering beams.

In the study of multiple-beam interferometers, this function $k(\psi)$ of (9.1) replaces $\cos\psi$ in the two-beam theory of Chapter 7. The function k represents both the instrumental functions for measuring spectral and spatial distributions and is the fringe profile in quasi-monochromatic radiation. In any particular interferometer, the appropriate function $k(\psi)$ is most simply found as the fringe profile given by a point source.

9.2. Airy function

The Fabry–Perot interferometer and related instruments make use of the multiple reflexions between two plane parallel surfaces. They are all characterized by the same function, the Airy function $A(\phi)$, a function of the increment of phase ϕ between successive beams. For two surfaces separated a distance d by a medium of refractive index n and a wave at an angle θ to their normal,

$$\left.\begin{aligned}\phi &= 4\pi\nu\frac{nd}{c}\cos\theta,\\ &= 2\pi\nu T,\end{aligned}\right\} \tag{9.5}$$

where $T=(2\pi nd\cos\theta)/c$ is the transit time for each double passage through the separating medium, normal to the surfaces. Let each surface be described by its complex reflexion and transmission coefficients for amplitudes. These are r and t for radiation incident from outside and r' and t' for radiation incident from the separating medium. Initially the two surfaces will be given different coefficients.

The amplitudes of successive waves are shown in fig. 72 for an incident plane wave of unit amplitude. The total transmitted amplitude for m waves is

$$\begin{aligned}U_{tm} &= t_1t'_2 + t_1t'_2r'_1r'_2\,e^{i\phi} + t_1t'_2r'^2_1r'^2_2\,e^{2i\phi} + \cdots\\ &\quad + t_1t'_2r'^{m-1}_1r'^{m-1}_2\,e^{i(m-1)\phi}, \qquad (9.6)\\ &= \frac{t_1t'_2(1 - r'^m_1r'^m_2\,e^{im\phi})}{1 - r'_1r'_2\,e^{i\phi}}.\end{aligned}$$

Fig. 72 Multiple reflexions between two parallel surfaces.

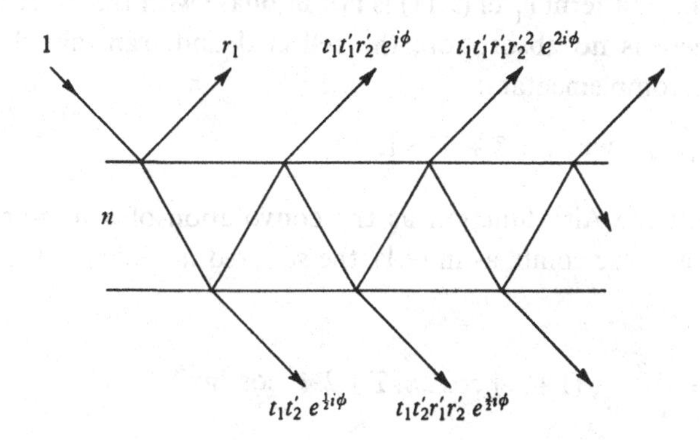

For an infinite number of waves, this tends to

$$U_t = \frac{\mathscr{T}}{1-\mathscr{R}\,e^{i\psi}}, \tag{9.7}$$

where

$$\mathscr{T} = |t_1 t_2'|, \quad \mathscr{R} = |r_1' r_2'|, \quad \psi = \phi + \chi, \tag{9.8}$$

χ being the correction for the phase changes on reflexion at the surfaces:

$$\chi = \arg r_1' + \arg r_2'.$$

The transmitted intensity is the Airy function

$$\begin{aligned} I_t(\nu) &= \frac{\mathscr{T}^2}{1-2\mathscr{R}\cos\psi + \mathscr{R}^2}, \\ &= \left(1 - \frac{\mathscr{A}}{1-\mathscr{R}}\right)^2 A(\psi), \end{aligned} \tag{9.9}$$

where $\mathscr{A} = 1 - \mathscr{R} - \mathscr{T}$, and $A(\psi)$ is the *Airy function*,

$$A(\psi) = \left[1 + \frac{4\mathscr{R}^2}{(1-\mathscr{R})^2}\sin^2 \tfrac{1}{2}\psi\right]^{-1}. \tag{9.10}$$

When the two surfaces are the same, $\mathscr{A}$, $\mathscr{R}$, and $\mathscr{T}$ are their (intensity) absorptance, reflectance, and transmittance and χ is twice the phase change on reflexion.

The total reflected amplitude is

$$\left.\begin{aligned} U_r &= r_1 + t_1 t_1' r_2' e^{i\phi} + t_1 t_1' r_1' r_2'^2 e^{2i\phi} + \cdots, \\ &= r_1 + \frac{t_1 t_1' r_2' e^{i\phi}}{1 - r_1' r_2' e^{i\phi}}, \end{aligned}\right\} \tag{9.11}$$

The reflected intensity does not reduce to as simple a form as (9.9) [Holden, 1949], since the first term r_1 of (9.11) is not in phase with the other terms. But, when there is no absorption, the reflected and transmitted interferograms are complementary:

$$I_r + I_t = 1 \quad \text{when} \quad \mathscr{R} + \mathscr{T} = 1.$$

To represent the Airy function as the convolution of a non-periodic function and a Dirac comb, as in (9.1), the squared modulus of the series (9.6) is used:

$$I_t(\nu) = \frac{\mathscr{T}^2}{1-\mathscr{R}^2}(1 + 2\mathscr{R}\cos 2\pi\nu T + 2\mathscr{R}^2\cos 4\pi\nu T + \cdots), \tag{9.12}$$

The transform $k(\tau)$ of this expression is, in the time domain, the impulse response of the interferometer. When the interferometer is used for spectroscopy, where measurements are made as a function of frequency, $k(\tau)$ is the transfer function for these. Since the transform of $\cos 2\pi N\nu T$ as a function of ν is $\frac{1}{2}\delta(u-N)+\frac{1}{2}\delta(u+N)$, where $u=\tau/T$, the transfer function is

$$k(\tau)=\frac{\mathscr{T}^2}{1-\mathscr{R}^2}\,\text{III}(u)\mathscr{R}^{|u|},\quad u=\tau/T,$$

and inverse transformation gives, from Table 2.1, the result given by Connes [1961],

$$\text{A}(\psi)=\frac{\mathscr{T}^2}{1-\mathscr{R}^2}\,\text{III}(\psi/2\pi) * \frac{2\rho}{\rho^2+\psi^2}, \tag{9.13}$$

where

$$\rho=-\ln\mathscr{R},$$

$$\psi=2\pi\nu T.$$

The function $a(\psi)$ of (9.1), which represents the profile of a single fringe, is the lorentzian function

$$a(\psi)=\frac{\mathscr{T}^2}{1-\mathscr{R}^2}\,\frac{2\rho}{\rho^2+\psi^2}. \tag{9.14}$$

The representation (9.13) is not unique, however, as any other function whose transform has the same values at the sampling points of $\text{III}(u)$ could be used instead of $a(\psi)$, for example, $I_t(\nu)\,\text{rect}\,(\nu T)$. But $a(\psi)$ represents a convenient function for mathematical manipulation. In terms of the order N and the finesse $\mathscr{F}$, defined below, it can be rewritten as

$$a(\psi)=\frac{\mathscr{T}^2}{1-\mathscr{R}^2}\,\frac{2\mathscr{F}}{\pi}\,\frac{1}{1+4\mathscr{F}^2N^2}, \tag{9.15}$$

a form given by Mertz.

The form of the Airy function is shown in fig. 73 for various values of $\mathscr{R}$ and no absorption. It represents an interference pattern of evenly spaced, narrow, bright fringes on a dark background. These fringes are repeated at $\Delta\psi=2\pi$ and become narrower as the reflectance $\mathscr{R}$ approaches unity, their width at half the peak intensity being given by (9.9), as

$$\delta\psi=2(1-\mathscr{R})\mathscr{R}^{-1/2}, \tag{9.16}$$

or by (9.14) as

$$\delta\psi=2\rho=-2\ln\mathscr{R}. \tag{9.17}$$

For $\mathscr{R}$ close to unity, these two expressions differ only by terms of the order $(1-\mathscr{R})^3$. From this half-width, the *finesse* $\mathscr{F}$ of the fringes is defined as the ratio of the *inter-order spacing* or repetition interval, $\Delta\psi = 2\pi$, to the half-width $\delta\psi$:

$$\left.\begin{aligned}\mathscr{F} &= \Delta\psi/\delta\psi,\\ &= \pi\mathscr{R}^{1/2}/(1-\mathscr{R}), \quad \text{or} \quad -\pi/\ln\mathscr{R}.\end{aligned}\right\} \tag{9.18}$$

Also of importance is the peak transmittance $\mathscr{P}$ of the interferometer at the centres of the bright fringes. This is given by

$$\mathscr{P} = \frac{\mathscr{T}^2}{(1-\mathscr{R})^2} = \left(1 - \frac{\mathscr{A}}{1-\mathscr{R}}\right)^2, \tag{9.19}$$

while the ratio of this to the minimum transmittance at the dark fringes is

$$\left[\frac{1+\mathscr{R}}{1-\mathscr{R}}\right]^2 = 1 + \left(\frac{2}{\pi}\mathscr{F}\right)^2. \tag{9.20}$$

When the interferometer is used as a filter, of more importance is the ratio to the total transmittance of the integrated transmittance of the bright fringe, out to twice its half width. This is the filtering efficiency $\mathscr{E}$ which can be defined in two ways, either from the Airy function itself, or from $a(\psi)$:

$$\left.\begin{aligned}\mathscr{E} &= \int_0^{\delta\psi} \mathrm{A}(\psi)\,\mathrm{d}\psi \Big/ \int_0^{\pi} \mathrm{A}(\psi)\,\mathrm{d}\psi,\\ &= \frac{2}{\pi}\arctan\left(\frac{1+\mathscr{R}}{1-\mathscr{R}}\tan\frac{\pi}{\mathscr{F}}\right),\end{aligned}\right\} \tag{9.21}$$

Fig. 73 Multi-beam fringes: the Airy function plotted against the phase difference ψ.

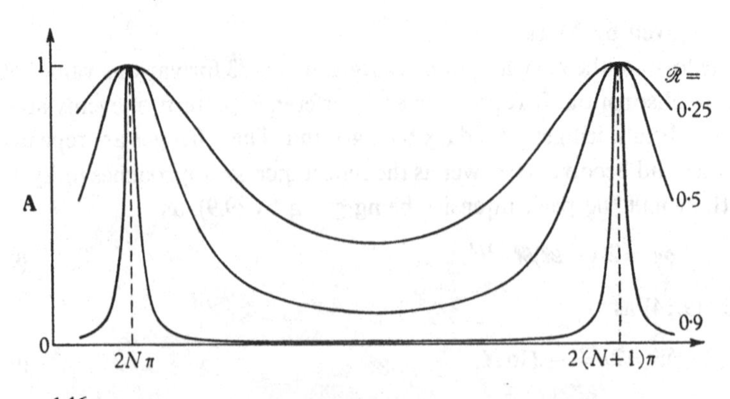

which tends to 0.70 as $\mathscr{R}\to1$; or

$$\left.\begin{aligned}\mathscr{E}&=\int_0^{\delta\psi}a(\psi)\,\mathrm{d}\psi\bigg/\int_0^{\infty}a(\psi)\,\mathrm{d}\psi,\\&=0.70.\end{aligned}\right\}\tag{9.22}$$

It should be noted that this filtering efficiency is not the same as the *filtrage* (facteur de filtrage intégral) of Chabbal [1958] which involves integrals out to the half-intensity points only.

For microwave interferometers and laser cavities, it is usual to speak of a *Q-factor* in place of the finesse. This factor is defined as the energy contained within the interferometer divided by the power dissipated per unit angular frequency. It reduces to

$$\begin{aligned}Q&=\nu/\delta\nu=\psi/\delta\psi,\\&=\mathscr{F}\nu T.\end{aligned}\tag{9.23}$$

It will be seen later that this is the same quantity as the resolving power R of an interferometer used for spectroscopy.

When there is no absorption at the surfaces, the reflected fringes (9.11) are the exact complement of the transmitted fringes, shown on fig. 73. Narrow, dark fringes appear on a uniform background. When absorption is present, it reduces the modulation, the departures of the intensity from the background value, more rapidly than it reduces the background intensity, and thus causes a loss of fringe visibility. For the transmitted fringes, although absorption reduces the intensity, the visibility is unaffected by it.

When the two reflecting surfaces are inclined to each other, the form of the fringes departs from the Airy function. Two effects cause this: the increment in phase between successive beams is no longer constant, so that higher-order waves become progressively out of phase; and successive waves make increasing angles to the normal, so that, for surfaces of finite size, the higher-order waves are lost from the edges, even when the incident radiation arrives normally to one surface. These effects combine to give fringes that are broader than the corresponding fringes for parallel surfaces, are unsymmetrical about their maxima, and have their maxima displaced from the points where the increment in phase between the first and second interfering beams is $2N\pi$.

An example of the intensity distribution in multiple-beam fringes between inclined surfaces has been calculated by Kinosita [1953] and is shown in fig. 74. The departures from the Airy function are clearly visible and these increase as either the angle between the surfaces or the mean separation is increased. When such fringes are used to compare the shape of a surface with some reference plane surface, it is desirable to keep the separation as well as the angle small, if the Airy formula is to be used to interpret the results. The condition given by Tolansky [1955] is

$$T \ll 3/(4v\theta^2 m^3), \tag{9.24}$$

where θ is the angle between the surfaces, T is again the increment of delay for each double passage, and m the number of beams contributing effectively to the interference (usually about 60).

9.3. The Fabry–Perot interferometer

The Fabry–Perot interferometer consists of two transparent plates with plane surfaces, the two inner surfaces being coated with partially transmitting coatings of high reflectance. These coated surfaces are set parallel, while to avoid multiple reflexions from the other surfaces the plates are usually made with a small wedge angle. The separation between the plates may be adjustable or fixed; in the later case the instrument is called a Fabry–Perot *étalon* or, when the separation is small, an *interference filter*. In the latter the separator is a dielectric layer, one half wavelength thick for order $N = 1$, deposited between the reflecting layers. The Fabry–Perot is the multiple-beam equivalent of the Michelson interferometer and, with an

Fig. 74 Multiple-beam fringes between inclined surfaces (full line) compared with the Airy function (broken line) [Kinosita, 1953].

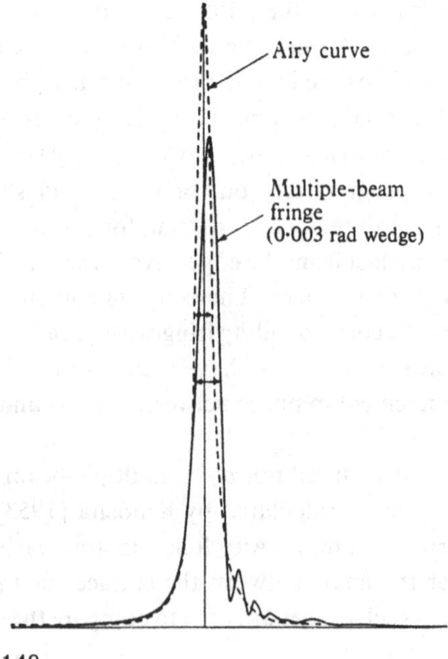

effective source at infinity, it can have tilt but no shear, and shift and delay but no lead. As it is a multiple-beam interferometer, these parameters all have a series of values. It has therefore similar applications to the Michelson interferometer: through its delay it measures spectral distributions, lengths, or refractive indices, while multiple-beam fringes are used to study phase variations. Through the shift, it is affected by source size.

The interferometer is characterized by the Airy function of fig. 73, which may be regarded as plotted as a function of ν, n, d, or $\cos\theta$. For constant ν and $\theta=0$, a point source of quasi-monochromatic light on axis, the Airy function would represent the intensity distribution in the fringes across the interferometer surfaces, the Fizeau fringes that show variations of either the spacing d of the surfaces or the index n of the medium between them. Tilts will introduce straight fringes, but, as explained above, the form of the fringes departs from the Airy function if the product of the tilt and the separation is too large. For an interferometer of constant optical path nd for axial rays, the Airy function represents, in terms of $\cos\theta$, the circular Haidinger fringes (sometimes called *Fabry–Perot fringes* for multiple-beam interferometers) given by an extended source. Finally, for polychromatic radiation, the function represents the transmittance of the interferometer for different frequencies ν.

The interferometer is described by its order of interference N, its peak transmittance $\mathscr{P}$, and its finesse $\mathscr{F}$. Usually these last two quantities should be as high as possible, and any increase in the reflectance of the surfaces, which increases the finesse, should not entail an increase of the absorptance and thus a reduction of the peak transmittance.

9.3.1. Reflecting surfaces

Both metal and multilayer dielectric coatings are used for the reflecting coatings. Of the former, silver is the most useful for visible radiation and aluminium for ultraviolet. The properties of silver coatings are given by Kuhn & Wilson [1950] and, for the ultraviolet, Bradley *et al.* [1965] have shown that aluminium is more suitable in some regions than even dielectric coatings. It should be evaporated rapidly, or in an ultra-high vacuum [Hutcheson, Hass & Coulter, 1971]. Multilayer dielectric coatings are the most efficient for visible radiation. These can now be made with reflectances very close to unity and the development of improved coatings has followed closely the development of lasers. The simplest coatings, designed for use at one frequency, consist of alternate layers of high and low refractive index, each having an optical thickness of one quarter the wavelength for which it is to be used. Modifications of this to give high reflectance over a greater range of frequencies are discussed in works on optical thin films, for example that of Macleod [1969].

For the far infrared, the reflectors used are metal grids, perforated metal plates, or multilayer stacks, consisting of quarter-wave dielectric sheets separated by quarter-wave airspaces. High reflectances, and hence high finesses are obtained more easily than with visible radiation, but diffraction losses are larger. Instruments have been described by Genzel [1965], and by Culshaw *et al.* [1960], and a microwave interferometer is shown in fig. 75. The use of Fabry–Perot interferometers at long wavelengths is reviewed by Chantry [1982], who also introduces two more detailed reviews.

9.3.2. Surface quality

The reflectance and absorptance of the coatings are not the only factors that limit the finesse and peak transmittance of an interferometer. They are limited also by departures of the surfaces and their coatings from flatness and parallelism, by any angular spread of the beam, and by diffraction at the edges of the interferometer reflecting surfaces. Traditionally the effects of diffraction have been ignored in the classical treatment of the Fabry–Perot interferometer, but they have become important now that this instrument has become widely used as a cavity for lasers.

The influence of surface defects has been studied by Dufour & Picca [1945] and Hill [1963]. If the variation of phase across the aperture of the interferometer is described by a distribution $D(\psi)$, where D represents the fraction of the aperture that has a phase between ψ and $\psi + \mathrm{d}\psi$, these authors describe the final interferogram by the convolution of this with the Airy function $\mathbf{A} * D$. If one surface of the interferometer is spherical with a maximum variation of phase from centre to edge of β, D has the form $\mathrm{rect}(\psi/\beta)$ and such a surface would limit the possible finesse of the interferometer to

$$\mathscr{F} = 2\pi/\beta = m/2, \tag{9.25}$$

Fig. 75 Fabry–Perot interferometer for microwaves [Culshaw, 1959].

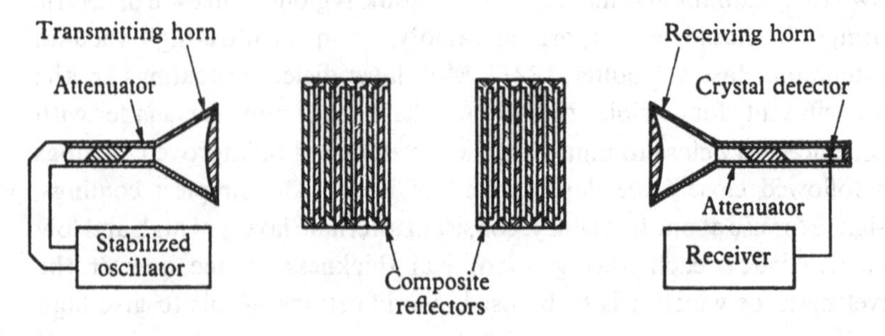

where $m=4\pi/\beta$ is a measure of the flatness of a surface that departs by λ/m from a plane. Optical flats can be made by the special technique, introduced by Otte [1965] and developed by Leistner [1975], to better than $\lambda/200$, and evaporated coatings can be deposited to the same uniformity [Ramsay, Netterfield & Mugridge, 1974]. For such flats, the limit to the finesse would be about 100, but for more usual flats and coatings it is about 30.

The result given above is the upper limit placed on the finesse by surface defects. An alternative treatment gives the effect of defects on the value of the finesse derived from the reflectance of the coatings by (9.18). This treatment follows the method used by Maréchal [1947] to derive tolerances for image-forming systems. It is based on a power-series expansion in terms of the small departures from flatness.

If the phase increment in the interferometer is written as $\psi + \phi(\mathbf{u})$, where ψ is the mean phase and ϕ its variation as a function of the position $\mathbf{u}$ at the aperture of the interferometer, the amplitude (9.7) can be expanded as

$$U_t(\mathbf{u})=\frac{\mathscr{T}}{1-\mathscr{R}\,\mathrm{e}^{\mathrm{i}\psi}}\left[1+\frac{\mathrm{i}\phi\mathscr{R}\,\mathrm{e}^{\mathrm{i}\psi}}{1-\mathscr{R}\,\mathrm{e}^{\mathrm{i}\psi}}-\tfrac{1}{2}\phi^2\mathscr{R}\,\mathrm{e}^{\mathrm{i}\psi}\frac{1+\mathscr{R}\,\mathrm{e}^{\mathrm{i}\psi}}{(1-\mathscr{R}\,\mathrm{e}^{\mathrm{i}\psi})^2}+\cdots\right], \tag{9.26}$$

and the intensity is the squared modulus of this. The integral of the intensity over the aperture of the interferometer gives the transmittance as

$$T(\psi,\phi)=\frac{\mathscr{T}^2}{(1-\mathscr{R})^2+4\mathscr{R}\sin^2\frac{1}{2}\psi}\times\left[1-\langle\phi^2\rangle\mathscr{R}\frac{1-12\sin^2\frac{1}{2}\psi}{\{(1-\mathscr{R})^2+4\mathscr{R}\sin^2\frac{1}{2}\psi\}^2}+\cdots\right] \tag{9.27}$$

where terms in ϕ have vanished since $\langle\phi\rangle=0$, and terms in $(1-\mathscr{R})^2\sin^2\frac{1}{2}\psi$ and $\sin^4\frac{1}{2}\psi$ have been omitted since they are much smaller than $\sin^2\frac{1}{2}\psi$.

The transmittance has one half its peak value at $\delta\psi$, obtained from

$$T(\tfrac{1}{2}\delta\psi,\phi)=\tfrac{1}{2}T(0,\phi).$$

This is found to be

$$\delta\psi=2\frac{1-\mathscr{R}}{\mathscr{R}^{1/2}}\left(1+\frac{3}{2}\frac{\mathscr{R}}{(1-\mathscr{R})^2}\langle\phi^2\rangle\right),$$

so that the finesse for the interferometer with defects is given by

$$\mathscr{F}'=\mathscr{F}_0\left(1-\frac{3}{2\pi^2}\mathscr{F}_0^2\langle\phi^2\rangle\right), \tag{9.28}$$

where $\mathscr{F}_0$ is the finesse found from the reflectance alone. Thus, for example, if the interferometer has one surface a sphere that departs from a flat by 0.012λ centre to edge, this will reduce a finesse of 30 to 24.

Equation (9.28) applies when the interferometer is used for spectroscopy. When it is used as a narrow-band filter to form a monochromatic image, for example, in solar astronomy [Ramsay, 1969], a tolerance based on the image quality is more appropriate. The central intensity of the transmitted image, the Strehl definition, is the squared modulus of the integrated amplitude $U_t(\mathbf{u})$ across the aperture, divided by $T(0, \phi)$ from (9.27). The central intensity so obtained is

$$I = \left| \int_{\mathscr{A}} U_t(\mathbf{u})\, \mathrm{d}\mathbf{u} \right|^2 \Big/ \int_{\mathscr{A}} |U_t(\mathbf{u})|^2\, \mathrm{d}\mathbf{u},$$

which, for a transmission band at $\psi = 2N\pi$, is

$$I_0 = 1 - \frac{\mathscr{R}^2 \langle \phi^2 \rangle}{(1-\mathscr{R})^2} + \cdots,$$

$$\approx 1 - \frac{\mathscr{F}^2}{\pi^2} \langle \phi^2 \rangle. \qquad (9.29)$$

For an image-forming system, the corresponding expression is $1 - \langle \phi^2 \rangle$. Thus the usual tolerances for such systems [Maréchal, 1947] should be divided by $\mathscr{F}/\pi$ for a Fabry–Perot interferometer. For a spherical error of a surface, which is equivalent to a defocusing in an imaging system, the tolerance is thus $\pi\lambda/4\mathscr{F}$. To give as good image quality as a mirror made to $\lambda/8$, the plates of a Fabry–Perot interferometer of finesse 30 should match to $\lambda/80$.

9.3.3. Source size

The effect of the finite angular spread of the beam corresponds to the effect of source size discussed for two-beam interferometers in §8.1. The size of the aperture that limits the effective source should be judged against the spacing of the Haidinger fringes. If these have a bright centre, a source that just covers this central fringe satisfies the multiple-beam analogue of Hansen's condition. Such a source would have a phase variation of $\delta\psi$ from centre to edge and an incoherent summation over the source would, following (9.25), give the limiting finesse due to a source of this size as

$$\mathscr{F} = 2\pi/\delta\psi$$

the same value as the finesse with zero source size. Since the instrumental function is the convolution of two functions of this finesse, the final finesse is reduced by a factor of approximately $1/\sqrt{2}$.

The angular size of the source across which there is a phase variation of $\delta\psi$ is derived from

$$\tfrac{1}{2}(\delta\theta)^2 + \theta\,\delta\theta = -\delta\psi/2\pi\nu T.$$

The solid angle subtended by this source is

$$\Omega = 2\pi/\mathscr{F}\nu T, \tag{9.30}$$

both for a central circular source, corresponding to a phase variation of $\delta\psi$ over the central fringe, and for an annulus with the same variation of phase across one of the outer fringes. This source size is equivalent to that given by (8.19) for a Michelson interferometer, since the finesse of a two-beam interferometer is two; a multiple-beam interferometer can be considered as having an effective delay of $\tfrac{1}{2}\mathscr{F}T$.

The final instrumental function is a convolution of the Airy function derived from the reflectance of the coatings, a function representing the surface defects, and a function representing the source size. This still does not take into account the finite size of the plates and the effects of diffraction at their edges and, for example, gives no indication of the intensity distribution across the beam as it leaves the interferometer. These diffraction effects are most important for microwave interferometers and are discussed by Culshaw [1960]. For 'passive' interferometers used with visible radiation they are relatively unimportant, but they are becoming more important also for visible radiation, when very accurate wavelength measurements are required or when the interferometer is used as a laser cavity.

9.3.4. Scanning

In contrast to an étalon with a fixed increment of phase ψ, a *scanning* interferometer is one in which ψ can be varied. This may be done in three ways: by changing the spacing d of the plates, by changing the refractive index n of the medium between them, or by tilting the interferometer to change θ. This last method is the least desirable as it causes a large loss of étendue, since the effective source becomes a small portion of one of the outer Haidinger fringes. In theory, however, scanning by a change of θ could be achieved by the use of an annular source of variable size, obtained for example as the image formed by an optical system of variable power.

A variation of the spacing d is obtained by mechanical displacement of one or both the plates. During this displacement the plates should remain parallel or a loss of finesse results, a tilted plate acting in the same way as a surface defect. The tolerance allowable on any tilts is very small, being that permissible for a two-beam interferometer divided by the relative finesse,

$\frac{1}{2}\mathscr{F}$. Many systems have been used or proposed for mechanical scanning and a number have been described in the proceedings of the two Bellevue Conferences [1958, 1967], and by Jacquinot [1960]. Slater, Betz & Henderson [1965] use magnetostriction and Clothier, Sloggett & Bairnsfather [1980] use an electromagnetic scan. The parallelism of the two interferometer surfaces can depend on accurate mechanical construction, a mounting that ensures parallel movement, or equal expansion of several supports. The mountings for parallel movement include elastic membranes, and elastically deformable parallelograms. A more recent approach is the use of servo systems to maintain parallelism. Ramsay [1962] has used a method that produces white-light superposition fringes [§9.6.1] between small regions on the opposite sides of the Fabry–Perot plates. When both are locked to the central white fringe, the plates have equal spacings at opposite sides. One such system controls the vertical separation of the edges of the plates, the other the horizontal. Bruce [1966] has used a system in which the peak transmittance over a finite region of plates is held at a maximum. This ensures parallelism at integral orders of the monitoring radiation.

If the index of the separating medium is changed by changing the air pressure, the problem of ensuring parallelism of moving plates is avoided, and pressure scanning was the most popular method before the introduction of parallelism control [Mack *et al.*, 1963]. Since the refractive index of air exceeds unity by about 3×10^{-4} for one atmosphere of pressure, an interferometer would need to have an order of interference $N > 3000$ to scan through one order with a one-atmosphere change of pressure. For interferometers with a lower order of interference, larger changes of pressure or gases of higher index have been used.

9.4. The spherical Fabry–Perot

A Fabry–Perot interferometer has been seen to have the same off-axis properties as a Michelson, its smaller allowable source size being due to the increased finesse. The field-widened version of the Fabry–Perot interferometer is the *spherical Fabry–Perot* described by Connes [1956], Hersher [1968] and Bradley & Mitchell [1969]. It consists (fig. 76(*a*)) of two spherical mirrors separated by the sum of their focal lengths and so is also known as a *confocal* Fabry–Perot. In its original form, only one half of each mirror was made partially reflecting, the rest being fully reflecting. One transmitted or one reflected wave thus occurred after two double passages of the interferometer and the increment in optical path between successive beams was four times the optical distance between the mirrors. Spherical Fabry–Perot interferometers, as now used for laser cavities, have the whole surface partially reflecting. An equivalent system is that shown in fig. 76(*b*), where one spherical mirror is replaced by a plane mirror at its focal point.

The étendue of the interferometer is limited by the spherical aberration of the mirrors, which, by limiting their usable apertures, limits both the area and the solid angle of the beam. In a plane Fabry–Perot only the solid angle is limited: the mirrors can be made as large as is technically possible. A comparison of the two interferometers must therefore bring in the diameter D of the mirrors of the plane interferometer and the result found by Connes is that the gain in étendue for the spherical Fabry–Perot is

$$G=2(d/D)^2, \tag{9.31}$$

where the plane interferometer has a spacing d and the spherical interferometer has mirrors of radius $\frac{1}{2}d$ these being separated by $\frac{1}{2}d$. The spherical Fabry–Perot thus gains at high orders of interference when d would be large compared with possible values for D, the mirror diameter.

If instead of having a common focus, the two mirrors of a spherical Fabry–Perot are set to be concentric, the interferometer is no longer field widened but has the properties of a plane Fabry–Perot with an effective source at the common centre rather than at infinity. Diffraction effects, however, are different and this system has some advantages as a microwave interferometer [Culshaw, 1961]. As a laser cavity this concentric system is just as difficult to adjust as a plane cavity. Again it has an alternative form with one spherical mirror replaced by a plane mirror through the centre of curvature of the remaining mirror; this is sometimes called a *hemispherical cavity*.

9.5. Fabry–Perot cavities

Most laser cavities are Fabry–Perot étalons with spherical, or flat reflectors. They are treated in detail in textbooks on lasers, for example, Maitland & Dunn [1969]. The form of the cavity is chosen to produce stable modes of oscillation. These are found by the method due to Fox & Li [1961]: a

Fig. 76 The spherical Fabry–Perot of Connes [1956]: (*a*) full form; (*b*) form with plane mirror.

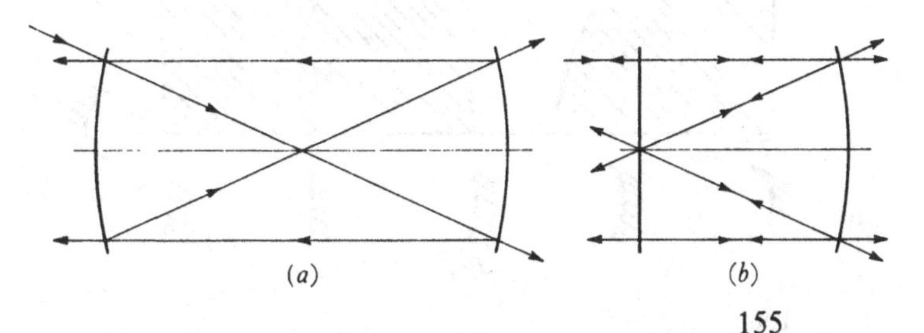

uniform plane wave starting from one of the mirrors is followed back and forward across the cavity, losing energy by diffraction on each transit. The distribution of amplitude across the wavefront changes, but may eventually settle down to a steady state that remains unchanged by further transits. It is then a stable mode of this cavity, and each such mode has a certain loss per transit. The final amplitude distribution has a gaussian profile, as has the beam leaving the laser.

A first indication of the stability of a cavity is given by ray optics. If successive reflexions bring rays in towards the axis, the cavity is stable. An unstable cavity reflects them progressively outward until they are lost outside the edge of the mirrors. The results obtained by Fox and Li, and Boyd & Kogelnik [1962] are shown in fig. 77. The regions of stability are those not shaded, and representative shapes of the mirrors are shown along the axes.

The forms considered previously, the plane, confocal, and concentric Fabry–Perot interferometers fall at the points *A*, *B*, and *C* respectively, while the systems equivalent to the last two, with one central plane mirror,

Fig. 77 Combinations of mirror radii that give stable resonators (unshaded regions).

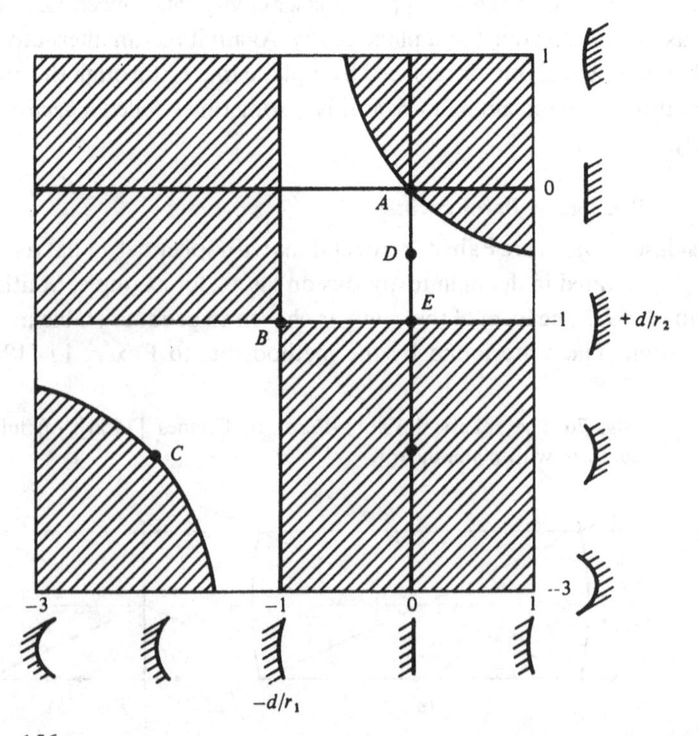

are at D and E. These are all systems that have obvious closed ray paths, and they lie on the boundaries between stable and unstable systems. Early lasers used a cavity with two plane surfaces, but this proved difficult to adjust and now the concentric and hemispherical cavities are more common. A confocal cavity is field-widened, and this means that it resonates in off-axis as well as in axial modes. It is used only when these are wanted. Other commonly used cavities lie in the region between A and B: two long-radius concave mirrors separated by less than half the sum of their radii. The rays in such a cavity do not have a simple closed path, but the wavefronts follow the shapes shown in fig. 78.

A laser oscillates in a certain mode when the gain in the active medium exceeds the loss in that mode. The loss depends partly on diffraction but more on the reflectance of the mirrors and the development of lasers has led to large technical improvements in the production of durable multilayer dielectric coatings of very high reflectance, 0.99 and greater.

A further factor in the choice of a suitable cavity is its dynamic stability, its ability when subject to small perturbations. This has been studied by Zhang [1981].

9.6. Fabry–Perot interferometers in series

When radiation passes through two Fabry–Perot interferometers in series, the instrumental function of the combination has large peaks where two peaks of the individual interferometers coincide and much smaller peaks where the peaks of one function occur against the background of the other. When the two interferometers are parallel, a third interferometer is formed and another set of peaks is produced, due to the multiple reflexions between the two separate interferometers. The combined system can be treated by the general Airy function (9.10) but now the transmission and reflexion coefficients in (9.8) represent those for the two interferometers and have the form (9.7) or (9.11).

Fig. 78 Representation of the shape of wavefronts in a laser cavity.

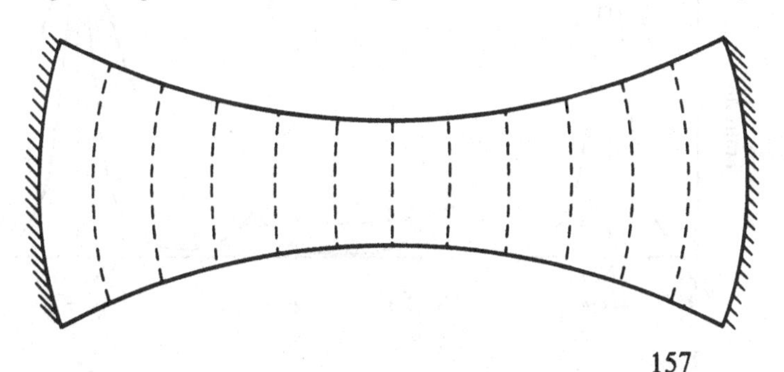

Usually such a full treatment is not required, for the two interferometers are normally decoupled so that these multiple reflexions do not occur. Methods are given by Schwider [1965]. The interferometers may be tilted, so that they are no longer parallel, or separated by a relatively large distance so that the fringes due to their interaction are finely spaced and not resolved; all that is seen is the averaged intensity over many fringes. A third method of decoupling two interferometers is to place absorbing material between them. This will reduce the general radiation, including the main peaks, by a fraction $1-\mathscr{A}$, but it will reduce the peaks due to the interaction to $[1-\mathscr{A}/(1-\mathscr{R})]^2$,, a much smaller value. In practice the Fresnel losses at the two uncoated surfaces of the inner interferometer plates are sufficient to reduce greatly any interaction.

When the two interferometers are effectively decoupled, their instrumental function is simply the product of the two instrumental functions for the separate interferometers. The form of this product is most easily understood when the two separate interferometers are perfect instruments described by the appropriate Airy function, if these are expressed in the form (9.14). The principal maxima of the product occur where peaks of the two Dirac combs coincide and, as these are convolved with the product of two single peaks, they are narrower than either peak. Smaller maxima occur at the unmatched peaks of either Dirac comb and these have the same shape as a peak of one of the single interferometers but with the peak transmittance greatly reduced. An example is shown in fig. 79, where the two interferometers have a low finesse, since this shows the reduction of the width of the peaks more clearly. The same type of reduction is also obtained by multiple passes through the same interferometer.

Fig. 79 Transmission of two Fabry–Perot interferometers in series with spacings in the ratio 3:1. The broken curve is the transmittance of the thinner interferometer.

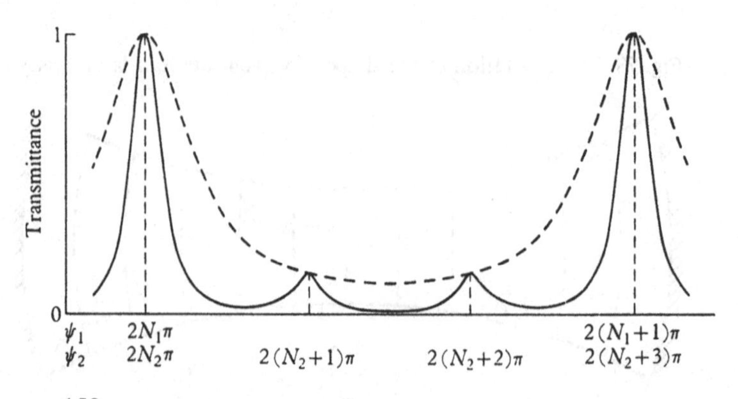

9.6. Fabry–Perot interferometers in series

Interferometers are used in series in spectroscopy to increase the free spectral range between maxima. An n-fold increase in the inter-order separation of one Fabry–Perot can be obtained either by the direct method of adding a second interferometer in series of $1/n$ times the spacing, or by a vernier method where the second spacing is $(n+1)/n$ times the first. In either case, the two Dirac combs coincide at an interval of n peaks of the first comb.

The theoretical investigation of the optimum ratio of spacings of a series of Fabry–Perot interferometers is a complicated problem when done in terms of the Airy function. After the preliminary work of Chabbal [1958], Mack *et al.* [1963] have given a computer analysis, while McNutt [1965] gives a theory in terms of integration in the complex plane. But, if the single-peak profile (9.14) is used, a very simple treatment is possible which provides approximate values of the half-width $\delta\psi$ and the filter efficiency $\mathscr{E}$ for different ratios of the spacings. These results are shown in fig. 80 for two interferometers. It is seen that a vernier method of increasing the free spectral range is preferable, as the greatest increase of efficiency occurs when the two interferometers have the same spacing; the use of two identical interferometers in series increases the filter efficiency from 0.70 to 0.88. This treatment can be readily extended to more than two interferometers in series, since products of more than two single-peak functions (9.14) can still be integrated simply.

Fig. 80 Half width $\delta\psi$ and efficiency $\mathscr{E}$ (fraction of light in the transmission band) for two Fabry–Perot interferometers in series, as a function of the ratio of spacings.

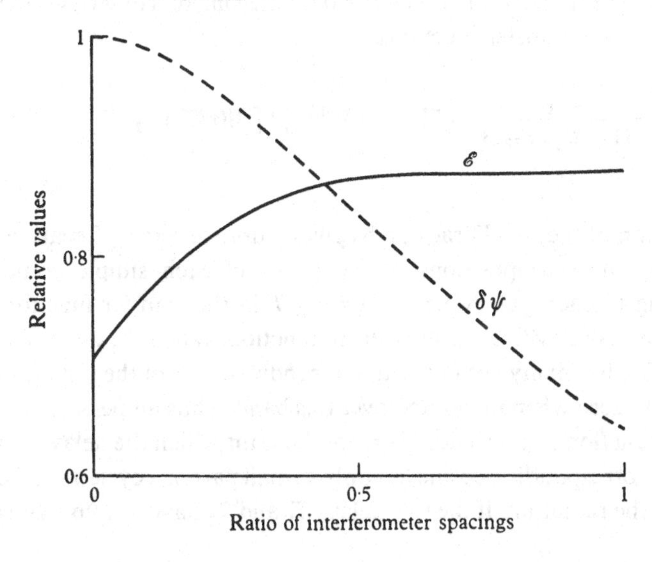

9.6.1. Fringes of superposition

The Airy function describes the intensity distribution in the fringes given by a Fabry–Perot interferometer with quasi-monochromatic light. For this interferometer, the effective delay is the product of the increment of delay T (for each double passage through the separating medium) and half the finesse and the light is quasi-monochromatic when its bandwidth is given by

$$\delta\nu < 2/\mathscr{F}T. \tag{9.32}$$

For radiation of larger bandwidth, the visibility of the fringes decreases and the fringes disappear when $\delta\nu$ is large compared with $2/\mathscr{F}T$; the radiation is polychromatic.

When two Fabry–Perot interferometers in series are illuminated by polychromatic radiation that would give no visible fringes with either interferometer separately, fringes may again be seen. These are *fringes of superposition* or *Brewster's fringes.* In a combination of two interferometers, a double series of different paths is possible, involving multiple reflexions in both interferometers. When the ratio of the two interferometer spacings is close to an integral fraction a/b, there is a series of pairs of paths that differ only by a small delay for all wavelengths, and hence interference fringes are obtained with broad-band radiation, which is, therefore, quasi-monochromatic for these paths.

The theory of fringes of superposition can be given in terms of the series (9.12), as in the original analysis by Benoît, Fabry & Perot, or in terms of the transfer function (9.13) which expresses the delays given by a single interferometer. The transfer function for two interferometers in series is the convolution of their transfer functions,

$$k(\tau)=\frac{\mathscr{T}_1^2\mathscr{T}_2^2}{(1-\mathscr{R}_1^2)(1-\mathscr{R}_2^2)}[\text{Ш}(\tau/T_1)\mathscr{R}_1{}^{|\tau/T_1|}] * [\text{Ш}(\tau/T_2)\mathscr{R}_2{}^{|\tau/T_2|}]. \tag{9.33}$$

The convolution of the two Dirac combs gives a double series of peaks at the sum and difference positions of the peaks of each simple comb. Corresponding to each pair of peaks at $\tau=\pm T$ in the transfer function, there is a cosine, $\cos 2\pi\nu T$, in the instrument function. When the delay T is large, this cosine has many periods with the bandwidth $\Delta\nu$ of the radiation and averages to zero when integrated over this band. Thus all peaks of the final transfer function can be ignored except those for which the delay is so small that the corresponding cosine has only a small part of a cycle with the bandwidth of the radiation. If the two delays T_1 and T_2 have a ratio that is close to the ratio of two integers a and b, that is

$$bT_2 - aT_1 = \delta T,$$

where δT is very small, the important peaks are the difference terms with delays that are small multiples of δT. For these, τ in (9.33) is given by

$$\tau = N\,\delta T = NbT_2 - NaT_1 \tag{9.34}$$

Thus the only terms of importance are

$$k(\tau) = \frac{\mathscr{T}_1^2 \mathscr{T}_2^2}{(1-\mathscr{R}_1^2)(1-\mathscr{R}_2^2)} \text{III}(\tau/\delta T)[\mathscr{R}_1^a \mathscr{R}_2^b]^{|\tau/\delta T|}. \tag{9.35}$$

To a constant factor, this is the transfer function for a simple Fabry–Perot interferometer in which the plates have a reflectance $\mathscr{R}_1^a \mathscr{R}_2^b$. This effective reflectance decreases as $a+b$ increases and hence the contrast of these fringes decreases the further a and b depart from unity. Like white light fringes in two-beam interferometers, these fringes have a white centre where $\delta T = 0$ and the surrounding fringes are coloured.

The above theory applies when both interferometers are used in transmission. Substantially different fringe shapes occur if one or both interferometers are used in reflexion; these cases are treated by Puntambeker [1973].

Applications of fringes of superposition are given by Cagnet [1954] and Schwider [1979]. They are particularly useful when an inaccessible separation is to be measured. By superposition fringes, this can be matched to the spacing of a separate interferometer, the spacing of which is accessible to measurement.

9.7. Gratings and radio interferometers

The Fabry–Perot interferometer is an example of a multiple-beam interferometer in which the component beams have a series of increasing delays. They also have increasing leads, when the circular Fabry–Perot fringes are used. Another class of multiple-beam interferometers are those with a series of shears. Since the shear is two dimensional these can be in one or two dimensions. Around visible frequencies these interferometers are represented by diffraction gratings, which are usually one dimensional. Although the optical diffraction grating is traditionally classed as an application of diffraction theory rather than interference, it could equally well be treated as a multiple-beam interferometer in which there is interference between many beams, one coming from each ruling of the grating. But the most interesting examples are the arrays of radio astronomy [Christiansen & Högbom, 1969].

A series of uniformly spaced aerials disposed along a line is analogous to a grating, yet in radio astronomy this is classed as an interferometer. Each vector separation of any pair of aerials represents one value of the shear, and corresponding to each value, such an interferometer is sensitive to a Fourier component of the object that has this spatial frequency. In the theory of optical image formation, the two-dimensional transforms of the object and its image are related by the optical transfer function of the image-forming system. The same theory applies when the object is a spatial distribution of radio emission and its image is merely a set of measurements. Further, as stated in (3.33), the transfer function is given by the auto-correlation of the pupil function. Pupils in visual instruments are usually single, fully transmitting apertures, but in radio astronomy a *dilute aperture* is commonly used, consisting of a series of relatively small aerials disposed in an array. The pupil function can be represented approximately by a series

Fig. 81 Simplified representation of three aerial arrays used in radio astronomy, and their transfer functions when used as either addition or multiplication interferometers.

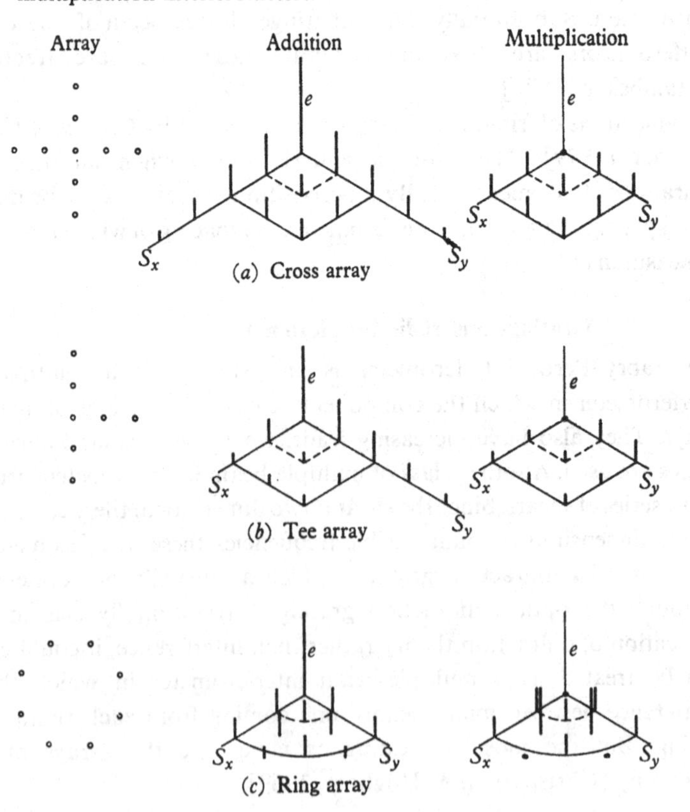

of points and the autocorrelation function has also a set of discrete values at the spatial frequencies that correspond to the separation of each pair of points. The Fourier transform of the object distribution is sampled at these discrete values and this leads to an instrumental function with a series of separate orders, called *grating responses*. These are caused by the repetition of the image at the reciprocal of twice the sampling interval.

If such an array were used as an addition interferometer in a conventional optical system, the transfer function would be weighted too heavily for zero spatial frequency (corresponding to the separation of each aerial from itself), the image having a large uniform background. Other spatial frequencies would receive contributions from no more than two pairs of aerials, except in the case of the commonly used cross or teė arrays where the vector separations in the two directions of the array are overemphasized. An ideal transfer function would give equal weight to all spatial frequencies or, even better, decrease the weighting approximately linearly with frequency so as to simulate the transfer function of a perfect optical instrument shown in fig. 16. Without this tapering, the image formed would show the diffraction rings of Gibbs' phenomenon.

As stated in §6.2, multiplication interferometers are commonly used at radio frequencies. If the signals from the two arms of a cross array are multiplied together instead of being added, the zero spatial frequency term and also those in the directions of the arms disappear. They can be restored, however, at the same level as other terms, by making the centre aerial common to both arms. As shown in fig. 81, such a multiplication interferometer corrects the uneven weighting found with an addition method. In the cross and the tee arrays, the two arms are multiplied together with the centre aerial common to both. For the circular array introduced by Wild [1961], the method of correcting the transfer function has been given by Wild [1965].

10

Measurement of mean phase

The largest field of application of interferometry is the measurement of phase. It is convenient to divide this into measurements of the mean phase, averaged over some field of observation, and measurements of the variations of phase across the field. This is not a clear-cut distinction and, over the history of interferometry, one type of measurement has adopted methods from the other, back and forward. But the applications are sufficiently different to make the division useful.

The total phase ψ has been given by (7.17) for a two-beam interferometer as

$$\psi = 2\pi\nu\tau - \beta_{12}(\tau_g) + \Delta\psi_u. \tag{10.1}$$

The mean phase is taken to be

$$\psi_0 = 2\pi\nu T - \beta_{12}(\tau_g), \tag{10.2}$$

and is either the phase averaged over some detector or its value at a reference point $\mathbf{u}_0$. The variations of phase $\Delta\psi_u$ have been given by (7.12) as

$$\Delta\psi_u = k(\Delta p_u + \Delta w_u); \tag{10.3}$$

they depend on the variations of optical path Δp_u due to the geometry of the interferometer and any variations Δw_u due to a specimen placed inside it.

Measurements of the mean phase then involve the term $2\pi\nu\tau$. Traditionally the quantity *order of interference* N is used, rather than the phase. It is given, when $\beta_{12} = 0$, by

$$N = \psi_0/2\pi = \nu\tau = \sum_{2-1} nd/\lambda_v, \tag{10.4}$$

the difference of optical path, $p = nd$, divided by the vacuum wavelength λ_v. Measurements of phase or order are thus used to obtain refractive indices, distances, or wavelengths.

The order can be written as

$$N = N_i + \varepsilon, \tag{10.5}$$

the *integral order* N_i plus the *excess fraction* ε. An interferogram in quasi-monochromatic light does not distinguish one fringe from another and so cannot give the integral order, but only the excess fraction. Integral orders are found by fringe counting, optical matching or multiplication, or the use of several different wavelengths.

In the expression (10.2), $\beta_{12}(\tau_g)$ is a correction term for the geometry of the interferometer and depends on the parameters shear and shift, shown in fig. 43; the other parameters tilt and lead affect Δp_u. Only when the interferometer is specially compensated does this term vanish. In the normal, uncompensated interferometer, the shift is directly proportional to the delay, and hence to the order of interference that is being measured, and a correction must be applied for the term β_{12}.

10.1. Measurement of fractional order

In the traditional method of phase measurement a tilt is introduced to give several fringes across the plane of observation. The mean phase is that at some reference point in this plane, and it is found by estimation of the position of the reference point relative to the fringe minima, while variations in phase are given by the departures of the fringes from straight lines. The unit for these measurements is a *fringe*, f, the spacing between successive fringe maxima or minima. For a single-passage interferometer, one fringe is equal to one wavelength, for double passage, half a wavelength of optical path.

10.1.1. Visual measurement

Visual estimation of fringe position is a simple but not very accurate method of measuring phases. It is possible to judge by eye the position of the sinusoidal fringes of a two-beam interferometer to about 0.05 f while the sharper multiple-beam fringes can be judged to perhaps 0.01 f.

Bottema [1960], and Hariharan & Sen [1960] have shown that considerably greater precision can be obtained by simple visual photometry. They divided the interferometer into two parts, one with the unknown path difference and the other with an adjustable compensator, graduated in path difference or phase. The two interferograms were presented side by side and tilts were eliminated so that these fields were uniformly illuminated. By visually matching the two intensities, they obtained a reproducibility of 0.001 f. Visual settings are also made with a source of white light. If the order of interference is close to zero, the fringes are coloured and the settings are made using the eye's colour discrimination rather than its ability to match intensities. The colours obtained with different interferometers are given by Kubota [1961].

One of the most important recent advances in interferometry is the increasing precision in measurements of phase. These advances are due largely to the replacement of older methods of judging the position of fringes by photo-electric setting on a maximum or minimum of the interferogram. Visual methods are now used chiefly for measuring phase variations.

10.1.2. Photo-electric measurements

Although fringes with a tilt present are used to measure phase variations by photo-electric methods, the mean phase is best obtained with no tilt and spread-out fringes. It is then a true mean, obtained by integration over the detector.

The direct method is photo-electric measurement of the intensity of the interferogram,

$$I = I_1 + I_2 + 2I_1^{1/2} I_2^{1/2} \cos\psi.$$

To obtain $\cos\psi$, the two intensities I_1 and I_2 must be measured separately, or derived from measurements of interferograms with known values of ψ. This method is used when it is not possible to separate the two interfering beams to apply the other methods given later. But its accuracy is limited by the usual problems that make it difficult to measure radiometric quantities to a relative accuracy of better than one per cent. This is not serious when the phase changes are small, when high sensitivity is obtained, even under difficult conditions. Thus Khanna & Leonard [1982] have measured vibrations of the basilar membrane in the inner ear of a living cat that have an amplitude of 10^{-5} f.

For high precision, methods that require measurements of intensities should be avoided. To estimate the ultimate (noise limited) precision of interferometry, the allied field of polarimetry [Dyson, 1963; Hopkinson, 1978] may be taken as an example of an optical measurement in which measuring techniques are highly developed and the ultimate precision is known. Polarimetry is the measurement of the rotation of the plane of polarization of linearly polarized light by an optically active medium. Such a medium is a *circular retarder*, having different refractive indices for the two circular polarizations, left and right handed. Any two-beam interferometer can, in principle, be converted into a circular retarder by giving the two beams opposite circular polarizations, and interferometry and polarimetry are equivalent measurements with one wavelength of path difference corresponding to π rotation of the plane of polarization.

The reproducibility of a modern photo-electric polarimeter is about 2×10^{-7} rad (0.05″ of arc) [King & Raine, 1981], which corresponds to 10^{-7} f in interferometry. While this may not be realizable in all interferometers, since they are more sensitive than a polarimeter to vibration and other sources of noise, it has been surpassed by some, for example, Zakharov *et al.* [1976], and Kwaaitaal, Luymes & van der Pijll [1978].

For some measurements, particularly those of length, the absolute reproducibility is of less interest than its value relative to the quantity measured. Thus the reproducibility can be represented by the root-mean-square fluctuations δN of the measurements of the order of interference N, and a measure of precision that increases as the precision gets higher is, following Hanes [1963], defined as

$$P = N/\delta N. \tag{10.6}$$

Hill & Bruce [1962, 1963] and Hanes have considered the quantum limit to the relative precision. For the quantity P, values of the order of 10^{10} should be possible with either Michelson or Fabry–Perot interferometers and thermal sources, and higher values with lasers.

10.1.3. Photo-electric setting

By the use of heterodyne interferometry, intensity measurements are replaced by measurements of electrical phase, which can be made more accurately [Tanaka & Ohtsuka, 1977]. The frequency of one or both of the interfering beams is changed as described in §4.7.2, and the phase of the signal obtained from their interference is compared with a reference phase. This is obtained from the component that changes the frequency, a rotating grating or $\lambda/2$ plate, or from the interference of the same two beams over a known delay.

Other methods of accurate phase measurement are null methods, in which, as in the visual method described above, the detector is used only to show when a compensator has brought the inteference to a fringe maximum or minimum. Usually the minimum is chosen, since there is less background light and hence less noise. To detect the minimum precisely, the phase difference between the two beams is modulated so that the detector receives an alternating signal. The fundamental component of this signal is zero at the fringe minima.

These methods require two extra components, a compensator and a modulator. Again the polarimeter provides examples of these. In many photo-electric polarimeters the modulator is a Faraday cell, which modulates the direction of linear polarization. This is equivalent to a modulation of the delay between two circularly polarized beams of opposite sense. The compensator is a second Faraday cell, now with direct current, or a rotatable analyser. The last also gives a phase difference to the two circular polarizations equal to twice the angle at which it is set. It is a very convenient form of compensator, since its angle gives the phase directly without special calibration.

The application of polarimetric methods to an inteferometer is illustrated in fig. 82. The light from the source is polarized, then enters the interferometer through a polarizing beam splitter; there are fewer phase problems within the interferometer if the beams are linearly polarized there. The beams are brought together at a second (or the same) polarizing beam splitter, and a $\lambda/4$ retarder then converts them into the two opposite circular polarizations, which combine to give a linearly polarized beam. The direction of polarization is at an angle that is one half the phase difference produced by the interferometer. A Faraday modulator modulates this direction and a second polarizer, the analyser, is rotated until the a.c. signal from the detector, at the modulation frequency, is zero.

Other components can be used to give similar results. Instead of the Faraday cell following the $\lambda/4$ retarder, a Pockels cell [Takasaki, 1966] before it, produces a varying retardation between the two linearly polarized beams. Polarization is not necessary; both the modulation and compensation can be obtained by movement of a mirror that will change the length of path. For the compensator, methods involving an electromagnetically moving coil are best [Bruce & Duffy, 1975], since piezoelectric transducers are not as linear and are less reproducible.

Modulation of the interferogram can also be obtained if the frequency of the laser source is modulated, either by modulation of the length of the cavity or by switching between the modes of the two-mode stabilized laser of §4.7.1. But this modulation is useful only under restricted conditions, when the delay is fixed and has a suitable, moderate value.

10.2. Measurements of integral order

The true integral order N_i applies to the fringe maxima. Settings are commonly made, however, at some other position on the fringes, so that it is the integral order plus some fixed fraction that is measured. For settings on the fringe minima, this function is one half. By electronic methods it is

Fig. 82 Polarimeter method of obtaining fractional order. S – source, P – polarizer, B – polarizing beam splitter, F – modulator (e.g. a Faraday cell).

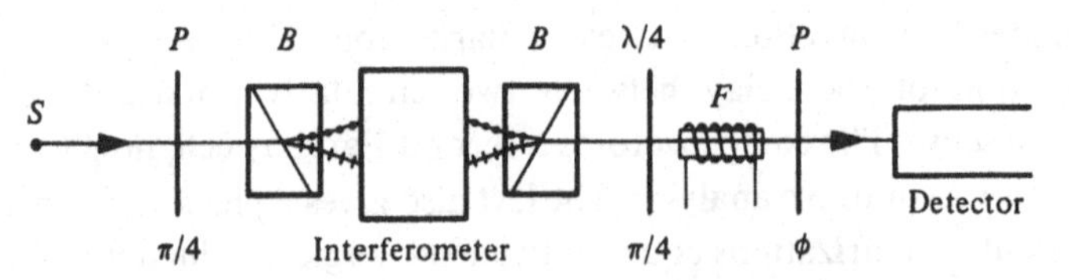

easiest to detect zero crossings of the interferogram and for these the fraction is $\pm 1/4$.

The obvious method of finding the integral order by counting fringes as the delay is increased from zero, or a known value, has become possible only since bright sources and fast reliable counters have been available. Counting can also be used with heterodyne interferometers, but it is now the beats that are counted between the signal with the varying delay and that from the same two beams with a fixed delay [Liu & Klinger, 1979].

When measurements were made visually, it was quite impossible to count fringes over large distances, and other methods of finding the integral order have been developed. The first of these combines the use of white-light fringes, to set two mirrors at the same optical distance, and *optical multiplication*, and was used in the early measurements of the wavelength of the cadmium red line. A series of étalons was used. For a Michelson interferometer these consisted of two parallel mirrors, held at a fixed spacing, and offset, so that a beam could strike each normally.

The method used by Michelson & Benoît [1895] has been given in detail by Candler [1951, p. 115]. Essentially, they stepped one étalon across the next longer one, using the position of the mirror in the other arm of the two-beam interferometer as a reference at each stage, and judging equality of optical paths by the position of the white-light fringe. Benoît, Fabry and Perot used Fabry–Perot étalons and, by placing two in series, found when their spacings were in an integral ratio (2:1) by Brewster's fringes. To adjust the ratio of spacings, one étalon could be tilted, or a third, wedged étalon used in series with the two being compared. Now such methods of optical multiplication are chiefly of interest for measuring distances greater than 1 m [Väisälä, 1954].

The second method is that of *exact fractions* in which the excess fraction is measured for several different, known wavelengths. An approximate value of the length is found by a mechanical measurement and, from this, an estimate of the integral order N_i for one of the wavelengths, λ_1; this may be in error by a small integer. A range of integers is therefore taken and the total order for this wavelength, $N_i + \varepsilon$, is multiplied by the ratios of wavelengths λ_1/λ_2, λ_1/λ_3, etc. to give the corresponding orders at the other wavelengths. If there is no phase dispersion, the correct set of orders is the one for which all the excess fractions agree with their measured values to the expected accuracy. Suitable sets of accurately determined wavelengths are given by the *Comité Consultatif pour la Définition du Mètre* for lines of ^{86}Kr, ^{198}Hg, and ^{114}Cd, and lines from stabilized lasers have since been added to these.

The method of exact fractions can be modified to give a direct result by the use of tunable lasers [Bien *et al.*, 1981]. The change in the number of orders, $N_2 - N_1$, is counted as the laser is tuned from ν_1 to ν_2. This difference is an integer and from it the delay τ can be obtained, since

$$N_2 - N_1 = (\nu_2 - \nu_1)\tau.$$

From this delay, the order N at any frequency ν is obtained as

$$N = \nu\tau = \frac{N_2 - N_1}{\nu_2 - \nu_1}. \tag{10.7}$$

Provided the uncertainty in the order so obtained is less than one half, its value at either ν_1 or ν_2 can be corrected to the known fractional value that was used for the setting and a better value of τ obtained from this. If a small frequency change is sufficient, the frequencies ν_2 and ν_1 can be measured by the beats between them and a stabilized laser at a nearby wavelength. For a greater frequency difference and greater precision, two stabilized lasers are used as reference [Goldsmith *et al.*, 1979].

10.3. Standard of length

Interferometer measurements of length can be regarded as either measurements of the wavelength of a spectral line, or measurements of length in terms of a known wavelength. Historically both viewpoints have had their day. When the metre was defined as the distance between two graduations on a certain platinum–iridium bar, interferometric comparisons with spectral wavelengths, particularly that of the cadmium red line, were made to measure these wavelengths absolutely and to study their reproducibility. When it was next defined in terms of the vacuum wavelength of an orange line of krypton 86, all lengths were referred back to this primary standard through interferometry. The same holds with the present definition of the metre, which bases it on the second by defining the speed of light to be

$$c = 299\,792\,458 \text{ ms}^{-1}. \tag{10.8}$$

Now any spectral line of which the frequency is known becomes a standard of length. In practice, such lines are provided by stabilized lasers.

The early history of the international standard of length is given by Baird & Howlett [1963]. Just after the adoption of the arbitrary line standard in 1889, Michelson and Benoît made the first measurement of the wavelength of the cadmium red line and demonstrated its precision as a standard of length. This experiment was repeated in six classical redeterminations and later, as pure isotopes became readily available, other lines were studied. In 1960, the metre was redefined as equal to 1 650 763.73 vacuum wavelengths of the radiation corresponding to the transition between the $2p_{10}$ and

$5d_5$levels of the krypton-86 atom, and means were recommended for realizing this new standard. With this line, the precision obtainable [Baird, 1963] is 10^8, or even 10^9 if special precautions are taken.

For a spectral line to be usable as a standard of length it had to be known to be readily reproducible without significant variations in wavelength. The quantum limit to the precision obtainable depends on the line's peak intensity; a narrow line is not inherently capable of higher precision than a broad one, but it has a greater coherence length, so that longer distances can be measured in one step. Finally, a standard line should have a symmetrical profile so that different methods of measurement will give the same value for the mean wavelength.

Although this standard of length has made possible a precision about 25 times higher than the old one it has some shortcomings. The line profile is slightly asymmetrical and the coherence length is less than one metre, so that a one-metre standard must be calibrated in two steps, or with a slightly more complicated interferometer [Bruce & Ciddor, 1967].

Meanwhile the good reproducibility of lasers stabilized on an absorption line (§4.7.1) had been demonstrated and, in a few cases, their frequencies had been measured by comparison with the caesium standard that defines the second. The value so obtained multiplied by the wavelength, found by an interferometer comparison with the krypton line, gave a precise value of the speed of light,

$$c = \nu\lambda_v.$$

This work showed the superiority of stabilized lasers over the krypton line, or any other emission line, both in reproducibility and, even more, in coherence length. A specific laser line has not been chosen to replace the krypton line, instead the value of c has been fixed at that given by these measurements. This implies no loss of accuracy, since the chief uncertainty in the measurements was the reproducibility of the krypton standard.

The frequency measurements involve a complex frequency chain through which the frequency of the laser is related to that of the caesium standard. Several such chains have been developed, leading to frequencies of lines in the mid infrared (28 THz, 10.6 μm), near infrared (88 THz, 3.4 μm), and visible (474 THz, 633 nm; 520 THz, 576 nm). These have been reviewed by Chebotayev [1980], and Deslattes [1980].

10.3.1. Standard wavelengths

A frequency chain up to a laser line in the visible or near infrared depends on a series of accidental near coincidences between the frequency of one laser and a harmonic of another, with sometimes a line from a third added.

The number of lines which can be measured and then used directly as length standards is therefore limited. Other lines for use as standards are measured by wavelength comparisons with these [Solomakha & Toropov, 1977].

A variety of interferometers is used for these comparisons, including Fabry–Perot étalons [Layer, Deslattes & Schweitzer, 1976; Woods, Shotton & Rowley, 1978], and Michelson interferometers, with plane mirrors [Dorenwendt, 1975], cat's eyes [Bouchareine & Janest, 1978], and with a plane mirror in one arm and a trihedral mirror in the other [Monchalin *et al.*, 1981]. This last has a rotational shear of π and is self aligning, since it will have fringes across the field unless the source is on axis. But it can be used only with lasers. All are used *in vacuo*.

For these measurements, a further method is available for obtaining the excess fraction. The interferometer delay is set to a fringe minimum for one laser and the frequency of the other laser is changed until it also gives a fringe minimum. The frequency change is measured by the beats between the radiation from the tunable laser and a laser of the same type, locked to an absorption line.

Frequency measurements can be made to about 10^{-13} and very precise interferometry is required to approach this accuracy. Corrections should be applied for diffraction, and these are largest when the wavelengths being compared are very different, for example, a visible line and the infrared line at 3.4 μm. They are smaller, the larger the aperture of the interferometer and differ for a beam directly from a laser, with a gaussian profile, and one that has been diffused to produce an effective source that is uniform over a small circle [Dorenwendt & Bönsch, 1976; Lyubimov, Shur & Etsin, 1978]. For the diffused source, a correction is also required for the phase term β_{12}. This correction is also needed when thermal sources are measured, and it is known as correction for *obliquity effects*.

10.3.2. Wavelength measurement

In addition to their use for the very precise comparison of standard wavelengths, interferometers are used for the routine measurements of laser wavelengths to a lower precision. These instruments have been given the name of *lambda meters*.

Many follow the designs used for Fourier spectrometers [Hall & Lee, 1976; Bennett & Gill, 1980], described in Chapter 13. An example of another type is that of Synder [1978], in which the Fizeau fringes from a fixed wedge fall on a diode array. From their positions a microprocessor computes the wavelength. Variations of this use the Fabry–Perot fringes from a coated wedge. This form of instrument can be used equally well with pulsed lasers.

10.3.3. Other standards

Interferometric measurements of length also enter into other standards. It is possible to relate electrical standards directly to the primary standard of length, since both capacitance and inductance can be calculated from the dimensions of a suitable capacitor or inductor. Such a calculable standard of capacitance, in which the effective dimension is measured by interferometry, has been described by Clothier [1965]. Clothier *et al.* [1980] have described an interferometric method for an absolute determination of the volt, and Avogadro's number, which determines the mole, has been measured by a combination of X-ray and visible interferometry [Deslattes, 1980]. For measurements of pressures, Terrien [1959], and Poulter & Nash [1979] have used interferometric manometers.

10.4. Interference comparators for length standards

Standards of length and their measurement have been reviewed by Giacomo [1980]. Two types of material standards of length are in common use: line standards whose length is the distance between two engraved lines, and end standards whose length is the distance between two optically flat, parallel ends. The form of end standards makes them particularly suited to the application of interferometry and interferometers have been used for calibrating gauge blocks for more than 50 years. While the primary standard was a line standard, it was more convenient to compare other line standards directly with it and the interferometric calibration of such standards is of recent development, although the principles have been used for absolute measurements of wavelength.

Present industrial and scientific requirements are for an accuracy of about one in 10^7. Since both a standard wavelength and a setting on the fringes [Ciddor, 1973] are reproducible to better than one part in 10^8, the main limitations are in the standard being calibrated, and the stability of the interferometer and its environment. Since it is inconvenient to make measurements *in vacuo*, the refractive index of the air needs to be known to the accuracy required of the measurement. To 5×10^{-8}, the refractive index can be calculated by the Edlén formula [Jones, 1981] from the measured temperature, pressure, and humidity, provided no unusual vapours are present. For higher accuracies, the refractive index of the air within the interferometer should be measured during the course of the length measurement by an interference refractometer. To obtain a uniform refractive index throughout, it is desirable to circulate the air and there are advantages in having the enclosure pressure tight. If measurements are made *in vacuo*, a correction must be applied for the expansion of the standard; this can be calculated from the elastic constants of the material.

The temperature of the standard would have to be measured to 10^{-3} K to correct to within 10^{-8} for any changes in length for most materials from which standards are made, while circulation of the air helps to keep the temperature uniform.

With the long path differences involved, corrections are required for the obliquity effect, β_{12}, when a thermal source or a diffused laser is used. While this effect could be eliminated by compensation of the interferometer, for example by the use of telescopes, as described in §8.3.1, this is not normally done. In other applications of interferometry, it is common to make the interferometer more complex if useful compensation is so obtained, but here the philosophy is that any complication is best avoided, as it could introduce additional instabilities and hence unexpected errors. The phase error β_{12} is corrected by the use of the theoretical value derived for a uniformly illuminated source of the shape used (usually circular). It is impossible, however, to correct for the loss of visibility that accompanies this phase error and a small source must be used.

10.4.1. Line standards

Baird [1963] has given an outline of the comparators for line standards used by different national standardizing laboratories. These interferometers are more complex than those for end standards, as they include a photo-electric microscope for setting on the graduations of the standard [Ciddor, 1982]. Either the scale or the reading microscope must be moved the distance to be calibrated, without the other moving. The instrument requires careful mechanical design to be extremely stable, and it should be mounted to be isolated from sources of vibration.

An example of an interferometer for line standards is shown in fig. 83. In this instrument, it is the scale that is moved, and it carries a mirror on its end to measure the movement. Two steps were required to calibrate a one-metre scale with ^{86}Kr radiation. In each step, the order of interference between the mirror on the scale and the transfer mirror is found by the method of exact fractions, the image of the reference mirror in the other arm of the interferometer being set to midway between these. Thus, when the scale is moved back 1 m from its position as shown in fig. 83, the reference mirror is moved back 0.5 m.

Much longer scales, such as surveyor's tapes, can be calibrated by a laser interferometer.

10.4.2. End standards

Interference comparators for calibrating or comparing end standards, or gauge blocks, are now highly developed; they are treated in detail by

Engelhard [1957]. Usually the gauges are wrung to a glass or metal optical flat and the distance between the top of the gauge and the flat measured interferometrically. The techniques used to obtain the integral order of interference are the same as those for line standards. For short gauges, a Fizeau interferometer or a Kösters interferometer is used. The latter, shown in fig. 84, is a Michelson interferometer with collimated light that

Fig. 83 An interferometer for one-metre line standards [Bruce & Ciddor, 1967].

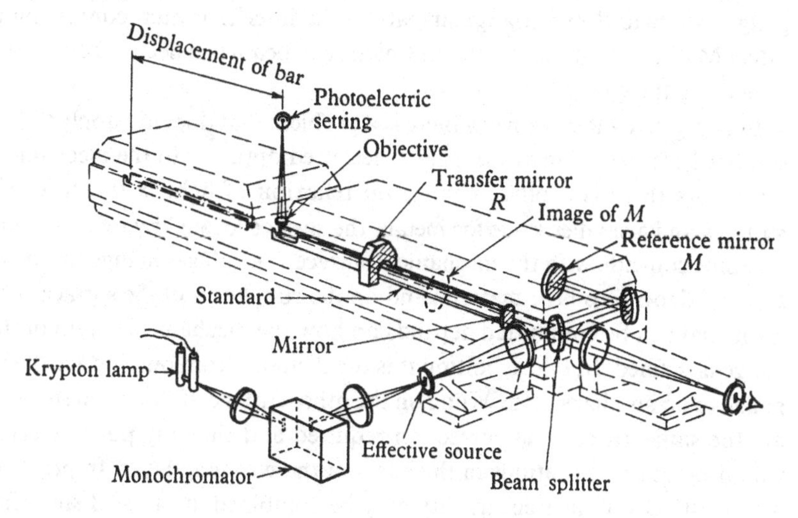

Fig. 84 Kösters interferometer for end standards.

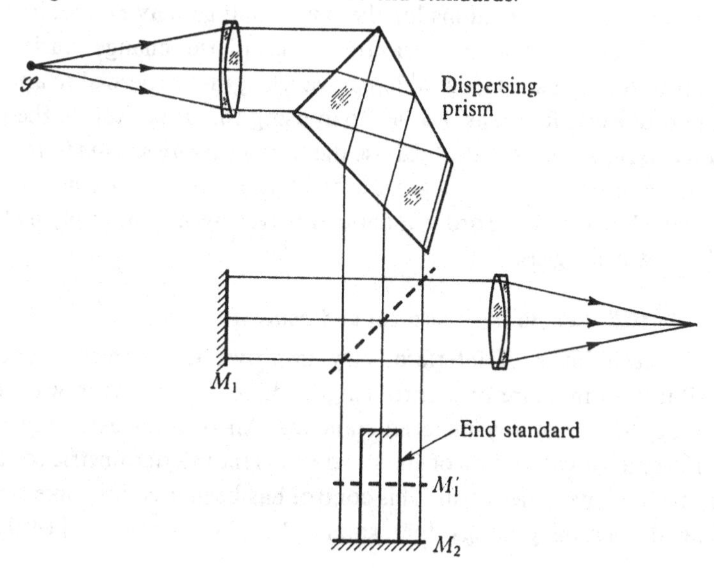

incorporates it own monochromator to separate out one spectral line from the source. A Michelson-type interferometer has the advantage that the reference mirror can be imaged midway along the gauge so that the maximum path difference and shift have half the values they would have in a Fizeau interferometer, and hence twice the range can be covered. For long gauges, a means of measuring the refractive index of the air should be included in the comparator, and such an instrument is described by Engelhard [1957]. An interferometer that compares end gauges without having them wrung on a flat is that of Dowell [Candler, 1951]. The two gauges are placed in a triangular-path cyclic interferometer, converting it into a Michelson interferometer in which each beam is reflected from a pair of ends of the gauges.

In end-gauge interferometry there is a problem that does not apply to line standards: the relation of the length measured optically to the mechanical length. As there is a phase change on reflexion of light, π for dielectric surfaces and a smaller value for metals, the surface located by reflected light does not coincide with the mechanical surface. The phase change for metal surfaces depends on the material and on the roughness of the surface. The error that can be committed depends on how the mechanical length of the gauge is defined. In one definition it is the distance from one surface of the gauge to an auxiliary plate wrung on the other surface. If these two surfaces are the same, there is no correction required and the only possible error would be due to variations in thickness of the wringing layer. In practice, metal surfaces of different quality may be combined, or a fused silica flat may be wrung to metal gauges, or ceramic gauges may be used. In all such cases, quite large corrections for the phase change may be required.

Except for transparent dielectrics, the phase change varies with wavelengths and its values at different wavelengths are needed to apply the method of exact fractions. Rather than using the dispersion of the phase change itself, workers in the field use the term *phase dispersion* to mean the change of order with wavelength, the difference between the true order and that calculated from the order at another wavelength by multiplying by the ratio of wavelengths.

10.5. Length measurement and control

As the techniques of interferometry improve, it becomes a practical possibility to measure or control lengths directly in terms of wavelength without the use of any material standard. An obvious extension of the interferometric calibration of line standards is the interferometric control of the machine that rules them. This control has been used for some time for ruling diffraction gratings, for example by Harrison *et al.* [1957] and

Babcock [1962], and is now being applied to the ruling of precision scales. To cover long distances, Mooney & Barlow [1965] narrowed the bandwidth of a spectral line by filtering it by a Fabry–Perot interferometer with plane reflectors. Sakurai & Tanaka [1965] controlled the pressure of the atmosphere inside the interferometer enclosure, so that the air wavelength of the 0.565 μm line of krypton 86 became a sub-multiple of 1 mm to one part in 10^8, thereby simplifying the control of the ruling engine.

Modern methods of controlling machine tools are based on programmed instructions. Frequently it is required that lengths be expressed and read as digits and this suggests the use of fringe counting. For most machines, a visible wavelength is too small a unit for the accuracy required, and an interferometer is needed in which one fringe corresponds to many wavelengths. This desensitizing of an interferometer can be done in several ways. If two spectral lines are used together of wavelengths in the ratio of two integers n and $n-1$, the beats between the two sets of fringes have n times the spacing of the fringes due to one line. Alternatively the interferometer can be made so that the direction of light is at an angle to the displacement to be measured, and the sensitivity is decreased by a cosine factor.

10.5.1. Moiré fringes

An important example of desensitized fringes is that of the *moiré fringes* produced by two gratings in series. Such an interferometer is very stable and therefore suitable for engineering applications; its use has been treated by Guild [1960] and Burch [1963].

Two gratings of the same pitch are used, one fixed to each of the two components whose relative motion is to be measured. One, the *measuring grating* is slightly longer than the distance to be measured, and a short *index grating* is moved along this, carrying with it an illuminating system and a photo-electric viewing system. When the two gratings have their rulings parallel, the moiré fringe is spread out to cover the full field but, if they are inclined at a small angle, a tilt is produced and moiré fringes cross the field perpendicular to the grating rulings. The two gratings should be close together, or a shift, shear, and delay will be introduced and these will reduce the fringe visibility when a large source of white light is used. When the index grating is advanced by one pitch of the ruling, one moiré fringe moves across the detection system. The effective spacing of moiré fringes is thus equal to the pitch of ruling. But the grating is not being used simply as a line scale, for, if the full length of the moiré fringe is read by the detector, the average is taken over this distance of any errors in the rulings.

Fine gratings, suitable for direct digital reading of length (50–250 lines per mm), can be made by the Merton–NPL process. A fine, accurate thread is turned, and a plastic replica of this is taken and opened out to form a flat grating. If required, the advantage, inherent in the moiré-fringe technique, of averaging the errors can be used by making the first grating control a machine that produces a second, better grating. Coarser gratings can be made photographically, either from an existing grating or as a photograph of interference fringes [Burch, 1960*a*]. This procedure was the forerunner of the method of making holographic gratings. The readings with moiré gratings need not be in multiples of a fringe, for the detecting system can read to fractions of a fringe, as in other interferometers. In fact the techniques of counting fringes have been developed chiefly to meet the needs of this application, and are discussed by Guild [1960]. Finally, the moiré-fringe technique is used also to measure rotations by means of radial gratings.

10.5.2. Small displacements

The measurement of very small displacements, such as those due to thermal expansion, is another application of interferometry. Some interferometric dilatometers have been described by Candler [1951], and Roberts [1975] has used an interferometer, of a type proposed by Dyson [1963*b*] as an interferometer suitable for use in a hostile environment. The principle is shown in fig. 85. Each beam goes across and back over the working space twice, one beam at the centre, the other beam at first one side, then the other. The interference shows the movement of the centre with respect to the edges, and is unaffected by either tilts or longitudinal displacement of the specimen as a whole. Polarization is used to tag the two beams.

The thicknesses of thin liquid films has been measured by Fisher, Parker & Sharples [1980] by a radiometric method, and by Israelachvili [1973], by multiple-beam interferometry. The latter method gives both the film thickness and its refractive index.

10.6. Measurement of refractive index

All the length measurements described can equally well be regarded as measurements of the refractive index of the medium within the

Fig. 85 Dyson's interferometer, as used to measure thermal expansions.

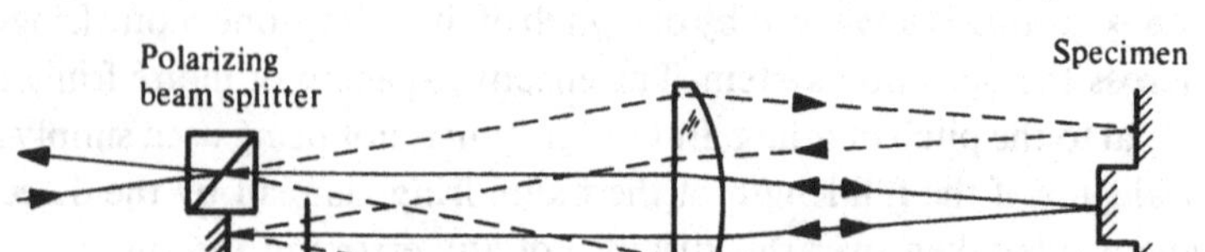

interferometer; either a length or an index must be known for the other to be measured. Normally an interferometer gives a comparative measurement, and absolute indices are found relative to a path *in vacuo*, usually of the same length. The ratio of the two indices compared is normally close to unity and interferometers are used to measure the absolute indices of gases or relative indices of liquids and solids. Absolute indices of the latter are usually found by a method involving the measurement of angles.

The refractive index commonly referred to in optics is the *phase index* n, the speed of light *in vacuo* divided by the phase speed in the medium. But, as stated in §7.4, when an interferometer makes use of radiation of a finite spectral band, it involves wave groups and measures the *group refractive indices* n_g, where

$$n_g = n + \nu \frac{\mathrm{d}n}{\mathrm{d}\nu} = n - \lambda \frac{\mathrm{d}n}{\mathrm{d}\lambda}. \tag{10.9}$$

The order of interference given by an interferometer with a source of quasi-monochromatic radiation is that derived from the phase index. The uncertainty of the whole number, the integral order, remains the same as in length interferometry and it may be resolved by two of the methods used there: fringe counting or the method of exact fractions. For gas refractometry, a measurement begins with the gas cell evacuated and fringes are counted as the gas is admitted to it. For liquids or solids, the thickness of the specimen would need to be variable. The method of exact fractions has the difficulty that, for a completely unknown specimen, the dispersion, and hence the different wavelengths within the specimen, are unknown. A bootstrap method is therefore used. Measurements are made initially on a thin specimen for which a change of unity in the order is relatively so large that it is impossible to mistake the correct integer. The approximate values, derived from this measurement, for the indices at the different wavelengths are used to apply the method of exact fractions to a larger specimen. The process is continued until results of sufficient accuracy are obtained.

Methods of interferometry based on judging when two optical paths are equal by the white-light fringe, or optical multiplication by the same criterion, measure the group order and give values of group indices. The frequency ν for which this index is measured is the mean frequency of the product of the spectral distribution of the source and the sensitivity curve of the detector. A more precise way of setting at this frequency is to use channelled spectra (§7.6) rather than the white-light fringe; the group paths are equal at the frequency for which the phase is stationary so that these fringes of equal chromatic order are parallel to the direction of dispersion. Channelled spectra are used to study the variations of refractive index in a region of anomalous dispersion. This is the *hook method* of Rozhdestvenskii [1951].

When the refractive index is required over a range of frequencies, the interference technique of Fourier spectroscopy is used. The index variations cause a variation of the phase of the complex transmittance of the specimen. The real part of this complex function is the usual intensity transmittance studied in spectroscopy. This type of Fourier spectroscopy is treated in §13.8.

For the older methods of deriving the refractive index from direct measurements of the order of interference, both two-beam and multiple-beam interferometers are used. These give comparative measurements and hence the ratio of the indices in two paths. To obtain absolute refractive indices, one of the paths is evacuated. An example of the measurement of refractive index in a Michelson interferometer is provided by the method of measuring the index of the air in an interference comparator, such as that described by Engelhard [1957]. The same interferometer is used that calibrates the gauges. An evacuated tube with oversize windows on its ends is placed in part of the beam. This provides a vacuum path through its centre and an air path around this. The difference in order for these two paths can be measured and, if the length of the tube is measured by the same interferometer, the index of the air can be found. If the tube has the same length as the gauge being measured, the difference in order can be used directly to correct the measurements on the gauge to vacuum orders.

The Fabry–Perot interferometer is widely used for gas refractometry, either as a single étalon or, more commonly, two étalons in series. Cagnet [1954] has used double passage through the same étalon. Again, if fringes are counted as gas is admitted into the étalon, the phase index is measured, while a comparison of orders by Brewster's fringes gives the group index.

Two interferometers have been specially developed for refractometry, the Rayleigh refractometer of fig. 2 and the Jamin interferometer, shown in fig. 86. Both are two-beam interferometers, but the Rayleigh uses division of wavefront, while the Jamin is equivalent to a Mach–Zehnder interferometer and uses division of amplitude. With the unavoidable shear associated with division of wavefront, the Rayleigh refractometer requires a narrow slit source and, as it has the same amount of tilt, the final fringes are finely spaced and require a large magnification for setting. The Jamin, however, is a compensated instrument that can be used with a large source. In spite of this inherent superiority of the Jamin refractometer, the Rayleigh is much more commonly used, possibly because of faults in the design of some Jamin refractometers [Kuhn, 1951]. Both instruments are used for the routine measurement of the refractive indices of gases or of the relative indices of two liquids.

10.6.1. Plasma diagnostics

Another important application of interferometer measurements of refractive index, at quite a different part of the spectrum, is the measurement of the electron density of a plasma. When the frequency of the radiation is greater than the *plasma frequency*, a function of the electron density, and both are much greater than the collision frequency, the plasma behaves as a dielectric and the radiation is transmitted with little attenuation. This radiation can then be used as a probe to study the plasma. Suitable radiation should also pass through the plasma without too great refraction: the refractive indices should not be too great. The radiation used ranges from millimetre microwaves, to the submillimetre infrared for dense plasmas. The electron density can be derived from measurements of the refractive index, found by interferometric methods, and the collision frequency is given by the attenuation in the plasma, either measured separately or derived from the visibility of the interference.

Interferometers are used to study relatively large laboratory plasmas; small plasmas can be investigated inside a microwave cavity. Forms of Michelson, Mach–Zehnder, and Fabry–Perot interferometers are all used, the different types being reviewed by Luhmann [1979], and Véron [1979]. For frequencies lower than about 50 GHz, the interferometers are made up of waveguides; at higher frequencies, free-propagation forms are used. Many plasmas approach radial symmetry, and refraction problems are less if the radiation is focused at the centre of the plasma rather than entering as a collimated beam.

Fig. 86 Jamin interferometer.

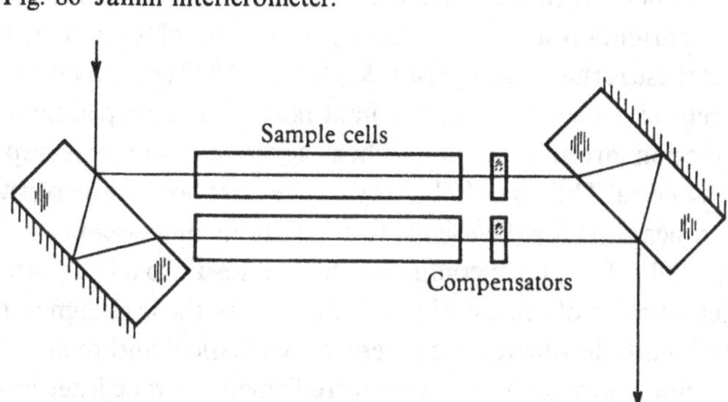

Fourier spectroscopy in the far infrared is also widely used in this field. It is treated in Chapter 13.

10.7. Speed of light

The remaining factor in the delay of an interferometer is the speed of light. It is now a fixed quantity, but previously its measurement was an important application of interferometry. Bergstrand [1956] has reviewed earlier methods, dividing them into direct methods, where the speed is obtained as the distance travelled divided by the time taken, and indirect methods, such as interferometric measurements of wavelength used in combination with a separate measurement of frequency. The latter were initially made with microwaves [Froome & Essen, 1969], since higher frequencies could not then be measured. But when these measurements became possible for the visible or near infrared, more precise values were obtained that led to the new definition of the metre. The same measurements are still made by interferometry, but they now have a different interpretation: the comparison of standard wavelengths.

Interferometry is still used to investigate any changes in the speed of light. Thus Fizeau, in 1851, used a two-beam interferometer to measure the change in the speed of light when the water in which it was travelling was moving. In a medium of refractive index n moving at a velocity v, the speed of light is given by the Lorentz formula

$$u = \frac{c}{n} \pm v\left(1 - \frac{1}{n^2} + \frac{\nu}{n}\frac{\mathrm{d}n}{\mathrm{d}\nu}\right), \tag{10.10}$$

where the factor in parentheses is Fresnel's *dragging coefficient.* Michelson & Morley [1886] repeated the experiments, again with water, while Zeeman carried out a series of measurements on both solids and liquids; his work is summarized by Zernike [1947].

Of more modern interest are the experiments, now called relativistic, which were originally aimed at detecting movement of the aether. The most famous of these is that of Michelson & Morley [1887] which aimed to show the difference between the speed of light parallel and perpendicular to the earth's motion around the sun. Their negative result was support for Einstein's Special Theory of Relativity. There have been many repetitions of this experiment and, by the use of lasers and frequency-locking techniques, Brillet & Hall [1979] have confirmed the null result to a few parts in 10^{15}.

A different class of relativistic experiments are those designed to detect rotations by interferometry. For these, both classical and relativity theory predict a fringe shift and this has been confirmed. A cyclic interferometer is used with as large as possible an area A enclosed by the beams. If it is rotated at an angular velocity ω about an axis at an angle θ to the normal to the plane of the interferometer, the delay due to the rotation is

$$\delta\tau = 4\omega A c^{-2} \cos\theta. \tag{10.11}$$

For the rotation of the earth, this effect was measured by Michelson & Gale [1925]. With an interferometer that enclosed an area of $2\times10^5\ \mathrm{m}^2$, a displacement of 0.230 f was observed, against a theoretical value of 0.236 f.

In the laboratory, Harress [1912] and Sagnac [1914] rotated the interferometer. The method used by Sagnac is shown in fig. 87: the light source and the photographic plate to record the fringes rotated with the interferometer. Various repetitions of this experiment have been reviewed by Post [1967]. Rosenthal [1962] has also proposed another interferometric method of sensing rotation, the *ring laser*, developed by Macek & Davis [1963]. This is shown in fig. 88. Up to four He–Ne gas tubes may be contained in the arms of a cyclic resonator. If this is rotated, the two outputs that traverse the resonator in opposite senses have their frequencies shifted by opposite amounts, producing a frequency difference

$$\begin{aligned}\delta\nu &= \nu\,\delta\tau/\tau\\ &= 4\nu\omega A p^{-1} c^{-1} \cos\theta,\end{aligned} \tag{10.12}$$

where p is the path around the ring. As it is desirable that the gain around the ring shall be just sufficient to produce oscillation, fewer than four gas tubes are normally used.

Fig. 87 Rotating cyclic interferometer used by Sagnac.

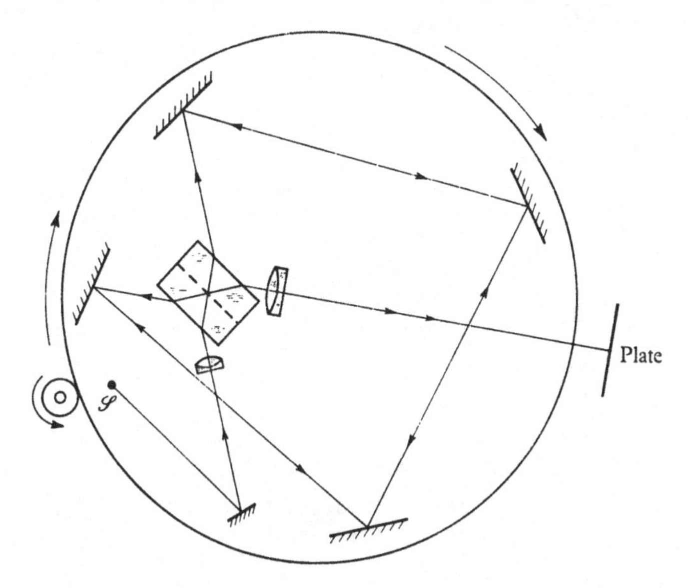

Fig. 88 Ring laser [Macek & Davis, 1963].

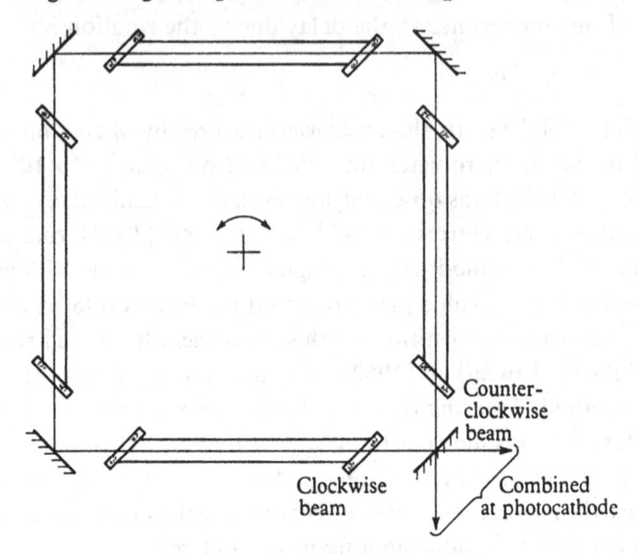

Both the cyclic interferometer and the ring laser are rotation sensors, and can be used as gyroscopes [Rowland & Agrawal, 1981]. The ring laser gives the stronger signal, but the two counter-running modes lock together at low speeds, and special methods of avoiding this must be added. They are reviewed by Aronowitz [1971]. The passive cyclic interferometer has not this problem, but needs a large enclosed area. It can still be compact if made from a fibre, with many turns around the enclosed area [Leeb, Schiffner & Scheiterer, 1979].

11

Measurements of phase variations

Ordinary optical images are records of intensity and, if an amplitude record that includes phase is desired, this is given by an interferometer with the object inside it. The results yield phase information of the type considered in the previous chapter: variations of length due to surface irregularities or variations of refractive index due to changes in composition. Records of both amplitude and phase are given by the method, due to Zernike [1948], of adding a *coherent background* to convert intensity to amplitude distributions, and include holograms. The form of interferometer used varies with the type of measurement, and surface irregularities are usually studied in reflexion by a double-pass interferometer, while a single-pass instrument is used for variations of index.

A review of the instruments used for testing optical components and optical systems is given by Malacara [1978].

Most of the methods used follow those described in Chapter 10, with visual examination of an extended field now more widely used when high precision is not required. The success of photo-electric settings for measuring the mean phase over small regions has led to the adoption of this method for phase variations also, the measurements being made in sequence with the detector scanning across the interference field (or the field across the detector). More recently, detectors that record the full image, Vidicons and charge-coupled arrays, have made measurements possible without mechanical scanning.

The general remarks on precision given in Chapter 10 apply, if the variations are not too rapid. For, as Ingelstam [1954] has shown for multiple-beam interferometry, there is an uncertainty relation connecting the accuracy of measuring a phase $\Delta\psi$ and the accuracy Δu of allocating it to a definite position u in the plane of observation:

$$\Delta\psi\,\Delta u \gtrsim N\lambda, \tag{11.1}$$

where N is the order of interference. Similar considerations apply to photo-electric measurements which require the collection of energy over a finite area to give high precision.

For nearly all measurements of phase variations, the source can be chosen to suit the interferometer, and traditional sources, such as the mercury-vapour lamp, are being displaced by lasers, with their greater power. But sodium and mercury lamps are still very useful when the delay, shear and shift are all small, as, for example, when optical surfaces are tested by Newton's rings.

11.1. Testing optical surfaces

The testing of optical components for surface quality is a good example of the application of interference methods. The most important example is that of optical flats, but the methods used for flat surfaces can be adapted simply to test spherical ones. This field has been reviewed by Schulz & Schwider [1976].

11.1.1. Flatness interferometers

The simplest method of comparing two flats is to place their surfaces in contact with a slight wedge of air between them. This gives a tilt and fringes are seen, located between these two surfaces, a form of Newton's rings. These have good contrast even when a diffuse source of relatively large bandwidth is used, since the system has very small shear, shift, and delay. For the fringes to give accurate half-wavelength contours of the space between the surfaces, they should be viewed from infinity (or from the centre of curvature, if the surfaces are spherical). At the same time it is desirable that the two surfaces being tested should be separated to avoid the risk of scratching. As a shift is introduced by this separation, the effective source should be small and at infinity.

An instrument that gives the source and viewing point at infinity is the *Fizeau interferometer* or *interferoscope*, shown in fig. 89. This instrument is also used, as mentioned in §10.4.2, to calibrate short end gauges. It has been drawn with a doublet collimating lens, but multi-component meniscus systems, described by Taylor [1957], are also used. The objective does not need to give a diffraction-limited image but only produce a beam that is sufficiently parallel so that all rays have the same optical path across the gap between the two flats, to the precision required for the measurement. The larger this gap, the better the correction needed. It is also desirable to correct the astigmatism of the beam splitter, as shown in §8.6.3.

The fringes are viewed through the standard flat, which should have a small wedge angle so that light reflected from its top surface does not reach the eye. The flat to be tested is held below the standard flat on an adjustable table. Two-beam fringes are obtained and visual examination of these gives the shape of the surface to about 0.03 f. The more complicated Twyman–

Green interferometer described later can also be used to test flats, and gives the same precision.

To increase the precision obtainable with a Fizeau interferometer, the two surfaces can be given high-reflectance coatings to produce multiple-beam fringes that can be judged to 0.01 f, or measured to 0.001 f by photo-electric methods [Koppelmann, Rudolph & Schrech, 1975]. Thin silver coatings are best, of about 0.9 reflectance, as dielectric multilayers of high reflectance would be much thicker and, unless very uniformly deposited, could change the shape of the surface. Normally the reflected multiple-beam fringes are observed, but for the multi-wavelength methods described by Herriott [1961], and Schulz & Schwider [1967] the transmitted fringes are used. These methods use as source a series of narrow spectral bands at equal wavelength intervals, obtained by passing broadband radiation through either a monochromator with many slits or another Fabry–Perot interferometer. Their spacing is adjusted so that, between successive orders of the fringes due to one band, there is an integral number of equally spaced fringes, one from each of the other bands. Although this modification does not increase the precision given by multiple-beam fringes, it gives fringes over a much larger portion of the surface, making a complete examination more rapid.

The need to coat a flat in order to test it is an inconvenience that can be avoided by the use of photo-electric setting on the two-beam fringes from uncoated surfaces. Methods for this have been given by Dew [1966], and

Fig. 89 Fizeau interferometer.

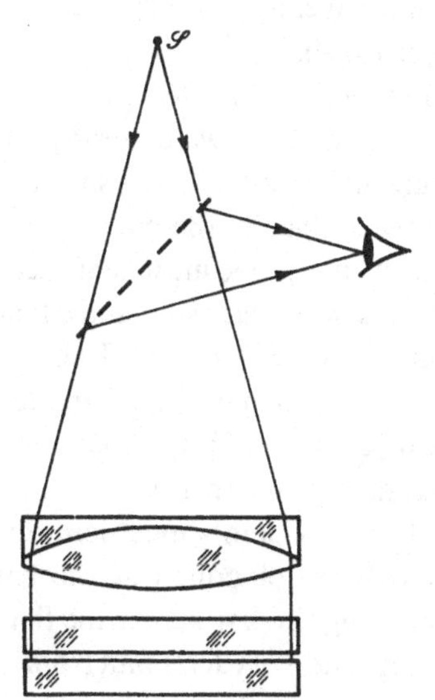

Roesler & Traub [1966]. The detector either views a spread-out fringe, or measures the deviation from a straight line of a fringe with a tilt present. By this latter method, Bruce & Sharples [1975] have obtained a precision of 10^{-3} f. With a spread-out fringe, a compensation method can be used. Since a Fizeau interferometer is not suitable for polarization methods, the compensation is obtained by movement of one flat. Modulation can be obtained by the same method or by modulation of the laser frequency, provided the gap between the flats is sufficiently large.

11.1.2. Standards of flatness

Optical flats are tested by comparing them with a standard. The standards themselves are either tested against an undisturbed liquid surface [Bünnagel, 1965] or three are intercompared and the results analysed to obtain the surface shape of each flat [Schulz & Schwider, 1976; Marioge, Bonino & Mullot, 1975]. Both methods have their difficulties.

If a liquid is to have a flat surface, it should be free from contamination and temperature variations, and well isolated from vibration. The liquids commonly used are mercury or a silicone oil. Being a good conductor, mercury has good temperature uniformity and is free from stray electrical charges, but it has the disadvantages of oxide contamination of the surface, sensitivity to vibration, and too high a reflectance to give fringes of good visibility against an uncoated glass surface. Clothier *et al.* [1980] give methods of avoiding ripples due to vibration. They slope the edge of the container inwards to eliminate the turn down of the meniscus, through which the ripples are coupled, and mount the container on thin vertical columns as an inverted pendulum. Bünnagel used a thin layer of mercury amalgamated to a silver-plated copper dish.

Silicone oils match better the reflectance of uncoated glass. An ideal oil has a low vapour pressure, high viscosity, and low water absorption [Dew, 1966]. To obtain better temperature uniformity, it is usual to place a perforated copper sheet just below the surface of the oil.

When three flats are intercompared by pairs, one must be turned over for one of the intercomparisons. This reverts the surface so that the final results apply only to one diameter; Schulz & Schwider [1976] show how to calibrate a series of diameters. But for the original intercomparison, three narrow bars can just as well be used [Dew, 1966], thick enough to have negligible sag, and one of the calibrated bars then stepped across the standard flat to calibrate it. The calibration points must join up to form triangles, and the preferred array is a set of points at the corners of equilateral triangles, obtained by stepping the bar across the flat in three directions at $\pi/3$ to each other. The departure of each point from the mean plane is found by a best-fit solution of the resulting overdetermined set of equations. The method is capable of 10^{-3} f; somewhat better than can be obtained from a liquid surface.

Calibrated flats are not necessarily stable. Dew [1974] has shown that they need recalibration periodically.

11.1.3. Spherical surfaces

Precision lenses are tested during manufacture against test plates of the required radius, which are made in pairs and tested against each other by Newton's rings. To avoid distortion, these should ideally be viewed from an image of the centre of the surface. A full calibration of a spherical surface requires four intercomparisons [Elssner, Grzanna & Schulz, 1980].

A range of radii can be tested in a spherical version of the Fizeau interferometer. The form of this that tests convex surfaces is the more useful, but the more complex, since it must focus to a distance in front greater than the radius to be tested. If the last surface of the objective has its centre at this focus, it reflects a reference wave. To cover a range of radii and numerical apertures, several different objectives are needed, or an objective with removable components. Path differences can be large, so a laser giving a single longitudinal mode is desirable as the source.

11.1.4. Ground-glass surfaces

During optical manufacture it is often desirable to test the shape of surfaces before they are polished. A rough flat surface can be tested by an interferometer that reflects a beam from it at grazing incidence, such as that described by Hariharan [1974] and shown in fig. 90, since it then acts as a specular reflector. If the roughness is not too coarse, the surface can also be tested with radiation of longer wavelength such as that from a carbon-

Fig. 90 Grazing-incidence interferometer for testing rough surfaces.

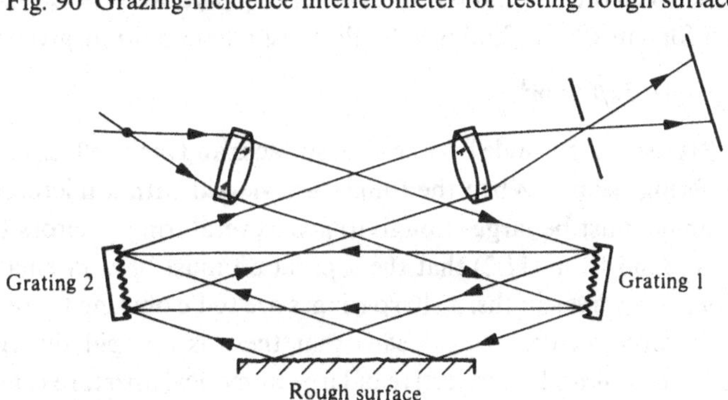

dioxide laser at $\nu = 28$ THz ($\lambda = 10.6$ μm). Both flat and spherical surfaces can be tested in an interferometer described by Kwon, Wyant & Hayslett [1980].

Either interferometer has a larger fringe interval than those that test polished surfaces with visible light. They are thus desensitized interferometers, of the type considered in §10.5, and these also have considerable use for measuring phase variations. In addition to the types mentioned there, there are the second-harmonic interferometers of Hopf & Cervantes [1982], and the contouring interferometers of §12.6.

11.2. Interference microtopography

The use of multiple-beam interferometry for the detailed study of surfaces has been developed by Tolansky [1960, 1970] and his co-workers, and applied to investigations such as the study of crystal surfaces and crystal growth. The technique combines microscopy with the testing of optical flats: a small reference flat is placed over the surface to be tested and the fringes viewed with a microscope. For two-beam fringes, the reference can be the top surface of a drop of liquid or clear cement placed on the specimen [Hariharan & Steel, 1974]. For tests with multiple-beam fringes, both surfaces are silvered. If the surface studied is that of a transparent specimen, transmission fringes can be used; for opaque specimens, it is necessary to observe the reflexion fringes. In the latter case the silvering of the flat is more critical, as any absorption in the silver reduces the visibility of the fringes. Even if the specimen is a metal, it is usual to silver it also, as this eliminates any confusion between surface irregularities and variations of phase change across crystal boundaries. Silver is used for the same reason when the thickness of an evaporated layer is measured. Part of the substrate is left free of the layer, both are coated with silver, and the step height at the edge of the layer is measured.

When a tilt is used to give several fringes, to be used as contours of the wedge between the flat and the specimen, the separation d between the two surfaces for a medium of index n should satisfy the condition, given in (9.24),

$$nd \ll 3\lambda/8\theta^2 m^3, \qquad (11.2)$$

where θ is the wedge angle between the surfaces and m the effective number of interfering beams. When the fringes are viewed with a microscope, the wedge angle must be large enough to give several fringes across the field, and it is found from (11.2) that the separation must be very small, of the order of a few wavelengths, if sharp fringes are to be obtained. Further, the relatively large wedge angle means that there is a rapid divergence of successive beams and an objective of large numerical aperture is needed to collect them. The same considerations apply as for microscopy in general: the higher the magnification, the greater the numerical aperture that should be used. Tolansky [1955] has found that good definition can be obtained with magnifications up to 250. A practical difficulty at high magnifications is the short working-distance of the microscope objective, which makes it difficult to fit a reference flat between the objective and the specimen. However, very thin flats of cleaved mica can be used.

The small separation necessary between the two surfaces means that the shift and delay are small. The light need not have a narrow spectral bandwidth and the effective source, at infinity, can subtend a few degrees at the specimen. When the reflexion fringes from an opaque specimen are observed, the vertical illuminator used in microscopy can be employed. The arrangement for observing reflexion fringes is shown in fig. 91.

When the specimen and reference surfaces are set accurately parallel, one fringe covers the whole field and variations in surface height show as variations of intensity. As multiple-beam fringes have steep edges, these variations of intensity are very sensitive to slight variations in distance, when the spacing is set for the steep part of the fringes. But it is not possible to discriminate between different orders by intensity alone. When the order of interference is close to zero, white light can be used to replace variations

Fig. 91 Arrangement for examining surfaces by multiple-beam interferometry.

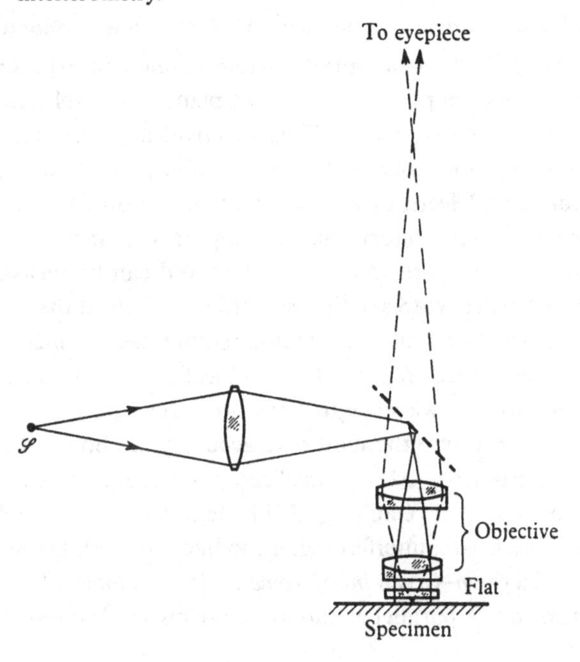

of intensity by variations of colour. In favourable circumstances, the eye may be more sensitive to these, and the colour also identifies the order.

Orders are most easily separated, however, by the narrow fringes obtained when a tilt is present. Tolansky [1955] has combined the two techniques to record in succession, on the same photograph, the fringe with no tilt and a set of narrow contours from the fringes with tilt. In some cases two sets of contours at right angles are useful. This method can be applied only to transmission fringes since the light background of the reflexion fringes would greatly reduce the contrast on successive exposures.

When it is not possible to satisfy the condition (11.2) and bring the surfaces close enough together to obtain wedge fringes of good visibility or to avoid the errors of asymmetry of these fringes, Tolansky [1955, 1960] has shown that narrow contours are given by the channelled spectra or fringes of equal chromatic order. Only a strip of the specimen can be examined at one time, and it is spread out into a channelled spectrum, the fringes of which show the variations of phase along the strip. The two surfaces can be parallel, so there are no wedge errors. The spacing of the fringes depends directly on the separation of the two surfaces and gives a direct indication of any change of order.

Surfaces are also studied by two-beam interferometry, particularly when it is not convenient to place a reference flat near the surface. The instrument used is an interference microscope, and it is discussed in §11.6.

11.3. Testing optical systems with Michelson interferometers

In terms of wave optics, an ideal optical system should convert a spherical (or plane) wave from each point in the object plane into a spherical wave centred on a point in the image plane. This is equivalent to the requirement that the optical paths along all rays between an object point and its image should be equal. Any defects, or *aberrations*, whether due to bad design, irregular surfaces, or non-uniform indices, appear as a variation $w(\mathbf{u})$ of optical path across the aperture of the system and can be measured by interferometry as the departures of the wave from a sphere at the exit pupil, centred at the image. A complete test would require measurements of this *wave aberration* for several representative object points. As aberrations small compared with the wavelength have little effect on image quality, optical testing requires only the moderate accuracy of about 0.1 f.

The use of interferometry to test optical components and optical systems was pioneered by Twyman [Candler, 1951]. He tested systems in double-passage, using a Michelson interferometer modified to work in collimated light. This is the *Twyman–Green interferometer*. In its simplest form it can test plane mirrors, plane-parallel windows, or prisms; the last test is shown in fig. 92. The *reference arm* of the interferometer provides a plane wave which interferes with the wave that has been twice through the specimen in the *test arm*, the departures from straightness of the fringes showing the variations in optical path caused by defects in the specimen.

When the source is a discharge lamp, the Twyman–Green interferometer has the advantage that the reference arm can be adjusted easily to give zero delay. An independent control is needed to correct the shift, such as those described in §8.3, but these corrections are not needed with a laser source. With them, Twyman–Green interferometers are still used, but the simpler Fizeau interferometer becomes an alternative [Forman, 1979].

Fig. 93 shows different arrangements of the test arm that can be used with either a Twyman–Green or a Fizeau interferometer. When a small telescope objective is to be tested, this is placed in the test arm and the plane mirror in this arm is replaced by a spherical *autostigmatizing* mirror with its centre at the image point of the lens. This sends light back through the focus so that it emerges from the lens again collimated. When the system is not a telescope objective but is intended for use with finite object and image distances, for example, a microscope objective, a *tube-length lens* must be added to the test

Fig. 92 The Twyman–Green interferometer, as used to test a prism.

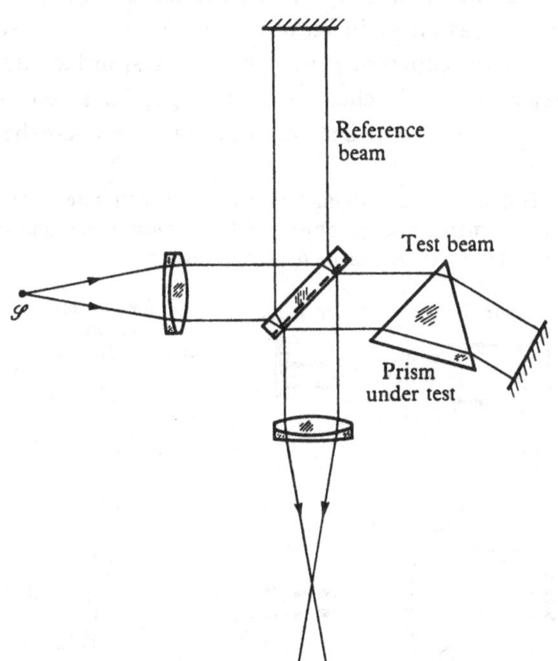

arm. This is a lens of negligible aberration, usually divergent, that converts the plane wave to a divergent wave from the object point. Finally, the test arm can include a nodal slide, so that lenses can also be tested off axis. A paraboloidal mirror is tested in the same way as a telescope objective; in the Fizeau interferometer, the spherical mirror can be supported on the reference flat. Finally, a larger objective can be tested in reverse, with a divergent lens and a plane mirror [R.E. Hopkins, 1962]. The further addition of correcting lenses, designed by Offner [Malacara, 1978], allows paraboloidal or hyperboloidal mirrors to be tested from their centres.

The radius of the autostigmatizing mirror used affects the shift [Hansen, 1942, 1955] and, more importantly, the position where the wave aberration is measured. The Fizeau fringes that show the aberration are localized at the end of the arms of the interferometer, that is, at the surface of the autostigmatizing mirror in the test arm. But the lens aberrations should be measured at the pupil of the lens. Thus an ideal autostigmatizing system is one in which the pupil of the lens is at the surface of the mirror, or is imaged to coincide with it. When the pupil is between the lens and the image, a convex mirror can be chosen to have its surface there, and a convex mirror close to the lens is a good approximation to the ideal system for use with a telescope objective, where the pupil is at the lens. But when the pupil is inaccessible, as in a microscope objective, a compound autostigmatizing system should be used, as shown in fig. 94. This consists of a hemispherical lens and a concave mirror concentric with the spherical surface of the lens. The ratio of the two radii can be chosen to suit the pupil position; as shown in fig. 94, the system is for use with a microscope objective where the pupil is

Fig. 93 Test arm of a Twyman–Green or Fizeau interferometer, for testing: (*a*) a small telescope objective; (*b*) a microscope objective; (*c*) a paraboloidal mirror; (*d*) a large telescope objective.

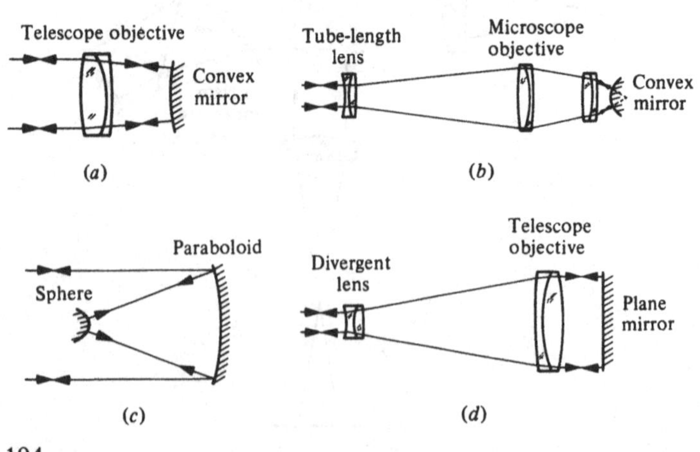

at infinity. These considerations of where the aberration is measured apply for both thermal and laser sources.

Three of the few cases where a Twyman–Green interferometer has tested a non-spherical surface are shown in fig. 95, the test arm only being drawn. The first shows a rough surface tested at grazing incidence, and the others are a hollow cylinder [Archbold, Burch & Ennos, 1967] and a screw. For the last two, an auxiliary component is needed to convert the parallel beam into beams normal to the surface being tested.

The Twyman interferometer requires a beam splitter larger than the lens being tested, unless a divergent lens is used, as in fig. 93(*d*). A simpler system, also suggested by Twyman and Green, and attributed by Burch [1940] to

Fig. 94 The perfect optical system of Dyson [1959], used as an autostigmatizing system.

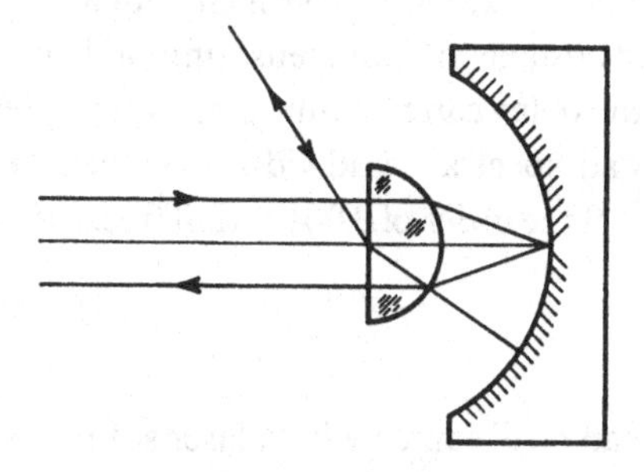

Fig. 95 Test arm for testing: (*a*) a rough surface; (*b*) an internal cylinder; (*c*) a screw [Burch, 1952].

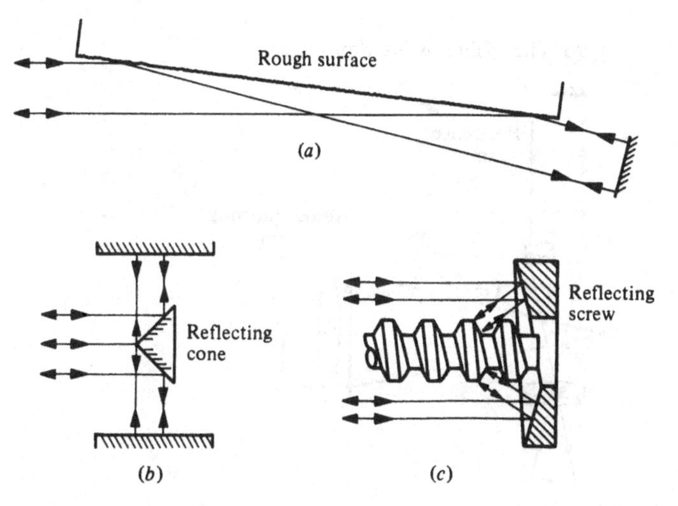

W.E. Williams, is shown in fig. 96. The beam splitter is no longer in parallel light and can be made quite small. While a Twyman interferometer can test lenses up to a maximum diameter, a Williams interferometer has a maximum angular aperture and the reference arm need not contain components as large as the lens being tested. Such a mismatch of the arms would lead to a shift unless compensated by the method shown in fig. 53.

The Twyman and Williams interferometers are both modified Michelson interferometers. They are very versatile, being also used for testing with infrared radiation [Corno, Lamare & Simon, 1977]. But this versatility means that they require considerable adjustment for every different test. They have two further disadvantages: reference components of high optical quality are needed, and the wide separation of the two beams means that any vibration of the instrument affects both beams differently and hence appears as a movement of the fringes. The complete elimination of vibration is very difficult and, while it can be done sufficiently to observe the fringes visually, it would blur a photograph taken with any but a very short exposure. This is then the advantage of compensating a lens-testing interferometer so that the shift and delay can be eliminated independently: a moderately large source of broad spectral bandwidth can then be used, permitting very short exposures, of the order of 0.001 s, that freeze the fringe pattern.

11.3.1. Laser sources

But now these short exposures can be obtained with a laser source, with no need for compensation. It also allows a further series of tests, shown in fig. 97. In each case the test arm shown produces a rotational or a folding shear, and no fringes would be seen with an extended source. The case 97(*c*) of a

Fig. 96 The Williams interferometer.

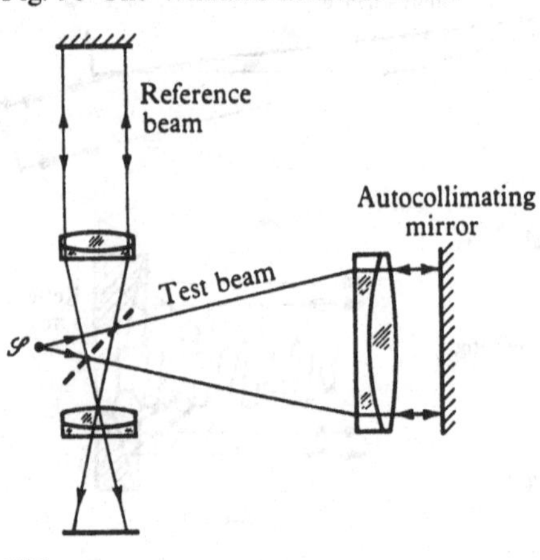

lens with a mirror at its focus shows the aberrations of even azimuthal order, that is, those terms in equation (3.27) for which m is even. This figure can also represent a cylindrical lens, which can also be tested in this way [Schnurr & Mann, 1981].

A laser used directly, or through a spatial filter, has the disadvantage that interference takes place between the main beams and all beams reflected from the surfaces of optical components, and all the unwanted fringes appear, superimposed on those due to the interference of the two main beams. The unwanted fringes can be eliminated by diffusing the light to give a larger effective source and correcting the shift of the interferometer; the delay, however, can be left uncorrected. If a ground glass is used as a diffuser, its diffraction pattern appears in the plane of observation unless it is moving. A colloidal diffuser also gives a moving pattern that is not seen.

A better solution is an interferometer designed specifically for use with a laser. To avoid surface reflexions, it should be as simple as possible, having a minimum of components, especially inside the interferometer itself, although double reflexions from components outside can still give unwanted fringes. The form used is generally that of a Williams interferometer. As the reference arm does not need to be adjusted in length, it can be built into the beam-splitting prism. Examples have been given by van Heel & Simons [1967], Houston, Buccini & O'Neill [1967], and Mahé & Marioge [1978]. Reflexions from surfaces, such as those of the plate supporting the beam splitter, are sent out of the beam by the use of oblique incidence or, since the radiation from a laser with Brewster windows inside the cavity is linearly polarized, the reflected component is eliminated entirely by using surfaces at the Brewster angle.

11.4. Common-path interferometers

Bates [1947] introduced the name 'wave-shearing' for an interferometer that tested a concave mirror without any reference system. It is based on the principle, considered by Mascart [Candler, 1951, p. 489], and demonstrated by Lenouvel & Lenouvel [1938], of comparing two

Fig. 97 Test arm for testing, with a laser source: (*a*) a cube corner; (*b*) a roof mirror; (*c*) a lens with a mirror at its focus.

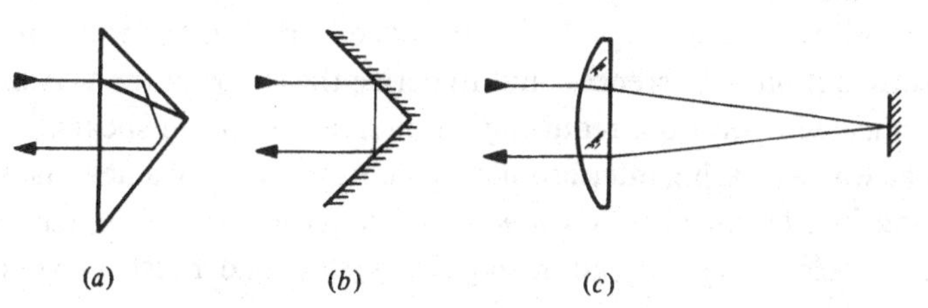

wavefronts from the concave mirror, one being sheared a distance **s** relative to the other. Essentially, they compare the mirror against itself. Thus if $w(\mathbf{u})$ is the variation of optical path caused by the aberrations of the mirror, the same disturbance occurs in both beams but at different positions and the final path difference is then given by (7.10) as

$$\Delta w_u = w(\mathbf{u} - \mathbf{s}) - w(\mathbf{u}). \tag{11.3}$$

The interferometer gives the difference in aberration at pairs of points separated by the shear. When this is small, the result is proportional to the gradient of the aberration,

$$\Delta w_u = |\mathbf{s}| \nabla w. \tag{11.4}$$

The constant of proportionality is thus vanishingly small, but this loss of sensitivity can be avoided in a second-harmonic interferometer designed by Hopf *et al.* [1981].

When used in single passage, a wave-shearing interferometer has a shear at the plane of observation and this limits the usable source to a slit that is narrow in the direction of shear; it can be made to have neither a shift nor a delay and these are not introduced when the system to be tested is put into the interferometer. When it is used in double passage, however, the shear cancels out at the final plane of observation, although it remains at the intermediate plane at which the specimen is placed. A large source can then be used. If double passage is undesirable for some reason, the same compensation can be obtained by having two interferometers with the same shear, one on each side of the specimen.

Three forms of this interferometer have been shown in fig. 1. The form used by Bates had a variable shear and tilt, but the Ronchi grating [1962] and the Wollaston prism have fixed values of these, unless used in pairs. Fig. 98 shows an interferometer consisting of two gratings due to Spornik & Yanichkin [1971]. The shear is changed by counterrotation of the gratings and a longitudinal displacement of the gratings as a pair with respect to the focal point changes the tilt. Like a Williams interferometer, a wave-shearing interferometer has a limited angular aperture up to which it can test, but the linear size of the system under test is unlimited. A different two-grating interferometer, due to Hariharan & Hegedus [1975], is used in a collimated beam. All wave-shearing interferometers do not give the wave aberration directly but only differences, and to derive the shape of the wavefront is a complicated operation, requiring an integration for all shears.

The wave-shearing interferometer is representative of a new class of lens-testing interferometers, *common-path interferometers*, in which both the test and reference beams go through the system under test. The two beams then have equal optical paths and for all tests the delay is automatically zero. While the shift is also zero in a wave-shearing interferometer, this may not be so in other types. To show aberrations, the two beams cannot have the same configurations but must have a shear or shift at the lens under test. This may be corrected, usually by double passage, at the final plane of observation so that a large source can be used. The effects of vibration are considerably reduced, since changes of optical path affect both beams. Only small differential movements, where the beams have different configurations, can affect the fringes.

In a wave-shearing interferometer, the beams are separated at the system under test by a lateral shear. It is possible to devise interferometers where there is a radial, rotational, or folding shear, or in which the separation is a shift, a *double-focus interferometer*. This and the *radial-shear interferometer* are the most useful for giving a picture of the full wave aberration, while the rotational- and folding-shear interferometers can show some types of aberration and suppress others.

11.4.1. Double-focus interferometers

When the reference beam passes through the system under test, the phase of the interferogram will be insensitive to the aberrations only if the system is at an image of the source. The aberration then plays the role of $w_1(\mathbf{x})$, discussed in §7.4, and affects the visibility of the test fringes but not their position. In the test beam, the system under test will, as usual, be at a

Fig. 98 Two-grating interferometer with independently variable shear and tilt.

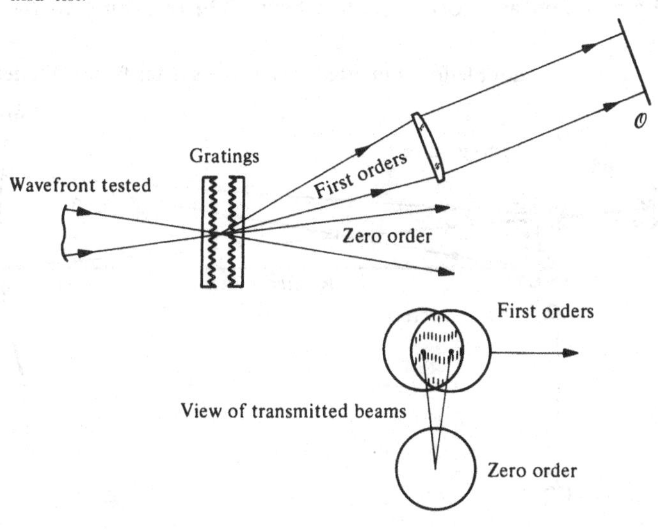

position conjugate with the plane of observation and its aberration acts as $w_2(\mathbf{u})$. This is the essence of the double-focus or *point-reference* interferometer. In practice a further lens is often needed at the complementary plane, conjugate with the source in the test beam and the plane of observation in the reference beam. This lens should be well corrected as its aberrations affect the reference beam: it is the comparison lens in this instrument, although appearing in series with the system under test, rather than in parallel.

With a single pass, a very small source would be needed. To permit a large source, the shift should be corrected before the final plane of observation. If this is done by the use of double passage, the two beams must be interchanged before their return passage through the interferometer. This is not necessary in shearing interferometers as lateral effects cancel on the return passage, but longitudinal effects add. To force each beam to take a definite path on the return passage, a polarizing beam splitter is desirable.

The first form of the interferometer was the scatter-fringe interferometer of Burch [1953], described further by Rubin [1980]. It is noteworthy for its simplicity and for the use of an unusual beam splitter: randomly diffracting surfaces. Unfortunately such beam splitters do not label the two beams as do polarizing beam splitters, and a bright false image needs to be removed by a suitable stop. The interferometer is shown in fig. 99. Two identical scatter screens are used, made as plastic replicas from the same ground surface. The direct image of the source through the first screen is focused at the centre of the system under test (a concave mirror in fig. 99), and this constitutes the reference beam, while the light diffracted by the screen fills the whole system and forms the test beam. On returning to the second

Fig. 99 A double-focus interferometer: the scatter-fringe interferometer.

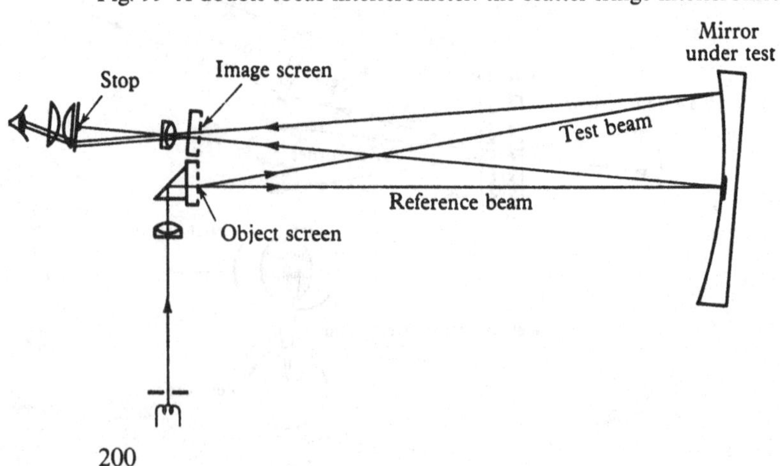

screen, the light diffracted by it from the reference beam interferes with the direct light from the test beam. The direct–direct light gives a bright false image, removed by the stop, while the doubly diffracted light is too weak to affect the interferogram. As the delay is zero, white light can be used.

Two polarization forms of this interferometer have been developed by Dyson [1957]. The first used a birefringent lens, a compound lens with a calcite component giving separate images for the two polarizations. The return beams had their polarizations interchanged by double passage through a quarter-wave plate, and so interchanged paths. No false image existed. The second form of double-focus interferometer used a polarizing beam splitter. A modification of this form that included a variable-power lens to give full correction for shift has been described by Steel [1965*b*]. Brown [1962] has described a further type under the name of 'exploded shear' interferometer.

In the double-focus interferometer, the source fringes have the same form as the test fringes in the plane of observation. When a tilt is introduced to increase the number of test fringes across the field of view, a shear is also produced that increases the number of source fringes. The size of the effective source then needs to be reduced to retain good visibility of the test fringes.

11.4.2. Radial-shear interferometers

Another method of reducing, though not eliminating, the effect of the aberrations on the reference beam is to pass this through the centre only of the system under test, where the aberrations are relatively small. The test beam would fill the system and both would see it at a plane conjugate with the plane of observation. As the two beams are separated by division of amplitude, they have the same étendue, so that the beam with the smaller spread must have a larger image of the source. The interferometer is a *double-magnification* system, forming in the same plane two images of the source of different sizes. The comparison between this method and the double-focus interferometer is shown in fig. 100.

If the ratio of the two spreads is r, the *shear ratio*, the disturbance due to the lens under test is seen by the two beams as the optical paths $w(\mathbf{u})$ and $w(\mathbf{u}/r)$, and the interferometer shows

$$\Delta w_u = w(\mathbf{u}) - w(\mathbf{u}/r). \tag{11.5}$$

In terms of the Zernike coefficients (3.27), the radial-shear interferometer measures

$$\Delta w_u = \sum_{l,m,n} b_{lnm}(1 - r^{-n})\eta^n \sigma^{2l+m} \cos m\phi; \tag{11.6}$$

each aberration term b_{lnm} is reduced by a factor $(1-r^{-n})$. The aberration of lowest order, $n=2$, is astigmatism, and even for $r=2$ when the two beams differ in aperture by a factor of two only, the relative error is only 0.25. If necessary, the results can be corrected for these errors, but for large shear ratios they are negligible.

The radial-shear interferometer was introduced by Brown [1962] and by Hariharan & Sen [1961*b*, 1962]. The latter form is shown in fig. 101. The two beams go around a cyclic interferometer in opposite directions. Two small telescope objectives of different focal lengths magnify one image and reduce the other, and the shear ratio is equal to the square of the ratio of the focal lengths. If the two objectives are separated by the sum of their focal lengths, the system has no shift for a point at infinity, and is thus suitable for testing a system that has a focal length large compared with the dimensions of the interferometer.

When used in single passage, a very small source is required and, as the shear increases radially, the visibility of the interferogram decreases from the centre. The shear can be corrected by using the interferometer in double-passage, but now polarizing beam splitters should be used. An interferometer of this type has been described by Steel [1965*b*]. In place of the two lenses of fig. 101, a single, low-power, microscope objective is used. Additional lenses are needed at the common images of the source to correct shift. It is also possible to make this interferometer with birefringent lenses and Steel [1965*a*] has described such an instrument, designed to test microscope objectives on axis. The principle is shown in fig. 102.

Fig. 100 A comparison of the double-focus and radial-shear interferometers.

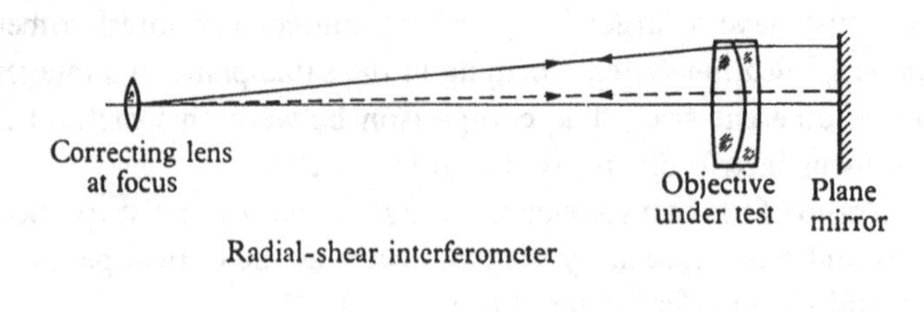

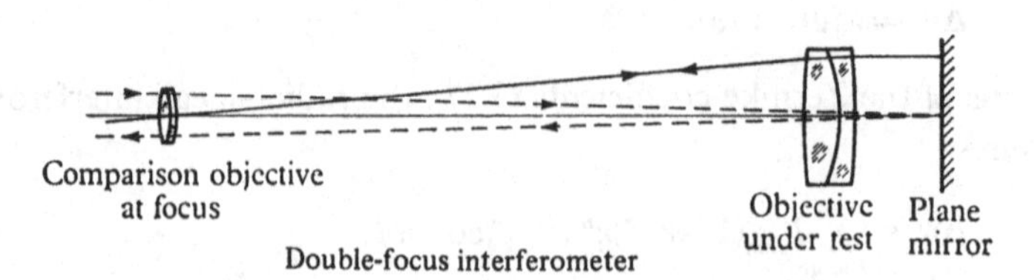

11.4.3. Rotational and folding shear

Interferometers with rotational or folding shears separate out the aberrations of different azimuthal order m in the series (3.27). A rotational shear of π gives

$$w(\phi-\pi)-w(\phi), \tag{11.7}$$

and a folding, or reverting shear about $\phi=\frac{1}{2}\pi$ gives

$$w(\pi-\phi)-w(\phi). \tag{11.8}$$

In both, terms even in m cancel to leave those with m odd: coma, distortion, etc. Hariharan & Sen [1961*a*] have used a double-passed Twyman interferometer to give separately the aberrations with m even and m odd. This and other double-passed interferometers are reviewed by Hariharan [Malacara, 1978].

Fig. 101 Cyclic form of radial-shear interferometer.

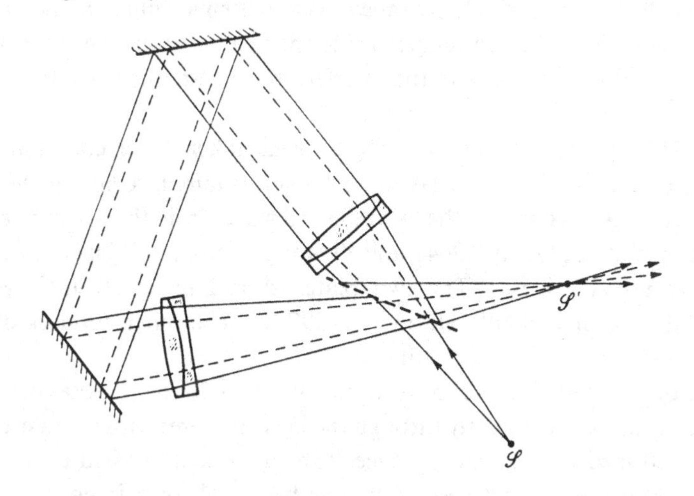

Fig. 102 A polarization form of the radial-shear interferometer.

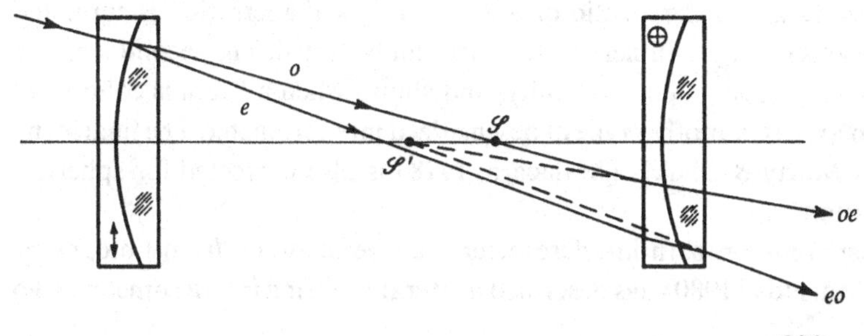

Folding and rotational shears can be obtained from a Twyman interferometer in which one or both mirrors are replaced by roof reflectors. One gives a reversion, and two sets of roof mirrors revert both beams, which can be rotated with respect to each other by rotation of one set of roof mirrors.

A classification of radial, rotational, and other shears is given by Bryngdahl & Lee [1974].

11.4.4. Common-path interferometers with lasers

Most of these common-path interferometers, and those reviewed by Murty [Malacara, 1978], have been designed for thermal sources. They have zero delay and, in some, the shear and shift are also corrected. Simpler instruments, without this correction, can be used with lasers.

A shear is most easily corrected by returning the light back through the interferometer and most tests of optical systems used to be done in double passage. With a laser, single-passage tests are often as suitable. The optical system being tested is no longer inside the interferometer, but forms an image of a point source, and the interferometer tests the wave from that image.

Fig. 103 shows four of the simple forms taken by a common-path interferometer when zero delay is no longer required. A lateral shear in collimated light is given by the two reflected waves from the two surfaces of an inclined plate [Murty, 1964], and Wyant & Smith [1975] have used this to test spectacle lenses. The two plates shown in fig. 103(*a*) give an adjustable shear and tilt [Hariharan, 1975*a*]. The outer surfaces of the plates have non-reflecting coatings.

The other interferometers have all radial shears. Two are lenses in which the wave that passes directly through the lens interferes with the first-order ghost, reflected once at each surface. The surfaces are coated to increase their reflectance, but not enough to give beams of equal intensity, or the higher-order ghosts would also be strong. The visibility, while not unity, is sufficient. The first lens has a shear ratio of -5 [Steel, 1975*a*]. If the design is changed to give a shear ratio of -2, the spherical aberration is corrected and beams of larger numerical aperture can be tested. The second lens, fig. 103(*c*) has a shear ratio near unity, and shows when a beam is collimated and on axis. It is another type of desensitized interferometer. The final form, due to Murty & Shukla [Malacara, 1978] is also corrected for spherical aberration.

Most common-path interferometers can be made up from holograms. Thus Matsuda [1980] has described a lateral-shearing interferometer. Two or more holograms are needed to produce a radial shear, as the polarization form of fig. 102 shows.

11.4.5. Point-diffraction interferometer

Even simpler than these common-path interferometers is the *point-diffraction* interferometer, or Smartt interferometer, first described by Linnik [1933*b*] and put into use by Smartt. It is described by Smartt & Steel [1975]. It is a direct example of Fourier optics. The addition of a plane wave to an aberrated wave to test it is on Fourier transformation the addition, to the aberrated image, of a δ-function, a point source coherent with it.

The interferometer is illustrated in fig. 104. It consists of an absorbing layer on a clear substrate and in this layer there is one small clear hole. This hole is small compared with the size of the Airy disk that corresponds to the numerical aperture at which the interferometer is being used, so that it acts as an unresolved point that diffracts light as a uniform spherical wave. This interferes with the aberrated wave that has passed through the absorbing layer, the transmittance of which is chosen to match the beams. A reflecting

Fig. 103 Common-path interferometers for a laser source: (*a*) lateral shear; (*b*), (*c*), and (*d*) radial shear.

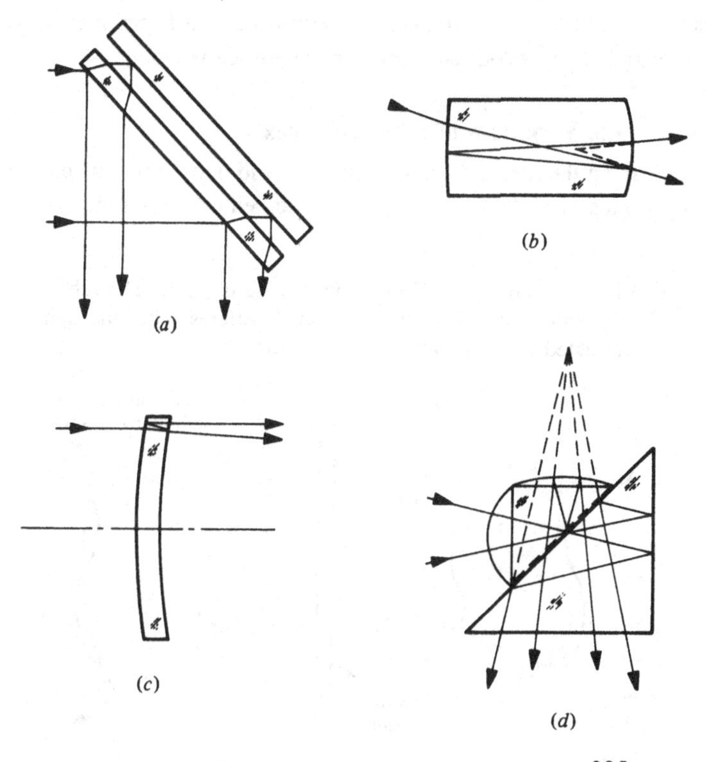

version is described by Speer *et al.* [1979], consisting of a diffracting particle on a reflector.

The tilt, and hence the number of fringes across the field, is changed by moving the pinhole laterally across the image. This changes the amount of light reaching the pinhole and hence the beam ratio. A selection of plates is therefore required, different pinhole sizes to suit different numerical apertures, and, for each, a range of transmittances to suit different tilts. Transmittances in the range 0.01 to 0.1 are usual. The plates are easily made as evaporated coatings in which a pinhole is formed by a very short exposure to the beam from a laser of sufficient power, focused through a microscope, the numerical aperture of which controls the size of hole.

By its construction, a point-diffraction interferometer has no delay. It requires an unresolved point source, but this need not be monochromatic. Fig. 105(*a*) shows an interferogram obtained from a telescope with a star as source. But, because the interferometer is wasteful of light, a laser is a convenient source for most tests, as shown in fig. 105(*b*).

The pinhole can be replaced by other diffracting apertures that have an autocorrelation function with a strong central peak, for example a narrow annulus; the relationship of this to the phase-contrast microscope is discussed by Smartt & Steel [1975]. Mori [1978] has used a line, but, since this diffracts in one direction only, it requires a small specimen surrounded by a clear field to produce a uniform reference wave.

11.5. Variations of refractive index

Surface irregularities lie on the surface and lens aberrations, although caused by a series of components, are specified on a reference sphere at the

Fig. 104 The point-diffraction interferometer. The aberrated wave, attenuated by an absorbing film, interferes with the spherical wave diffracted by a discontinuity in the film.

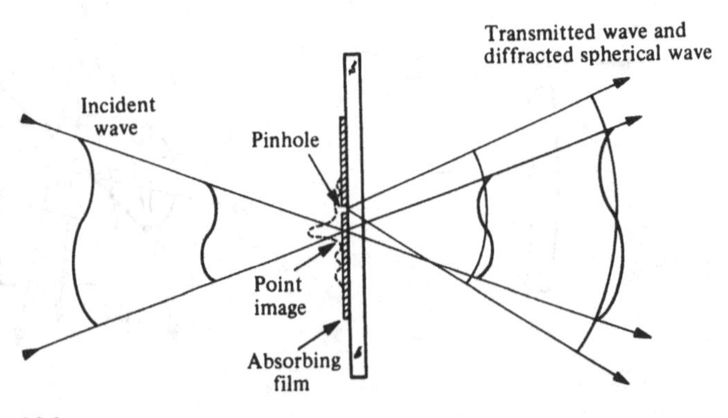

exit pupil. But no such localization applies to the variations of refractive index in the thick specimens on which studies of index variations are made: mixing liquids, air flow in wind tunnels, and flames. Even when the variations of refractive index are small, so that ray paths can be taken to be straight lines, more than one view is needed to locate these variations. But when they are large, rays are refracted so that they are no longer straight, and the derivation of the index distribution from a set of interferograms, taken from several directions, is more difficult. The problem has been reviewed by Baltes [1978], and Vest [1979], and it occurs also in ultrasonic tomography. An iterative technique for its solution is given by Cha & Vest [1981].

The refraction of rays in the test beam also influences the design of the interferometer. Single passage through the test space is required, and components following this space should not have aberrations, since the test and reference beams can go through different parts. When a Mach–Zehnder interferometer is used, the test beam should be reflected at the second beam splitter, not transmitted.

Variations of refractive indices in solids are most commonly studied to test whether a material is suitable for optical use. For a specimen to warrant testing, any variations of index should be small and, as for other optical components, a double-pass test in a Twyman–Green interferometer can be used. When the specimen is thick, it introduces a shift and a small source must be used. The bandwidth of the source is also limited, but not only by the usual first-order effect. When the group delay is made zero, the first derivative of phase with respect to frequency is zero at the centre of the

Fig. 105 Point-diffraction interferograms taken by Smartt [Smartt & Steel, 1975]: (*a*) large telescope with a star as source; (*b*) small telescope with laser source.

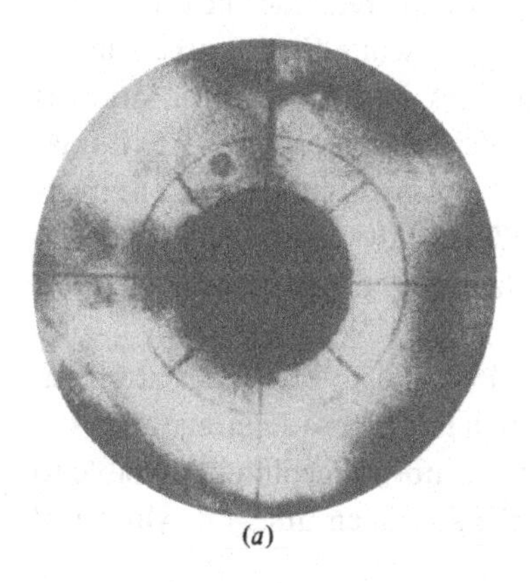

(*a*)

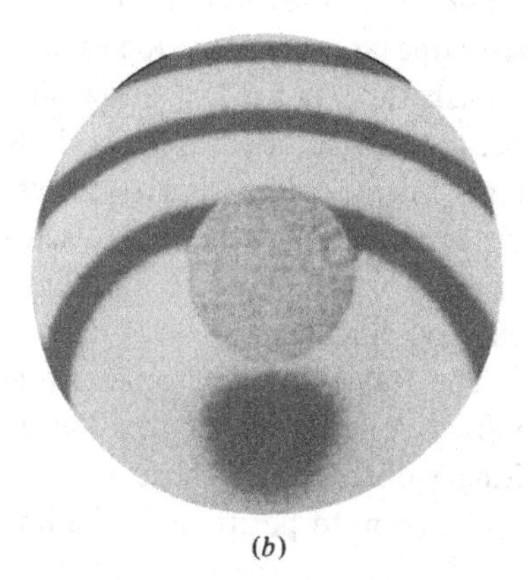

(*b*)

band of radiation used. The dispersion of the medium, however, leads to a high second derivative and the phase can vary significantly within the bandwidth unless a narrow spectral line is used. A second test applied to isotropic materials is that for freedom from birefringence, and this is essentially an equivalent test, performed with a polarization interferometer.

Most applications are to liquids and gases. Variations of refractive index denote variations in density and these can be related to changes in temperature, pressure, or relative concentration of different components. Thus interferometry is used for studies of diffusion in liquids, combustion in flames, compressible flow in wind tunnels, electron density in plasmas, etc. It has the advantage, common to other optical methods, of making measurements without disturbing significantly the phenomenon studied, an important advantage in aerodynamics where probes seriously disturb the flow pattern. Other optical methods of studying variations of refractive index are those, like the schlieren method, that measure the deviation of radiation produced by these variations. While direct interferometry gives the distribution of index, deviation measurements give the gradient of the distribution, a property shared, however, with a shearing interferometer. When the gradient is small, interferometer methods are more convenient; when large, deviation measurements are better.

The interferometer most commonly used is the Mach–Zehnder interferometer, and its use in this field is treated in detail by Weinberg [1963]. The forms based on diffraction gratings, shown in fig. 69, are one type used. In this application of interferometry, and particularly in aerodynamics, a large field of view is often required. The conventional Mach–Zehnder interferometer described in §8.4 has the light collimated outside the interferometer and, for a large field, would need very large beam splitters and mirrors. The beam splitters can be made small, however, if the collimating systems are inside the interferometer, and Johnson & Scholes [1948] have shown that it is not necessary for the reference beam to have the same large cross-section as the test beam. If a white-light source, such as a flash tube, is used to allow short exposures, suitable compensation is needed. The interferometer is shown in principle in fig. 106 but, in its final form, mirror collimators have been used. The compensation gives the high-order correction of the delay discussed in §7.6 and the compensators consist of glass plates, of crown and flint glass, that can be tilted to change their effective thicknesses.

The Rayleigh refractometer (fig. 2) is also used to study variations in refractive index, particularly in liquids. It has the disadvantage that the fringes observed are those at infinity and it is not immediately possible to relate them to positions in the image of the specimen unless a cylindrical lens is used. This focuses in the same plane the vertical fringes and the specimen, the latter being assumed to be uniform horizontally. This and other modifications to the interferometer that enable it to measure gradients are described by Weinberg [1963]. However, gradients are probably better measured by shearing interferometers.

Common-path interferometers have similar advantages and disadvantages to those discussed for optical testing. They are simpler and easier to adjust but measure gradients rather than direct distributions of index. A lateral shear is normally used. This and other applications of shearing interferometers have been reviewed by Bryngdahl [1965]. Dyson, Williams & Young [1962] have used a double-focus interferometer to study plasmas.

11.6. Interference microscopes

The methods of microtopography discussed in §11.2 involve the separate use of a microscope to observe the fringes given by an interferometer. An interference microscope combines the two functions of interferometer and microscope into a single instrument. As with ordinary microscopes, different designs exist for examining opaque (reflecting) objects and transparent (transmitting) ones and those for use with reflecting objects suffer more from the problem of scattered light. Since a microscope combines a high numerical aperture, to give resolving power, with a finite field of view, it requires a beam of finite étendue, and a laser is not normally a suitable source, unless diffused.

The interference microscopes for examining reflecting surfaces are modifications of the Michelson interferometer. Thus, if the mirrors at the

Fig. 106 A Mach–Zehnder interferometer with arms of different cross-section [from Johnson & Scholes, 1948].

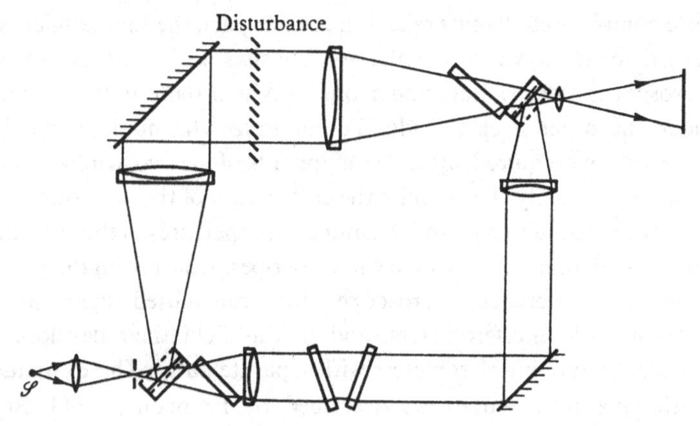

ends of the arms of a Twyman interferometer are replaced by a pair of microscope objectives with plane mirrors at their foci, as has been shown in fig. 42, the Linnik [1933*a*] microscope results. One of these mirrors is the surface being examined, the other is a reference mirror, chosen to have about the same reflectance as the specimen so that the fringes have good visibility. The interferometer is compensated by the use of matched objectives in the two arms; when the magnification is to be altered, both objectives must be changed. In comparison with the multiple-beam methods discussed earlier, this microscope suffers by being a two-beam instrument with less sharp fringes, but it is much more versatile, being capable for example of studying inclined surfaces such as a diffraction grating set to reflect back the first (or a higher) order.

A simpler instrument can be made if the interferometer is fitted in front of the microscope objective, and forms due to Sagnac and to Krug and Lau are described by Françon [1961]. Although the need to fit a beam splitter in front of the objective limits the numerical aperture obtainable and hence the useful magnification, systems of moderate power are now manufactured. Alternatively, instead of being at the usual angle of $\pi/4$, the beam splitter and reference mirror can both be normal to the axis of the microscope, the latter obscuring a small part at the centre of the beam. Françon [1961] and Smith [1956] describe a form due to Mirau, while Dyson [1953] has made a separate interferometer unit that fits in front of any objective. As it is a re-imaging system, a large numerical aperture can be retained. Lateral-shearing interferometers are also used for opaque specimens but these are fairly obvious modifications of the forms for transmitted light.

While the basic instrument for studying reflecting specimens is the Michelson interferometer, that for transmission is the Mach–Zehnder. A microscope based on the latter has been described by Horn [1958], and is more complicated than its reflexion counterpart, the Linnik microscope. It has the form shown in fig. 107, and consists essentially of two identical microscopes side by side, under one of which there is the specimen and under the other a clear slide. To minimize the number of adjustable compensators required, optical components of the two arms are selected to be matched closely. To simplify the calculation of the variations in optical path from the interferogram, the numerical apertures of the condensers are much less than used in ordinary microscopes. Apart from this instrument, however, interference microscopes for transmitted light are usually common-path interferometers, and in this field their development has preceded that of interferometers with separate paths. These common-path interference microscopes are described by Françon [1961], Bryngdahl

[1965], and Krug, Rienitz & Schulz [1964]; the last authors give a survey of all types of interference microscopes.

The two principles most commonly used are lateral shear and double focus. The lateral shear may either be small, to give gradients of path difference, or sufficiently large for the reference beam to pass through a clear area of the microscope slide to one side of the specimen; the direct distribution of path is then observed. The shear is produced usually by a birefringent component. Considered as a rotation of a spherical wavefront about its centre, the angular shear should be centred in the pupil of the microscope. In the object space surrounding the specimen itself, the pupil is at infinity (telecentric stop) and the shear is a linear displacement that can be produced by a crystal plate or Savart polariscope, of one of the forms shown in fig. 67, placed near the object. Behind the objective, the shear can be produced by a Wollaston prism, shown in fig. 66, with its centre at the exit pupil of the objective. If this pupil is not accessible, the modified Wollaston prism of fig. 66(*b*) has the point of intersection of the sheared

Fig. 107 Interference microscope for transmitting objects, with the two beams completely separated.

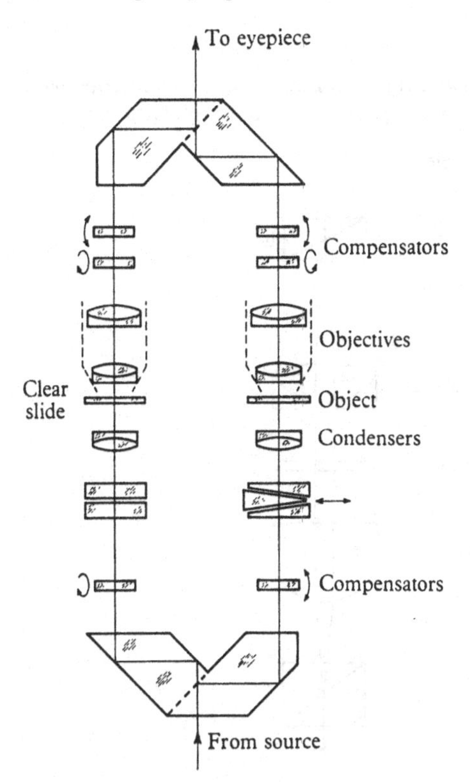

rays external to the prism. Positions can also be found in the eyepiece where a Savart polariscope can be placed to form an interference eyepiece that will convert an ordinary microscope into an interference microscope.

The two alternative systems are illustrated in fig. 108. If a single shear only is used, the illumination must come from a slit source, but an extended source can be used if a compensating shear is used before the object, as shown in this figure. At the same time the shift and delay can be made zero and a white-light source used.

In double-focus interference microscopes, there is a shift at the object, which may, however, be compensated by a second double-focus system. The instruments used seldom attain the ideal conditions, illustrated in optical testing, where the final fringes lie in a plane that is conjugate with the object for the test beam and with the pupil for the reference beam. Usually the reference beam is only slightly defocused from the object, so that its phase corresponds to the phase averaged over small regions of the object. In polarization interferometers, the double focus can be obtained either with a plate of birefringent material, cut parallel to the axis, or with lenses of birefringent material; strictly speaking, the first method gives astigmatism as well as a defocusing. Both of these have been developed by Smith [1956],

Fig. 108 Alternative forms of an interference microscope with a lateral shear, with either Savart polariscopes *S*, or Wollaston prisms *W*.

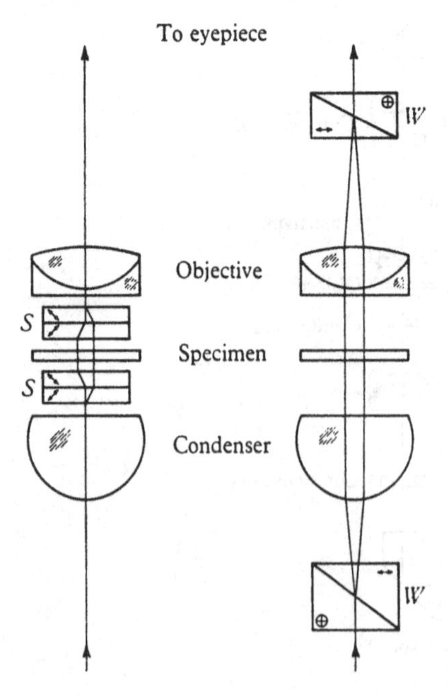

and Françon [1961], and one is illustrated in fig. 109. If the second double-focus lens below the object is omitted, the source must be circular and small.

Double-focus interferometers without birefringent elements have been made by Dyson [1949], and Philpot [1952]. The latter has developed several cyclic interference microscopes which may have a shift, lateral shear, or rotational shear at the object. A related instrument is described by Rienitz, Minor & Urbach [1962]. Since microscope specimens do not tend to follow as simple a phase relation as lens aberrations, radial or rotational shears are not used on their own in microscopy.

The interferogram given by an interference microscope can be related to variations of refractive index in the specimen. For quantitative work, corrections must be made for the finite angle covered by both the source and the aperture of the viewing system. These are discussed by Ingelstam & Johansson [1958], Françon [1961], and Guillard [1963–64]. Similar corrections are required for reflecting specimens and are given by Gingell, Todd & Heavens [1982].

11.7. Alignment by interferometry

A plane in space can be defined as the locus of points equidistant from two reference points while a point on a line is equidistant from three. Alignment may therefore be tested by an interferometer that measures departures from equality of these paths. Usually two beams are used to define a plane, while a line is defined by the intersection of two planes.

The unlocalized fringes of Young's experiment define a plane as the locus of the central white fringe. This method of alignment has been applied by van Heel [1961] to a variety of problems with only the modification that two slits are used instead of pinholes. To align in two dimensions, two sets of measurements are made, or the two slits are replaced by a round hole surrounded by one or more annular holes. The line is then defined by the central maximum of the diffraction pattern produced.

An interferometer with a rotational shear of π defines a straight line. When the source is on axis, its two images coincide and no fringes are formed. If it is off axis, straight fringes are produced at right angles to the

Fig. 109 Double-focus interference microscope with birefringent lenses *L*.

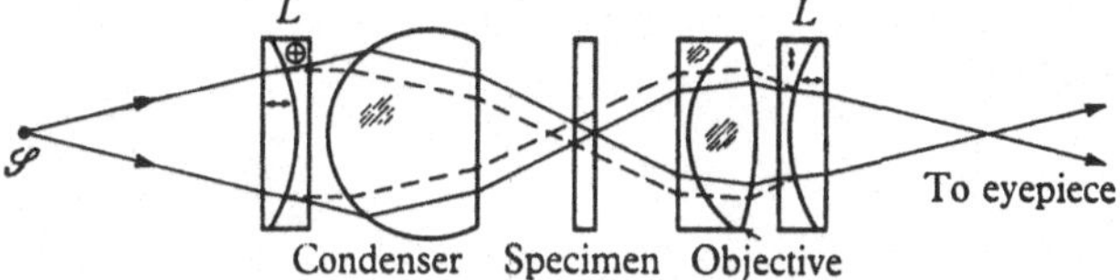

direction it is off axis. If such an interferometer is moved between the source and a screen so that no fringes appear, the interferometer traces a straight line [Gates & Bennett, 1968]. A convenient form of such an interferometer is a confocal Fabry–Perot étalon with partially reflecting surfaces. The beam reflected once at each surface is inverted with respect to the directly transmitted beam.

If the two beams have no relative inversion, but instead a shift, as produced by a plane Fabry–Perot étalon, it produces circular fringes centred about a line normal to the étalon surfaces that passes through the source. As this interferometer is moved, it shows departures from squareness.

A different method of testing interferometrically for departures from a straight line has been described by Dyson [Strong, 1958]. The line is defined by setting up two spherical mirrors with their foci coincident. A shearing interferometer is moved between them and any lateral movements of this off the line joining the two centres causes a movement of the fringes.

12

Hologram and speckle interferometry

Lasers have made a larger impression on the techniques of interferometry than the ability to use simpler instruments, described in Chapter 11. The new techniques of hologram and speckle interferometry have extended interferometry to studies of rough surfaces and of changes that occur with time. They have now become major methods of non-destructive testing. Detailed treatments are given by Ostrovsky, Butusov & Ostrovskaya [1980], Vest [1979], and in conference reports, such as Robertson [1976].

Hologram interferometry involves the interference of waves, at least one of which is a reconstruction from a hologram. It is the most precise of the techniques treated in this chapter, but covers the smallest range. Simpler, but less precise techniques, covering a larger range of displacements, involve records of the speckle produced by the object. These are *speckle interferometry* and *speckle photography*. Both are included under *speckle metrology*, a name that distinguishes this speckle interferometry from the technique of the same name used in astronomy. This is described in Chapter 14.

12.1. Classification of methods

Because holograms can store waves and reconstruct them later, they introduce a new variable, time, into interferometry. This ability is shared by the speckle techniques.

12.1.1. Treatment of the time variable

Hologram interferometry can measure changes with time in three ways. If the object is sufficiently stable, *real-time* interference can be used. A hologram of the object is made and processed, either *in situ* or in a mount that can be relocated accurately. It reconstructs a wave that can interfere later with the wave from the object to show changes as they are taking place. The number and shape of the fringes across the field can be altered by fine adjustments to the hologram mounting or to the reference beam. This method is said to produce *live fringes*.

When the object is not sufficiently stable for this method, or when the change to be measured cannot be delayed until a hologram is made and processed, two holograms are made in succession on the same plate. A double-pulsed laser is used for rapidly changing objects, such as living people. When this hologram is reconstructed, it produces *frozen fringes*, which show the change between the two exposures. It records only a single time difference, but a multi-aperture method can be used [Hariharan & Hegedus, 1973] to record a series of these on one plate.

In a simple frozen-fringe hologram, the number of fringes across the field cannot be altered by movements of the hologram or the reference beam. To obtain adjustable fringes, two techniques are available. One is Abramson's sandwich hologram [Abramson, 1976; Abramson & Bjelkhagen, 1979], the other is the use of separate reference beams for each exposure and its reconstruction. The latter method also allows the introduction of phase or frequency differences between the two reconstructions.

Vibrating objects are studied by *time-averaged* fringes. A simple harmonic variation of the path difference, $p \cos \omega t$, when averaged over times long compared with its period $2\pi/\omega$, gives a fringe modulation

$$J_0(2\pi p/\lambda). \tag{12.1}$$

The brightest fringe follows the nodes, where $p=0$, and the visibility decreases with increasing fringe order.

12.1.2. Objects studied

Classical interferometers measure phase variations through transmitting specimens or across reflecting surfaces that give specular reflexions. A rough surface must be viewed at grazing incidence. If one or both mirrors of a Michelson interferometer are replaced by rough surfaces, the fringes produced are a very fine set of contours of this roughness. Usually they are not resolved and only a speckle pattern is seen. But if two interfering waves come from the same rough surface, their irregularities match and the interference between them gives broad fringes that show the relative displacement. This must be small compared with the scale of the roughness, or the detailed structure of the two waves will no longer match and the fringes will lose visibility.

Only normal displacements of a specular surface can be measured, since a displacement of a mirror in its own plane does not affect the wave reflected from it. But hologram interferometry of a rough surface can measure both normal and in-plane displacements. It is essentially a more complex technique, with a more elaborate theory needed to relate the fringes observed to surface displacements. Fuller treatments than the outline given here are provided by Schumann & Dubas [1979], and Stetson, for example [1979].

12.2. Phase variations of classical interferometry

Hologram interferometry can be used in place of traditional interferometric methods for measuring phase variations. Usually it simplifies the equipment and makes possible the separation of the desired variation from other unwanted phase irregularities.

12.2.1. Transmitting objects

When the specimen studied is transparent and has small phase variations due to variation of refractive index, interference is obtained between the wave with no specimen present and the wave through the specimen. Often both waves include the effects of the chamber or cell that contains the specimen, so that any phase variations in this cancel out and only the phenomenon being studied affects the final interferogram. To obtain this cancellation in classical interferometry, two interferograms must be recorded, with and without the specimen, and their difference found. This difference can be obtained as the moiré fringes formed between photographs of the two interferograms.

An example of hologram interferometry is shown in fig. 110. A hologram has been taken of a cold electric lamp. A wave through the hot lamp interferes with the reconstructed wave from this hologram to give live fringes, which show the variations in optical path in the heated gas above the filament. The path variations due to the glass bulb do not appear. This same cancellation allows wind-tunnel interferometry to be carried out

Fig. 110 Hologram interferogram of an electric lamp with: (*a*) direct illumination; (*b*) diffused illumination.

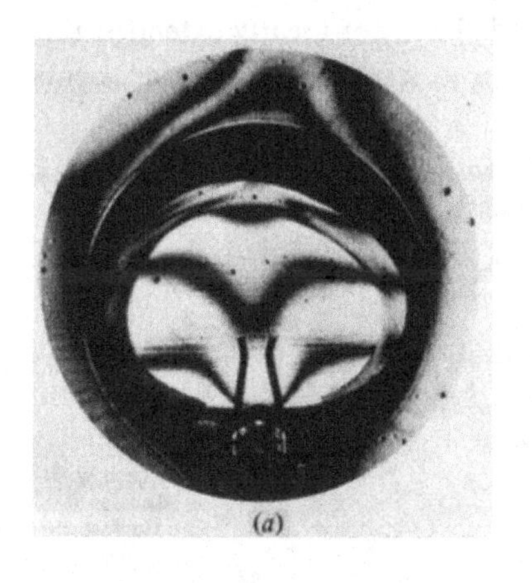
(*a*)

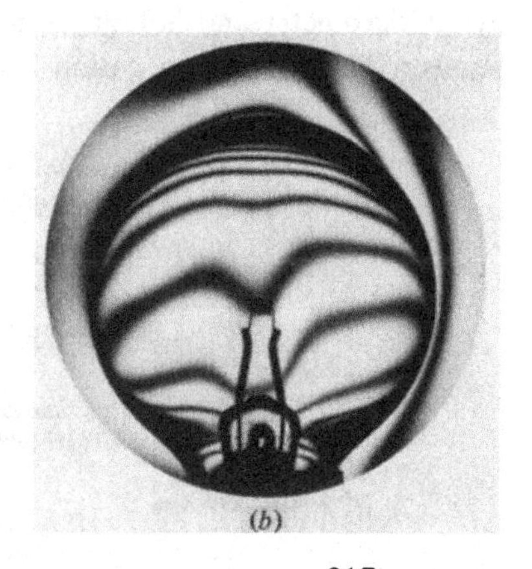
(*b*)

through windows of poor optical quality, and is the basis of the two-wavelength methods of plasma diagnostics. For the latter, the two holograms can be taken in the two different wavelengths with reference beams at different angles, chosen so that a single reconstructing beam gives the two reconstructions in the same direction. They interfere and the interferogram shows the variations of the dispersion of the refractive index, which depends on the electron concentration, but is almost unaffected by the variations of the index itself, which show the atomic concentration [Ostrovsky *et al.*, 1980].

When the phase variations are large, as for example in flames or liquids, hologram interferometry has the same problems and limitations in resolving very fine fringes and relating these back to a three-dimensional distribution of refractive index that are discussed in §11.5. But hologram methods have led to a new attack on the problem, for example that of Schulz [1976]. Also, frozen fringes from holograms taken a very short time apart will often allow rapid changes to be resolved.

12.2.2. Reflecting objects

The application of hologram interferometry to testing reflecting objects dates from Archbold *et al.* [1967], who compared cylinders against a hologram of a master in the arrangement shown in fig. 111. This illustrates the main requirements of such measurements, which are a special system to illuminate the object at grazing incidence, here a conical lens, and a spatial filter that passes only the specular component of the reflected light. When the fringes are viewed by eye or photographed, this filter must be the entrance pupil of the imaging system. In fig. 111 the interferogram is obtained on an annular image and requires interpretation in terms of bodily displacements and surface errors of the object under test.

As mentioned in Chapter 11, not many surface shapes have been tested by interferometry and hologram methods have not greatly extended this number. Their chief application has been to testing optical components

Fig. 111 Optical system for comparing cylinders by hologram interferometry [Archbold *et al.*, 1967].

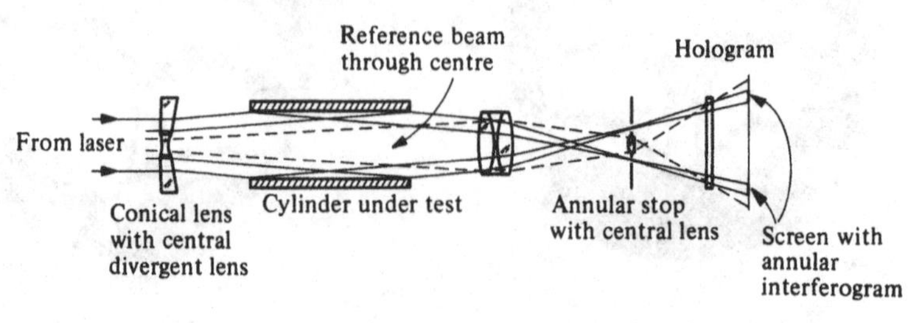

with aspherical surfaces [Hildebrand, 1976; Schulz & Schwider, 1976; Malacara, 1978]. These are smooth enough to give a specular reflexion at normal incidence and no spatial filter is needed, except where in-line holograms are used [Mercier & Lowenthal, 1980]. In classical interferometry a corrector system is used that converts a spherical wave to one that meets the surface normally and then converts the reflected wave back to spherical. Such a corrector is not easily made as a recorded hologram but it can be computed [Ichioka & Lohmann, 1972; Schwider *et al.*, 1980]. But an unsymmetrical test is possible. From a perfect specimen of the surface a hologram is made that converts the irregular wave reflected from the surface illuminated by a spherical wave to another spherical wave. This hologram can then be used to test other specimens of the same surface. Also, in reverse, it converts a spherical wave into one of suitable form for contouring the surface by the methods of §12.6.

12.2.3. Diffuse illumination

When a transmitting specimen has large index variations or a reflecting surface departs considerably from its standard, the wave produced is so distorted that it does not focus within the pupil of the imaging system used to record the fringes. The complete object cannot then be seen without some parts being distorted or in shadow, as are the outer regions of the lamp bulb in fig. 110(*a*).

The use of a diffuser to scatter the laser light overcomes this problem, as is shown in fig. 110(*b*). It avoids also the need for special correctors. But the characteristics of the interference then change from unlocalized fringes to the localized fringes obtained with an extended source or from the scattering objects of the next section.

12.3. Hologram interferometry of rough surfaces

The measurements by hologram interferometry on objects with diffusely reflecting surfaces can give the vector displacement for each point on the surface. This requires at least three interferograms to give three non-coplanar components of this vector.

12.3.1. Theory

The measurement of the displacement of a point P on a rough surface is shown in fig. 112. The point has the positions P_1 and P_2 before and after a vector displacement $\mathbf{d}$. The surface is illuminated from a point S and observed from the point O. Since the displacement $|\mathbf{d}|$ is very small compared with the distances from P to S and O, the ray directions from S to P_1 and P_2, can be taken to be the same, in the direction of the unit vector $\mathbf{r}$.

Similarly the viewing direction can be taken as $\mathbf{r}'$ from both positions of P to O.

The optical path difference between rays from S to O via P_2 and those via P_1 is

$$p = n[P_1M - P_1N].$$

If account is taken of the direction of the ray vectors, this can be expressed as

$$\begin{aligned} p &= n\mathbf{d}\cdot(\mathbf{r}-\mathbf{r}'), \\ &= \mathbf{d}\cdot\boldsymbol{\sigma}, \end{aligned} \tag{12.2}$$

where $\boldsymbol{\sigma}$ is the *sensitivity vector* for a measurement made with these positions of S and O. It is defined by

$$\boldsymbol{\sigma} = n(\mathbf{r}-\mathbf{r}'); \tag{12.3}$$

its direction bisects the two ray directions from P to S and to O and its magnitude is

$$|\boldsymbol{\sigma}| = 2n\cos\tfrac{1}{2}\theta \tag{12.4}$$

where θ is the angle between these two rays. The sensitivity vector is either known from the arrangement used to make the holograms or derived from the holograms later [Pryputniewicz & Stetson, 1980].

To find the displacement vector $\mathbf{d}$ at any point on the object, at least three measurements are required of the path difference p for three non-coplanar sensitivity vectors $\boldsymbol{\sigma}$. Then the displacement is given by the usual expression

Fig. 112 The relationship between the vector displacement $\mathbf{d}$ and the change of optical path: the derivation of the sensitivity vector $\boldsymbol{\sigma}$.

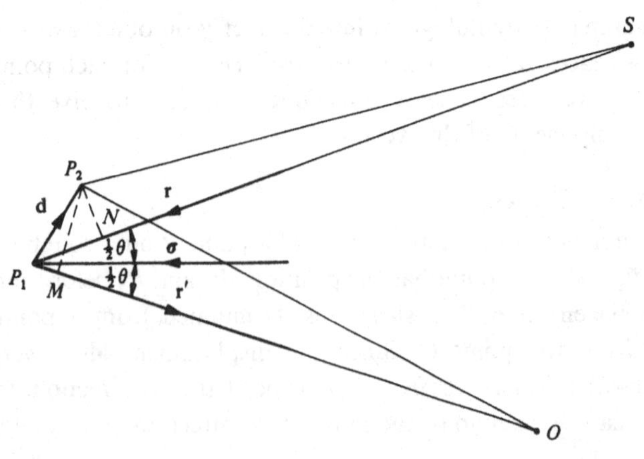

for a vector whose components are given along three general vectors;

$$\mathbf{d}=p_1\boldsymbol{\sigma}_1'+p_2\boldsymbol{\sigma}_2'+p_3\boldsymbol{\sigma}_3', \tag{12.5}$$

where $\boldsymbol{\sigma}_1'$, $\boldsymbol{\sigma}_2'$, $\boldsymbol{\sigma}_3'$ are the reciprocal vectors of $\boldsymbol{\sigma}_1$, $\boldsymbol{\sigma}_2$, $\boldsymbol{\sigma}_3$ defined by

$$\boldsymbol{\sigma}_1'=\frac{\boldsymbol{\sigma}_2\times\boldsymbol{\sigma}_3}{[\boldsymbol{\sigma}_1\boldsymbol{\sigma}_2\boldsymbol{\sigma}_3]}, \text{ etc.,} \tag{12.6}$$

$[\boldsymbol{\sigma}_1\boldsymbol{\sigma}_2\boldsymbol{\sigma}_3]$ being the scalar triple product.

The different sensitivity vectors can be obtained by varying the direction of illumination, the direction of view, or both. Measuring techniques can be divided into those that measure the fringes in a plane conjugate with the object and those that measure the alternative interferogram to this in a plane away from the object. Two examples are given here; Briers [1976] has listed these and other methods.

The surface strain in an object depends on the gradient of the vector field $\mathbf{d}(\mathbf{x})$ and the flexure on second derivatives. This gradient is called the *fringe vector* by Stetson [1974], and its use to derive strains is given by Pryputniewicz & Stetson [1976]. To find it from the displacements by finite differences requires accurate measurement, since accuracy is lost in the process. A knowledge of the surface shape is needed and this can be found by the contouring methods given in §12.6.

12.3.2. Fringes at the object

If the detection system used, whether the eye or a camera, is focused on the object, the fringe positions can be related to each point on the object surface. Different sensitivity vectors can be obtained by the use of three different directions of illumination. The observing directions can all be the same, when the three holograms are either superimposed on the same plate or made on different plates in the same position, or they may differ slightly so that the three holograms are recorded on different parts of the same plate. The method, due to Ennos [1968], is illustrated in fig. 113.

There is the usual uncertainty of integral order of the fringes but often this does not matter if only relative displacements are required. When absolute values are needed, these can be found either by having a point known to be stationary on the object, or attached to it by a coupling strip, or by the use of several wavelengths, as discussed in §10.2. Also, the sign of the path difference and hence of the displacement is not indicated but must be inferred from other information.

12.3.3. Fringes at the hologram

Although methods based on measurement of the fringes at the object are most commonly used, it is also possible to derive the displacement vectors from measurements away from the object [Tsujiuchi, Takeya & Matsuda, 1969]. The most convenient position to make these is at the hologram itself. Then the different directions of the sensitivity vector are provided by the different directions of view from different parts of the hologram. To relate the observed interference to a point at the object, a small aperture is placed at an image of the object and the interferogram viewed through it. It is moved across the image to study each object point in turn. If fringes are counted as it moves, differences of integral order can be found.

The magnitude and direction of each displacement vector is derived from the interferogram by a method similar to that given in §4.2.1. The theory is slightly more complicated, since now the direction of illumination must be included.

As indicated in §6.7, the fringes in hologram interferometry are localized. Both methods of analysis have required measurements of interferograms in planes which are not necessarily the plane of localization and so a limiting aperture is required, small enough to give fringes of good visibility. In the method using fringes at the hologram this aperture is the one that defines the region at the object for which the results apply; for the case of fringes at the object, the aperture should be at the hologram.

12.3.4. Experimental methods

While it is possible to photograph the interferograms and measure them subsequently to obtain the fringe order at any point, this is time consuming

Fig. 113 Optical arrangement for hologram interferometry based on fringes at the object. At least one more direction of illumination is required, inclined to the plane of the two shown.

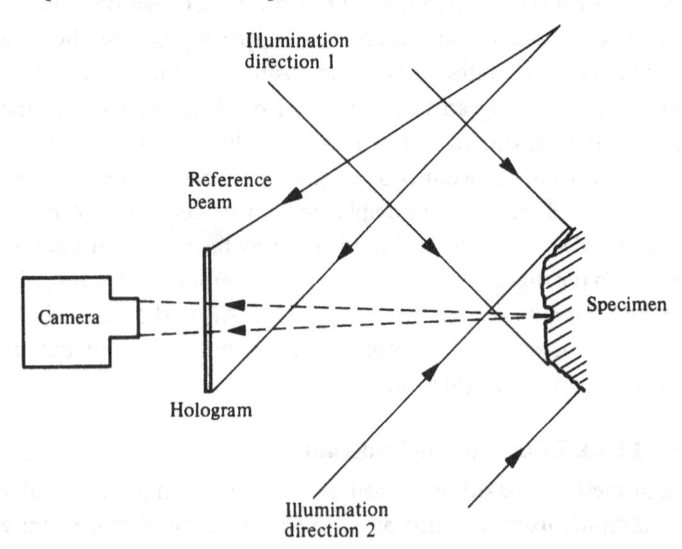

and not very accurate. When the fringes are not uniformly spaced, interpolation does not give an accurate value of the fractional order.

Photo-electric recording of the fringes followed by electronic processing is being increasingly used. The methods follow those given in Chapters 10 and 11. For frozen fringes, or very slowly changing objects, a scanning method has been used by Dändliker [1980]. Fractional orders are obtained by the heterodyne method. To introduce the required frequency shift to one reconstruction in the frozen-fringe method, two separate reference beams must be used.

Two-dimensional detector arrays can be used. These are integrating systems and, rather than the continuous phase variation of the heterodyne method, phase changes in steps of $\pi/3$ have been used by Hariharan *et al.* [1982] to obtain fringe orders to 0.002λ. The separate interferograms are analysed for each point of the array and stored, and the displacement vectors computed from these results. A similar method of analysing classical interferograms is given by Dörband [1982].

Usually hologram interferometry employs transmission holograms. But Boone [1975] has shown that the use of reflexion holograms can reduce greatly the stability requirements, since they can be held on the object itself.

12.3.5. Partial analyses

The theory of hologram interferometry developed from earlier methods, which were studies of part only of the displacement. These methods are still useful when the full displacement vector is not needed.

For certain views, that is, certain directions of the sensitivity vector, one interferogram usually gives an ambiguous answer. Uniformly spaced straight fringes are obtained both from a tilt of the object about an axis in its surface and from an in-plane translation. But these can be distinguished by the position of localization of the fringes. Tilting the object produces a tilt, while a translation produces a shear. So, on a qualitative argument, the first gives fringes localized at the object while the second gives them localized away from it.

For a more accurate analysis, the position of localization, if it exists, can be found by a construction, which is an extension of that given in §7.5.2. The concept of homologous rays, introduced in §6.6.1, is now applied. A ray from the source is scattered by the object, before and after displacement, into pairs of homologous rays. If these intersect, there is localization at the point of intersection. The form of the fringes produced and the localization has been given by Briers [1976], and Vest [1979] for different object displacements.

Many partial analyses require the choice of a suitable sensitivity vector to show the desired component of motion but not those normal to it. The choice of a layout that achieves this is assisted by Abramson's *holodiagram* [1970, 1972]. This is shown in fig. 114.

The ellipses represent the three-dimensional ellipsoids of constant path from a source S to a point of observation O via some object. Instead of all the ellipses at increments of path of λ, fig. 114 shows about every ten-thousandth ellipse. Movements of the object point along an ellipse do not change the path and so do not show in the interferogram. The maximum change of path is given by movements normal to the ellipses but the sensitivity of these varies, being inversely proportional to the spacing between adjacent ellipses.

Abramson [1972] has summarized the many applications of this diagram. In particular, it can be used to plan a layout for the desired sensitivity vector and to derive fringe patterns by a moiré method [Abramson, 1971]. If the ellipses are drawn as broad fringes and two such patterns are superimposed, displaced by the amount of the object displacement, they give a moiré pattern that shows, where it intersects the object, the interferogram that the displacement will produce.

The holodiagram shows that it is easy to obtain a sensitivity vector normal to the object but not possible to have one in the surface to measure in-plane displacements only. For these measurements, speckle methods are better. Indeed for any displacements they are often more suitable for partial analyses than holographic methods, particularly when great accuracy is not needed.

Fig. 114 Abramson's holodiagram: curves of constant optical path from a source S to an observer O via an object at P.

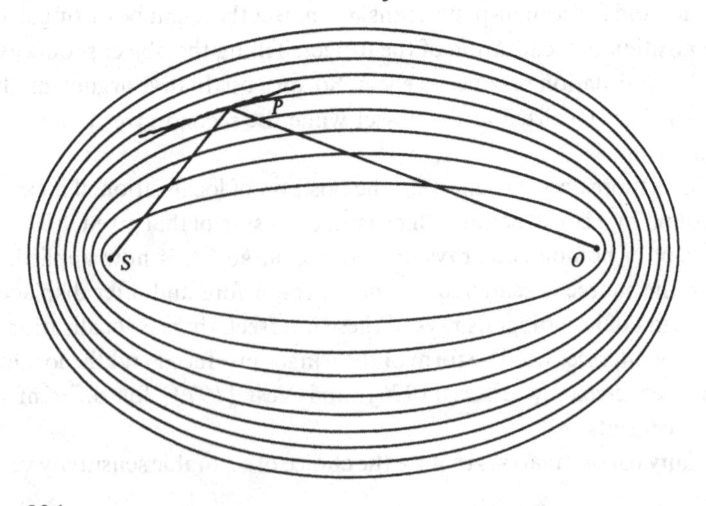

12.4. Speckle interferometry

Hologram interferometry is based on the use of holograms, which contain more information about the object shape and reflectance than is needed to measure only displacements. These holograms require a recording medium with a high resolving power and this, in practice, means a low sensitivity. Sufficient information for obtaining displacements is contained in the speckle pattern produced by the object and this can be recorded with lower resolving power. Speckle methods are given in more detail by Ennos [1978], Erf [1978], and Vest [1979].

The two waves in a Michelson interferometer in which one or both mirrors are replaced by a flat diffuser still have the same average phase difference as they had from the mirrors, but on this is superimposed the random phase of the surface roughness. They interfere to produce a speckle pattern that changes if one surface is moved. If the two patterns, before and after the change, are compared, the phase change caused by the movement can be derived. For every $2N\pi$ change of phase the speckle returns to its original form so that the correlation pattern of the speckle before and after the change appears as an interferogram that gives contours of changes of $2N\pi$.

12.4.1. Visual speckle interferometer

If the speckles are large enough, interference effects can be seen by eye. The method, developed by Archbold *et al.* [1969] and refined by Stetson [1978], is to view the surface with a low-power telescope, shown in fig. 115. Light from the laser source that illuminates the object is added by a beam splitter inside the telescope to give interference between a uniform wave and the wave scattered by the object. An iris diaphragm at the front controls the speckle size.

If the surface moves slowly, the speckles twinkle as each goes through a

Fig. 115 Optical system of the visual speckle interferometer [Ennos, 1978; Stetson, 1978].

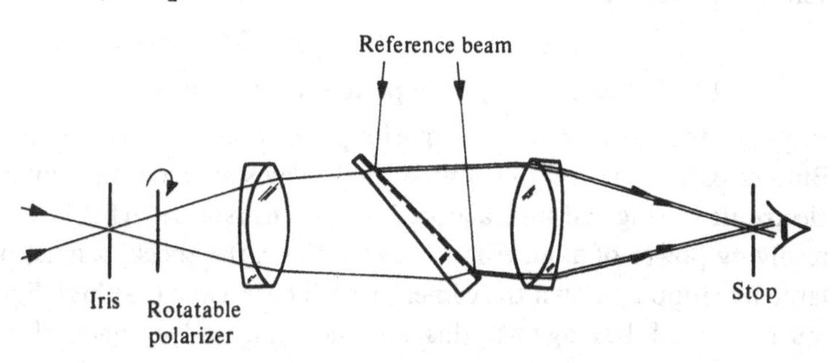

cycle of phase difference. If it vibrates rapidly the speckle is blurred, and the stationary nodes stand out against this blur as regions of higher contrast. This visual system is useful for looking at objects set up for holography to see that conditions are stable, and for adjusting vibration modes to resonance before studying them by quantitative methods.

12.4.2. Photographic methods

Quantitative speckle interferometry is carried out by correlating photographs of two speckle patterns formed by the interference between the wave scattered by the object and a reference wave. Several methods are used to obtain this correlation [Ennos, 1978].

In the first, a negative of the first pattern is superimposed on a positive of the second. Where the patterns are correlated a bright part of the speckle is covered by a black part of the negative and no light is transmitted. Where they are not correlated, some light is passed. Instead of a positive, the actual speckle pattern can be used to give real-time correlation.

In another method, the two records are made on the same plate with a small displacement between them. When this plate is illuminated by a collimated beam from a laser, the far-field diffraction pattern shows fringes, similar to Young's fringes, due to this displacement. A spatial filter placed at the centre of one or more dark fringes will give an image of the object that is dark where the speckles are correlated, and lighter where they are not.

If there is no shift of the plate between the two exposures, the correlation can still be seen. Where the patterns are correlated, two identical speckle patterns add and the result has the same statistics as each pattern. In terms of fig. 36, the probability distribution follows curve 1 when the reference wave is also from a rough surface and curve 2 if it is a uniform wave. But where they differ in phase difference by π, the two speckle patterns add in the same way as two incoherent patterns and so follow the distribution of curve 3, which has a lower contrast. The difference can be increased by non-linear effects in photography, which converts the difference in contrast to a different average density.

12.4.3. Electronic speckle pattern interferometry

Probably the most useful technique for performing the correlation is that of Butters & Leendertz [1971] who used television recording followed by electronic storing and subtraction. The system is shown in fig. 116. Since the resolving power of a television camera is low, the speckles must be made large by stopping down the camera lens. This means that high light levels must be used but against this are the many advantages of avoiding photographic processing and of obtaining almost immediate results on the monitor.

ESPI as it is usually called is now the most widely used method of speckle interferometry [Erf, 1978]. It can observe vibrating surfaces directly to give quantitative results that cannot be obtained by the visual interferometer, and it can be employed with any of the special systems described below.

12.4.4. Special speckle interferometers

Several different arrangements have been developed for speckle interferometry that measure movements in a specified direction.

Normal displacements are measured in an interferometer arranged as a Michelson interferometer with the object in place of one mirror. The system is insensitive to in-plane movements of the object, provided the illumination and viewing are both normal.

In-plane displacements are measured in the arrangement of fig. 117, due to Leendertz [1970]. The object is illuminated by two beams from the same

Fig. 116 Electronic speckle pattern interferometry.

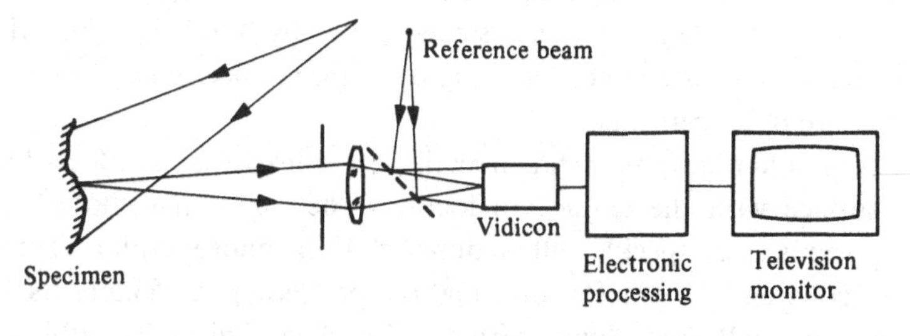

Fig. 117 Speckle interferometry for measuring in-plane displacement.

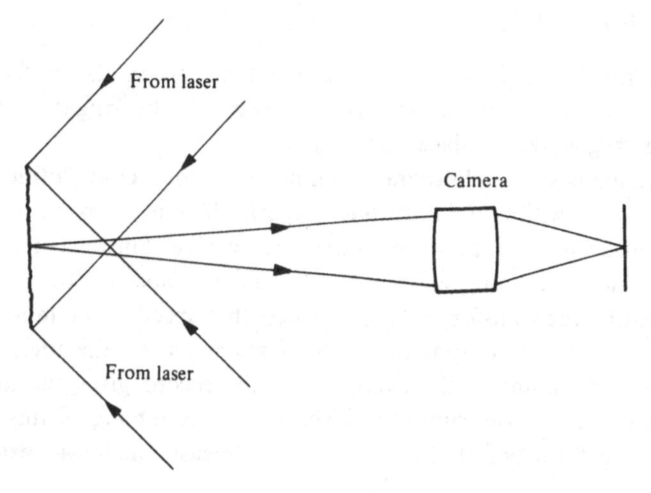

laser, these being at equal angles on opposite sides of the normal and lying in the plane in which the movement is to be measured. The interference takes place between two waves from the same surface, one from each illuminating beam. For a normal displacement, each of these waves receives the same phase change and the interference is unaffected. But in-plane displacements change the phase of each wave by opposite amounts. To measure displacements in the plane at right angles to that of the illuminating beams, further illuminating beams are needed.

Shearing interferometers are used to measure changes to the gradient of the phase across the surface, that is, flexure of the object. If the surface is illuminated and viewed normally, the results are unaffected by in-plane movements. The lateral shear is produced by any of the classical shearing interferometers, which is placed in front of the camera that photographs the speckle.

12.5. Speckle photography

Speckle interferometry makes use of two interference patterns between two coherent waves. But the simple speckle pattern produced by a single wave scattered by a rough surface has been shown by Burch & Tokarski [1968] to contain information on the position of the surface, which can be used to measure displacements.

If two photographs are taken on the same film of the speckle pattern from a surface with the surface in focus, if the object has moved between exposures, each speckle will be doubled. If the photograph is examined in the coherent system for information processing of §3.5.1, its Fourier transform will show Young's fringes [Ineichen, Eglin & Dändliker, 1980]. For a displacement d' on the film, their angular spacing is given by

$$\sin u = \lambda / d'. \tag{12.7}$$

If the movement is not the same at all points on the surface, the photograph can be scanned by a narrow beam and the fringes formed by each region give the displacement there.

The doubling of the photograph can be regarded as convolution of the speckle pattern with two δ-functions d' apart. The transform then consists of the transform of the speckle multiplied by a cosine. If instead of a double exposure, a continuous exposure is made, the speckle transform is multiplied by the transform of the function that specifies the motion.

If the speckle photographs are taken not at an image of the object but at the back focal point of the camera lens, the record gives the angular distribution of speckle from the object and any doubling of this image denotes a tilt of the object. These two different measurements are examples of the general result obtained by Stetson [1976] that, for any direction of view, a pair of double-exposed speckle photographs, taken at different focal settings, gives information on both the in-plane and normal displacements. Speckle photography has, therefore, the same capabilities as hologram interferometry, without requiring the stable conditions needed for interferometry [Ennos, 1980]. It is less precise but covers a larger range of displacements, ranging from just greater than the speckle diameter upward. A limit to the precision is the quality of the camera lens [Stetson, 1977].

12.5.1. White-light speckle

This name is used for the technique of Burch & Forno [1975] and Boone & de Backer [1976]. The object is now characterized, not by its speckle, but by a photograph of its surface irregularities taken in white light. A smooth surface can be coated to produce suitable irregularities. Correlation fringes showing the movement are obtained by the same methods used for speckle photographs.

Burch and Forno applied a fine rectangular grid as surface coating and the correlation fringes can be regarded as a moiré between the two images of this grid. To enhance the response of the camera lens to the spatial frequency of this grid, they added to it an aperture made up of two pairs of slits at right angles. This aperture also produces a grid image of an object with random surface irregularities and so improves the contrast of the correlation fringes.

12.6. Object contouring

Interference methods can be used to put contours on the surface of a solid object or on its holographic reconstruction. This can be done directly, if the object is placed in the standing waves produced by two interfering beams [Tsuruta & Itoh, 1969; Welford, 1969]. But, because the beams are at close to grazing incidence, large portions of the object are in shadow. Hologram and speckle methods, in which the illumination is closer to normal, avoid this.

12.6.1. Hologram methods

If two holograms of the same object, taken on the same plate, have phase differences that are in a constant ratio to each other, their reconstructions interfere to produce contour fringes across the surface of the reconstructed object. The same result applies if one reconstruction is replaced by the object itself.

True contours that follow plane sections of the object require a sensitivity vector of constant magnitude and direction over the whole object. From the object, both the source and point of observation should appear to be at infinity, that is, they should be at the focus of collimating lenses.

The phase difference in a hologram is $2\pi p/\lambda$, where p is the path difference between the object and reference beams. To change the scale of the phase requires a change of λ, the wavelength of the light in the medium in front of the object. This can be done by a change of the frequency of the light (and hence its vacuum wavelength) or a change of the refractive index of the medium. In either case the effective wavelength, which determines the contour interval, is

$$\Lambda = \lambda_1 \lambda_2 / \Delta\lambda. \tag{12.8}$$

The actual interval depends on the angles of the incidence and view, being $\frac{1}{2}\Lambda$ when both are normal.

For the two-frequency methods, two superimposed holograms made with light of different frequencies are reconstructed with the same light. The optical system must be arranged so that, with the change of frequency, the reconstructed images still coincide and the same ratio of phases applies to the whole hologram. When a hologram is reconstructed with light of a different frequency, the reconstructed wave is changed in direction and curvature. This can be derived from a modification of equation (5.3). There is thus both a lateral and a longitudinal displacement of the image that increases with its distance from the hologram. The optical system for two-frequency contouring should be arranged so that this displacement is negligible. The ratio of the phase differences should also be the same across the whole hologram. Two of the methods used are shown in fig. 118. In the first, an image hologram ensures coincidence of reconstructions, while the

Fig. 118 Systems for contouring by two-frequency holography: (*a*) image hologram, formed through a small aperture; (*b*) reconstruction at infinity (at the principal focus of a lens).

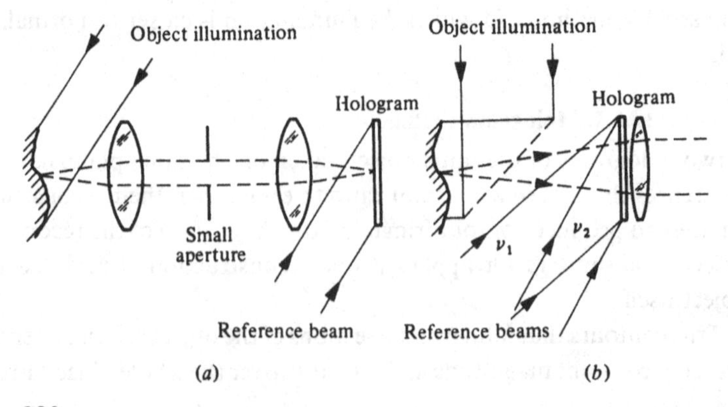

ratio of phases is constant provided the illuminating and reference beams are at the same angle. In the second, normal illumination and viewing is used and two different reference beams must be used at angles proportional to the two wavelengths.

For the two-index method of Shiotake *et al.* [1968], the change of wavelength applies only to the immediate surround of the object, which is placed in a suitable cell, and the system is simplified. In the arrangement of fig. 118*b*, one reference-beam angle is sufficient. Large index differences can be obtained by immersion of the object in two different liquids; smaller differences, and hence large contour intervals, in two different gases.

12.6.2. Speckle methods

Speckle interferometry can be used for generating contours in the same way as holographic methods. The most usual display is electronic [Denby, Quintanilla & Butters, 1976], and again, either two frequencies or two refractive indices can be used.

12.6.3. Shadow moiré

For the coarsest contour intervals, the method of shadow moiré is used. The object is illuminated through a grid with a small enough source so that light and dark bands are projected on its surface. It is then viewed through another grid from a different angle. The grids are so arranged that each would project to the same spacing, from the source for the first, from the eye or camera for the second, on a plane object. Such a plane will then appear light when the grids are in register, dark when out of register.

When the object is not plane, the projection of the first grid no longer appears as straight lines from the viewing direction. With the second grid these produce a moiré pattern that contours the object. The technique, introduced by Weller & Shepard [1948], has been described by Takasaki [1979].

13

Interference spectroscopy

Spectroscopy is the study of the distribution of electromagnetic radiation with frequency, and interference methods are used in those spectral regions where power detectors must be employed, $\nu > 0.5$ THz or $\lambda < 0.6$ mm. Different names are used for instruments according to the detector used: a spectroscope is a visual instrument, a spectrograph is photographic, and a spectrophotometer or spectroradiometer is now usually electronic. A common name for the last two instruments is *spectrometer* and it is used here; it should be remembered, however, that this name has also an older use for an instrument that measures angles and hence the refractive indices of prisms.

13.1. General theory of spectroscopy

Any spectroscopic instrument measures the product of three spectral distributions: the radiation studied $I(\nu)$, the detector sensitivity $d(\nu)$, and its own transmittance $\mathcal{T}(\nu)$. This product will be denoted by $g(\nu)$, where

$$g(\nu) = \mathcal{T}(\nu) I(\nu) d(\nu). \tag{13.1}$$

A spectrometer is a linear system and the recorded distribution $b(\nu)$ is the convolution of $g(\nu)$ with $k(\nu)$, the *scanning function* or *instrumental profile* of the equipment,

$$b(\nu) = \int_{-\infty}^{+\infty} k(\nu')g(\nu - \nu')\,\mathrm{d}\nu'. \tag{13.2}$$

Fourier transformation gives

$$B(\tau) = K(\tau)G(\tau), \tag{13.3}$$

where $K(\tau)$ is the transfer function of the equipment, a function of a time difference τ. For most instruments, B and G have only the meanings of mathematically derived transforms of b and g. But (6.14) and (6.15) show that $G(\tau)$ is a physical entity for a two-beam interferometer, it is the *interferogram* or the intensity variation expressed as a function of the delay

τ. Two-beam interferometers are used for *Fourier spectroscopy*, where the transform is recorded rather than the spectrum itself as in direct methods.

13.1.1. Resolution

The narrower the scanning function of a spectrometer, the smaller its limit of resolution $\delta\nu$ and the greater its *resolving power* or ability to separate close spectral lines. The resolving power is defined as

$$R = \nu/\delta\nu = \lambda/\delta\lambda. \tag{13.4}$$

The uncertainty relation (6.2) has shown that the width of the scanning function is reciprocally related to the maximum delay T in the interferometer:

$$\begin{aligned} T\,\delta\nu &= 1/(2\varepsilon) \\ &\approx 1, \end{aligned} \tag{13.5}$$

where ε is a constant, of order one or less. It is the ratio of the actual resolving power to its theoretically possible value, and depends on the shape of the scanning function and how its width is defined. The resolving power is then given by

$$R = 2\varepsilon\nu T = 2\varepsilon\sigma P, \tag{13.6}$$

the second form being in terms of the wavenumber σ and the maximum path difference P.

The same relation (13.6) holds also for classical spectrometers. For a prism it should be remembered that group time is involved and, by Fermat's principle, refraction through a prism gives a constant phase time across the wavefront. The difference between phase and group times gives the classical expression for the resolving power of a prism of base b,

$$R = b\frac{\mathrm{d}n}{\mathrm{d}\lambda}.$$

The scanning function is the output of the spectrometer for an input of one frequency, a δ-function. Classical spectrometers disperse the light by a prism or grating and use slits to separate different frequencies. For an input of one frequency, the instrument forms an image of the entrance slit, and this is scanned by the exit slit to give a scanning function that is the convolution of the two rectangular functions that represent the slits, convolved again with the spread function of the optical system. For infinitely narrow slits, the spread function and scanning function are the same: a sinc^2 function for a rectangular prism or grating. At the other limit of geometrical optics, where diffraction is ignored, the scanning function is triangular.

13.1.2. Apodization

The scanning function of some interferometers is approximately a sinc function. This has much larger oscillations than the scanning functions of classical instruments and, around any rapid changes in the spectral intensity, oscillations appear that can be mistaken for weak spectral lines. A test spectrum and its convolutions with sinc and sinc^2 scanning functions are shown in fig. 119. The false lines produced by the first scanning function are evident.

The technique of changing the scanning function to avoid such false lines has been named *apodization* or 'removal of the feet' [Jacquinot & Roizen-Dossier, 1964]. The transfer function of a spectrometer represents the relative use made of each delay present, and Fourier theory shows that apodization is equivalent to tapering the transfer function so that the weighting of the higher delays is progressively reduced, and any discontinuity at $\tau = T$ removed. A two-beam interferometer that uses all delays with equal weight has its transfer and scanning functions given by

$$K(\tau) = \text{rect}(\tau/2T), \quad k(\nu) = \text{sinc}\, 2\nu T. \tag{13.7}$$

If this is given a linear taper, the functions become

$$K(\tau) = \text{tri}(\tau/T), \quad k(\nu) = \text{sinc}^2\, \nu T. \tag{13.8}$$

The taper has apodized the scanning function but has doubled its width. It is for this reason that the sinc^2 function of double width is included in fig. 119 as this would be obtained if the sinc function were apodized. The sinc^2

Fig. 119 A test spectrum (*a*) and its convolution with: (*b*) an unapodized scanning function (sinc); (*c*) an apodized sinc^2 function of twice width (from the same spectrometer); (*d*) a sinc^2 function of the same width as (*b*).

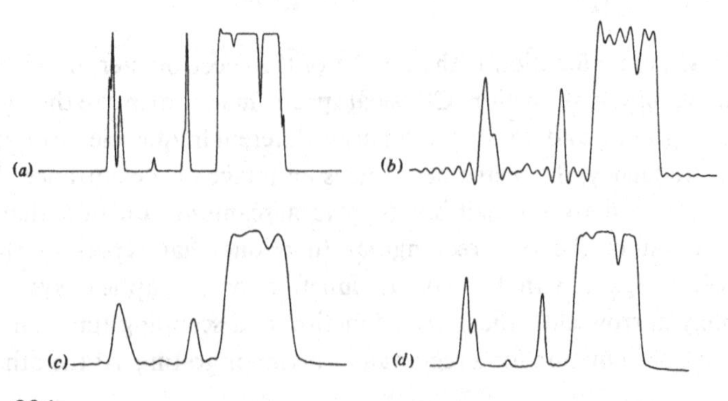

function of the same width, fig. 119(d), requires a spectrometer with twice the maximum delay. This effect of apodization is included in the factor ε of (13.5) and (13.6) and, if the width of the scanning function is taken to be the distance from its centre to the first zero, $\varepsilon=1$ for an (unapodized) sinc function and $\varepsilon=\frac{1}{2}$ for the (apodized) sinc^2. Other functions suitable as tapers and the corresponding scanning functions have been shown in fig. 8, and given by Vanasse & Sakai [1967], and by Norton & Beer [1976]. It is seen that, as the apodization is increased, the width of the scanning function increases further and ε becomes less than $\frac{1}{2}$.

Apodization is obtained instrumentally by building the tapered transmittance into the spectrometer. Or it can be done mathematically by convolution of an unapodized spectrum and an apodized scanning function [Filler, 1964].

13.1.3. Sampling

The observed spectral distribution will extend over some range of frequencies $\Delta\nu$. The number of times that the width of the scanning function fits into this range is the number of spectral elements m that are resolved, given by

$$m=\Delta\nu/\delta\nu, \tag{13.9}$$

$$=2\varepsilon T\,\Delta\nu. \tag{13.10}$$

As $\Delta\nu$ is finite, $B(\tau)$ is a band-limited function and, in Fourier spectroscopy, it need only be sampled at a finite interval $\delta\tau$. If the spectral range extends from $\nu=0$, (2.36) shows that

$$\delta\tau=1/(2\Delta\nu), \tag{13.11}$$

the factor one half entering because the same range $\Delta\nu$ must be considered as applying to negative frequencies. For a sinc scanning function ($\varepsilon=1$) there are then $m+1$ sampling points required, $2m+1$ for sinc^2 ($\varepsilon=\frac{1}{2}$). If the range $\Delta\nu$ extends up to a maximum frequency ν_m, Connes [1961] has shown that the sampling interval may often be increased and that it is given generally by

$$\delta\tau=N/2\nu_m\leqslant 1/2\Delta\nu, \tag{13.12}$$

where N is the integral part of the ratio $\nu_m/\Delta\nu$. Such a sampling interval gives repetitions of the spectrum that do not overlap, analogous to the various orders given by a grating, which in its N-th order samples the time difference at $\delta\tau=N/\nu$. A prism with continuous sampling gives a single spectrum.

The spectrum itself is also band limited since its transform $G(\tau)$ exists as $B(\tau)$ only for $\tau \leqslant T$. Thus, in direct spectroscopy, the spectrum needs to be sampled only at intervals

$$\delta v \leqslant 1/2T, \tag{13.13}$$

while in Fourier spectroscopy it can be calculated at this interval and intervening points found by convolution, as shown by (2.36). For this application, (2.35) and (2.36) become

$$B(\tau) = [B(\tau) * \text{Ш}(\tau/2T)] \operatorname{rect}(\tau/2T),$$
$$b(v) = [b(v)\text{Ш}(2vT)] * \operatorname{sinc} 2vT,$$

and convolution of the sampled spectrum with sinc $2vT$ gives intervening points of the (unapodized) spectrum. To apodize the spectrum, its transform should be multiplied by a tapering function $A(\tau)$. This function must vanish for $\tau > T$, so we can write

$$A(\tau) \operatorname{rect}(\tau/2T) = A(\tau)$$

and (2.35) gives

$$G(\tau)A(\tau) = B(\tau)A(\tau) = [B(\tau) * \text{Ш}(\tau/2T)]A(\tau), \tag{13.14}$$

since, with no tapering, $G(\tau)$ and $B(\tau)$ are the same up to $\tau = T$. The last equation leads to the modified sampling theorem due to Jacquinot & Roizen-Dossier [1964] and Filler [1964],

$$g(v) * a(v) = [b(v)\text{Ш}(2vT)] * a(v); \tag{13.15}$$

an apodized spectrum can be obtained if the same sampled spectrum is convolved with an apodized scanning function $a(v)$, provided this is band limited to $\tau \leqslant T$. A similar technique can be used to change the apodization if the original spectrum is already apodized. In this case the convolution is taken with the transform of the ratio of the two tapering functions.

The total number of sampling points in the spectrum is then $2T\,\Delta v + 1$, or $m + 1$, one more than the number of spectral elements. As shown by (13.12), the number of sampling points M in the interferogram is not less than m/ε. Noise considerations discussed later show that this is the minimum number of sampling points and that it may be desirable to use more points of the spectrum in direct spectroscopy, to improve the signal-to-noise ratio. But in Fourier spectroscopy, where the number of sampling points should be kept to a minimum, if the signal-to-noise ratio needs to be improved, more time can be spent on obtaining the measurement at each point.

13.1.4. Noise

The relation (13.6) suggests that resolving power can be increased without limit by increasing the maximum delay in the instrument. But such an increase will spread the available energy over more spectral elements giving less signal for each. This limits the resolution possible since the signal must be detected against a background of noise.

The signal obtained from the spectral distribution (13.1) in a time t is proportional to

$$q(\nu) = Etg(\nu), \tag{13.16}$$

where E is the étendue of the spectrometer. Superimposed on this signal is *noise*, random variations of q characterized by their root-mean-square value Δq. Two cases will be distinguished. The first is that of photon noise and shot noise, which follow the statistical variations of the radiation and have a noise power $(\Delta q)^2$ that is proportional to q. The signal-to-noise ratio is then

$$(q/\Delta q)_p \propto [Etg(\nu)]^{1/2}. \tag{13.17}$$

This expression holds for noise that is due either to the signal radiation itself or to some background radiation and will be called the *photon-noise* case.

The most important type of noise that is independent of the signal is that in the detector itself. *Detector noise* has a power that is proportional to the area of the detector and hence to the étendue of the system, provided the detector is being used most efficiently over the largest practical solid angle. When this noise is the more important,

$$(q/\Delta q)_d \propto (Et)^{1/2} g(\nu); \tag{13.18}$$

the change from (13.17) is the increased importance of $g(\nu)$, which includes the transmittance of the instrument.

The effects of atmospheric turbulence are sometimes treated as a third class of noise, scintillation noise, that depends directly on the signal q, so that the noise power depends on q^2.

Photo-emissive cells and photomultipliers, used as detectors of visual and ultraviolet radiation, have very low detector noise and photon noise is the limit in these regions. In the infrared, detector noise is the more important, although, as the techniques of making infrared detectors improve [Kimmit, 1977], the region over which detector noise predominates is shrinking. But these new detectors are not as linear and thermal (noisy) detectors are still used when radiometric accuracy is required.

The cross-section area of a spectrometer is limited only by practical considerations, but there are theoretical limits to the solid angle that limit the étendue. They depend directly on the shift h and, in uncompensated interferometers, where h is directly proportional to the delay, they are inversely proportional to the resolving power R. Hence, the signal-to-noise ratio is inversely proportional to the square root of the resolving power

$$q/\Delta q \propto R^{-1/2}. \tag{13.19}$$

The same relation holds for a dispersing instrument in which resolving power is increased by decreasing the slit width and hence the étendue. It also applies in Fourier spectroscopy if different lengths of a recorded interferogram are used to give spectra at different resolutions. If the spectrum does not consist of one single line, most of the signal is concentrated around $\tau = 0$ in the interferogram, and a reduction in the length used reduces the noise but not the signal.

The case of a spectrum consisting only of one line narrower than the scanning function must be distinguished. In the argument above, it is seen that the signal then depends directly on the amount of the interferogram that is used and

$$q/\Delta q \propto R^{1/2}. \tag{13.20}$$

The same result follows for dispersive instruments, for narrowing the slit does not reduce the signal received from a line until it appears as wide as the slit. Then the relation (13.20) changes over to (13.19).

The signal-to-noise ratio is slightly reduced by apodization. Connes [1961] has considered this effect in Fourier spectroscopy and has shown that there is less loss with mathematical apodization than when, for example, an absorbing screen is used in the instrument and some signal is lost.

13.2. Comparison of methods

There are two important comparisons to be made: that of interference and classical methods of spectroscopy, and that of direct and Fourier methods of interference spectroscopy. Comparisons of this type have been given by Jacquinot [1958] and Kiselyov & Parshin [1964]. Fourier and Hadamard spectroscopy are compared by Tai & Harwit [1976, 1977].

13.2.1. Etendue advantage

A classical spectrometer that disperses the radiation and separates different frequencies by their different positions uses slits, which reduce its solid angle in one dimension. To obtain the required resolution, the slits must be narrow in the direction of the dispersion although they can be long at right angles to this.

Jacquinot & Dufour [1948] have shown that an interferometer with division of amplitude has circular symmetry and has an acceptance angle that, in all azimuths, is comparable with the angle subtended by the long dimension of the slit of a classical instrument. If the length of the slit subtends an angle β, the gain in solid angle is by a factor of approximately $2\pi/\beta$ and, if the two instruments have the same area of cross section, there is the same gain in étendue, and the signal-to-noise ratio is improved by $(2\pi/\beta)^{1/2}$. But not all interference spectrometers are used symmetrically: a Fabry–Perot interferometer used off axis does not have this gain, nor does an interferometer with division of wavefront.

It has recently been shown by Girard [1965] that a dispersing system can be used without slits. He has used entrance and exit grilles that subtend large solid angles; these have a hyperbolic form, but other shapes are possible. The scanning function is the autocorrelation function of the grilles and, for a grille to be suitable, this function should have a narrow peak that may be superimposed on a uniform background; the Wiener spectrum of a grille should be a constant. The background can be eliminated by subtracting, from the results obtained with an identical pair of grilles, those given by a pair one of which is the complement of the other; however, its presence in the scanning function leads to a large increase in photon noise. The Girard instrument is not an interference method, but it has shown that the étendue of interferometers can be equalled by dispersing instruments.

Instead of being regarded as a means of increasing the total energy available, the increase in étendue of an interference spectrometer can be used to increase the field of view and to study simultaneously the spectral distribution of radiation at different points of the field. This corresponds to the use of the interferometer as a *tunable filter*, through which an image is formed. An application is the observation of the sun in monochromatic light of narrow bandwidth for which birefringent filters or Fabry–Perot interferometers are used. The dispersion instrument used for the same studies, the classical spectroheliograph, cannot image the whole sun simultaneously. The image must be scanned across the slit and built up line by line.

The limit of $2\pi/R$ on the solid angle accepted by an interferometer means that, when it is used as a filter, the frequency transmitted varies by the half width of the transmission band from the centre to the edge of the field of view. This is normally too large a variation, and, to reduce it by a factor α, the angular radius of the field should be reduced by the factor $\alpha^{1/2}$.

13.2.2. Multiplex advantage

A spectrograph recording on a photographic plate uses the whole exposure time t to record each spectral element, but a scanning spectrometer must divide t between the m elements present so that each is observed for a time t/m only. Other effects being equal there is a reduction of the signal-to-noise ratio by a factor of $m^{-1/2}$ compared with the spectrograph. To regain this factor, m separate detectors with separate channels would be needed.

In electronics, many signals can be transmitted simultaneously over the same channel if they are suitably coded so that they can be separated at the output end. This is a *multiplex system*, and Fellgett [1951, 1958] has shown that Fourier spectroscopy is a *multiplex method*, in which each optical frequency is coded as a cosine signal of a lower electrical frequency. The signals may be decoded by Fourier analysis. The time for which each spectral element is recorded is increased by a factor m compared with direct methods, and, for regions limited by detector noise, the signal-to-noise ratio is increased by a factor of the order $m^{1/2}$, if the transmittance and étendue are the same. Fleming [1977] gives the factor as $m^{1/2}/2\sqrt{2}$. But this gain does not apply to spectral regions where the main noise is photon noise; the increase in the recording time increases also the noise. In fact, since the bandwidth of the detector $\Delta\nu_d$ may be larger than that of the spectrum being examined, the number of spectral elements contributing noise, $m_d = \Delta\nu_d/\delta\nu$, may exceed the number m in the signal, and there is a loss in signal-to-noise by the factor

$$(m/m_d)^{1/2} = (\Delta\nu/\Delta\nu_d)^{1/2}. \tag{13.21}$$

For scintillation noise, the multiplex effect operates in reverse as a disadvantage. While Fourier spectroscopy is used in the visible and ultraviolet [Luc & Gerstenkorn, 1978], for this radiation it should be inferior to other interference methods.

In addition, the distribution of photon noise is different in the two instruments. In a direct method of spectroscopy, the noise is proportional to the signal and is strongest on the peaks of lines and least between them. This method is thus the better for looking for weak emission lines. In Fourier spectroscopy, the process of taking a transform redistributes the noise evenly across the spectrum. This and the relatively poor efficiency of a single Fabry–Perot interferometer mean that Fourier spectroscopy may be preferable for studying absorption spectra. Fourier methods are, however, most useful in the infrared, where detector noise is a limitation and energy is at a premium. They have both the étendue and the multiplex advantage and have been successful in giving spectra of planets [Connes & Michel, 1975], the upper atmosphere [Baker, Steed & Stair, 1981], and weak laboratory sources [Connes *et al.*, 1970; Guelachvili, 1972]. Resolving powers of 10^6 have been obtained over 10^6 spectral elements, and these resolving powers have only recently been exceeded by laser spectroscopy.

13.2.3. Effects of errors

As indicated in §6.4.1, errors in interferometers are more serious than in other optical systems. If, in the latter, a certain fraction of the intensity is misdirected, the same effect produces the square root of this fraction as an error in an interferometer. For spectroscopy this means that the drive that changes the delay of a Fourier spectrometer must be more accurate than the drive of an engine for ruling gratings. Limits on the errors of an interferometer drive are given by Sakai [1971], and Bell & Sanderson [1972]. They show that, to have errors less than 0.01 of the intensity, the drive has to be good to about 0.01 of the wavelength.

13.3. Earlier methods of interference spectroscopy

Direct methods of interference spectroscopy, where there is no multiplex effect and the result obtained is a measurement of the spectrum itself, are more suitable for visible and ultraviolet radiation. Both detectors and dispersing spectrometers are well developed for these regions, and interferometers are required only for problems involving high resolving power where the maximum possible étendue is desirable. The resolving power of prisms and gratings is limited by practical considerations of size; until recently 10^5 was considered high but now values up to 2×10^6 are possible. Above this, interferometers are required.

Another spectroscopic field in which interference methods are now used almost entirely is the precise measurement of spectral wavelengths. This application has been considered in Chapter 10.

The most important direct spectroscopic instrument is the diffraction grating. This is now regarded as a classical instrument rather than an interferometer and will not be considered here except for comparison purposes. A grating with n rulings used in its N-th order has a resolving power

$$R = nN, \qquad (13.22)$$

and the spectrum is repeated in the next order at $\Delta\nu = \nu/N$. Normally gratings are used in low orders and have a larger spectral range $\Delta\nu$ than interferometers. Errors in the ruling of the grating cause radiation to be lost from the central maximum of the scanning function to outer regions. If the errors are periodic, this appears concentrated as a satellite peak or *ghost*; if they are random, it is spread evenly. There have been two important advances in the theory and production of diffraction gratings. One is the introduction of holographic gratings, described in §5.6.1. The other is the development of their electromagnetic theory [Petit, 1980] to the stage where it can now explain satisfactorily the distribution of radiation between different orders. R.W. Wood first showed, in 1910, how gratings should be blazed to concentrate a large amount of the radiation into one order, but it required this new theory to explain completely the effects of groove shape and surface material.

Wood called a rather coarse grating blazed for a high order an *echelette* grating and Harrison [1949] extended this idea to produce *echelle* gratings with $n \sim 10^3$ and $N \sim 10^3$. Belonging to this same family of instruments is one that is traditionally regarded as an interferometer, the *echelon* of Michelson. This is not a ruling but consists of 20 to 40 plates of fused silica, all of closely the same thickness, joined in optical contact to give facets of equal width, as shown in fig. 120. These dimensions should be correct to better than $\lambda/20$ for a reflexion echelon.

Another classical instrument of interference spectroscopy, also difficult to construct to the required precision and now seldom used, is the Lummer–Gehrcke plate. It is also shown in fig. 120. Initially, it was the most widely used method of interference spectroscopy. As it employs division of

Fig. 120 (*a*) The reflexion echolon; (*b*) the Lummer–Gehrcke plate.

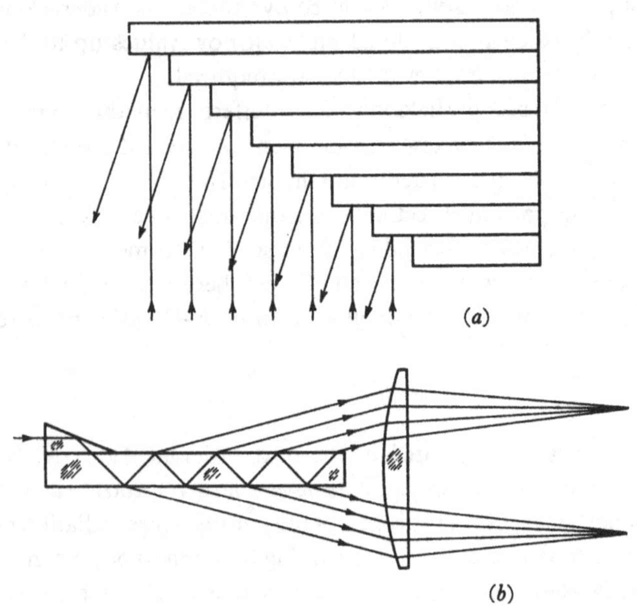

amplitude, it does not require a slit source, as does the echelon, for all beams to be coherent, but as these beams leave with increasing shears, it has some of the properties of an interferometer with division of wavefront, namely, a lack both of circular symmetry and of the large étendue that goes with this. Although often considered an obsolete instrument, it has been used by Warde [1976], with some modification, for the spectroscopy of thin dielectric films.

Fuller descriptions of these two instruments are given by Candler [1951], Tolansky [1955], and Born & Wolf [1959].

13.4. Fabry–Perot interferometers

The two instruments discussed above have been largely supplanted by the interferometer, first introduced by Boulouch [Duffieux, 1969], and developed by Fabry & Perot [1897]. For two plane parallel reflecting surfaces separated a distance d by a medium of index n, a ray incident at an angle θ to the normal has an increment of phase

$$\begin{aligned}\psi &= 4\pi n\sigma d\cos\theta \\ &= 2\pi\nu T,\end{aligned} \tag{13.23}$$

for each double passage, where T is the transit time $2nd\cos\theta/c$; strictly this expression also should include the phase changes on reflexion, given in (9.8), but these can be regarded as a correction to the separation d. The scanning function is given by the Airy function (9.10)

$$k(\nu) = \frac{\mathscr{T}^2}{1+\mathscr{R}^2-2\mathscr{R}\cos\psi} = \frac{\mathscr{T}^2}{(1-\mathscr{R})^2+4\mathscr{R}\sin^2\frac{1}{2}\psi}, \tag{13.24}$$

where $\mathscr{R}$ and $\mathscr{T}$ are the reflectance and transmittance of the reflecting surfaces. This function has been shown in fig. 73; it has narrow peaks at

$$\nu T = 2n\sigma d\cos\theta = N \quad (N \text{ integral}) \tag{13.25}$$

of half-width

$$\delta\psi = 2\pi/\mathscr{F}, \quad \delta\nu = 1/\mathscr{F}T, \tag{13.26}$$

where $\mathscr{F}$ is the finesse, given by (9.18).

An interferometer with a fixed spacing, a Fabry–Perot étalon, has a radial dispersion and transmits radiation of frequency ν at angles θ given by (13.25), the different orders N representing successive interference fringes. The fringes for different spectral lines can be separated by using the étalon in combination with a spectrograph, as shown schematically in fig. 121, so that each spectral line is broken up into a narrow strip of this circular fringe pattern [Tolansky, 1947]. The photographic recording of a spectrograph enables a large amount of information to be recorded in one experiment. This technique is used chiefly for the study of hyperfine structure and can also give accurate values of wavelengths by the methods described by Harrison, Lord & Loofbourow [1948].

The method of use in fig. 121 resembles that for the echelon or the Lummer–Gehrcke plate and does not take advantage of the radial symmetry and consequent large étendue of the Fabry–Perot. The importance of radial symmetry was shown by Jacquinot & Dufour [1948] who introduced the Fabry–Perot spectrometer to take advantage of its symmetry. The interferometer acts as a monochromator or adjustable filter, and is followed by a detector, light being collected over one of the circular fringes. Although it is most convenient to use the central fringes, all fringes subtend the same solid angle, given by (9.30) as

$$\Omega = 2\pi/\mathscr{F}\nu T. \tag{13.27}$$

Special masks or annular mirrors have been used to collect the radiation from more than one fringe, either to increase the amount of energy obtained at one frequency or to form a multi-channel system, as described by Hirschberg & Platz [1965], that collects several frequencies simultaneously. As, by (13.26), the scanning function has a width $\delta\nu = 1/\mathscr{F}T$, the resolving power is

$$R = \nu/\delta\nu = \mathscr{F}\nu T. \tag{13.28}$$

This expression may be combined with (13.27) to give the maximum solid angle as

$$\Omega = 2\pi/R; \tag{13.29}$$

when the interferometer is used as a filter, this is the solid angle over which the mean frequency transmitted varies by $\delta\nu$. As will be shown later, this same limit (13.29) applies to other important interference spectrometers.

It has been shown in Chapter 9 that the transmittance of a Fabry–Perot interferometer, and hence its scanning function as a spectrometer, is a convolution of three functions. The first is the Airy function (13.24), which

Fig. 121 The Fabry–Perot étalon combined with a spectrograph.

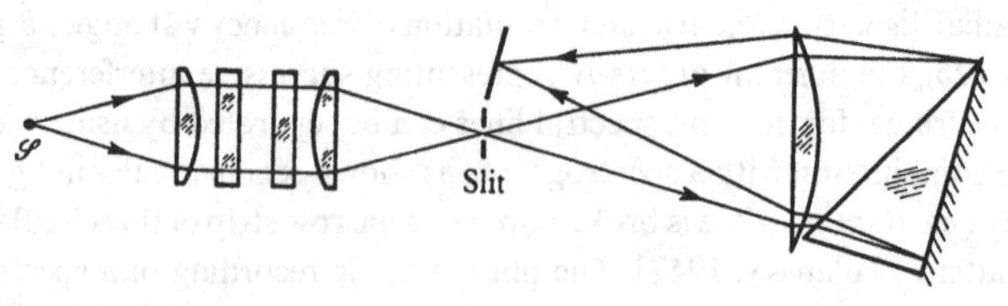

has a width dependent on the properties of the coatings of the reflecting surfaces. Although a finesse of between 20 and 40 is considered reasonable, modern multilayer dielectric coatings are now good enough to give a finesse of 10^3. This is approached in practice only over the small apertures used for laser cavities. For larger apertures, the second term of the convolution that depends on the flatness of the mirror surfaces is the limit to the available finesse, although improvements in optical techniques for producing flat surfaces are slowly raising this limit. The third function in the convolution is that denoting the size of the source.

The interferometer is scanned to transmit different frequencies by changing ψ, by a change of either the spacing d or the index n of the intervening medium. Scanning methods have been discussed in §9.3.4. The spacing of the different orders,

$$\Delta\nu = 1/T = \mathscr{F}\,\delta\nu, \tag{13.30}$$

is relatively close. A Fabry–Perot spectrometer may be used on its own to study the hyperfine structure of a single spectral line, but to cover a broader spectrum, it must be used with a pre-monochromator that can select one of the orders transmitted by the interferometer. This may be a dispersing instrument, and then the arrangement follows that in fig. 121 with the spectrograph replaced by a scanning monochromator. The matching conditions are now more critical, for the width of the slit of the monochromator should be at least equal to the usable source-size of the interferometer. This is possible only with a monochromator made with a large grating and, in common arrangements, the monochromator limits the étendue of the combination even though it has the lower resolving power. If, instead of a monochromator, a spectrograph is used, the Fabry–Perot interferometer becomes the multi-channel instrument, described by Chabbal & Pelletier [1965] under the name *Simac* (spectromètre interférentiel multi-canal). The spectrum is recorded either on a simple photographic plate or by a Lallemand electronic camera. If the plate is moved in synchronism with the scan of the interferometer, a very large amount of information can be recorded on a single plate, each separated order of the interferometer being spread out as a spectrum.

Another method of separating orders is to use several Fabry–Perot interferometers in series. This method has been developed by Chabbal [1957] who suggested that each interferometer in the series should decrease in spacing by a factor of about 10. Later work by McNutt [1965] has shown that it is preferable to use non-integral ratios, such as 1.0000:0.8831:0.7244, for three spacings, rather than a simple geometric series; the results in fig. 79 have also shown the advantages of vernier-type spacings. One difficulty of this method is that all the interferometers must be scanned together. Pressure scanning may be used with all the interferometers at the same pressure. Equal pressure, however, is only a special case; Mack *et al.* [1963] have shown that, to keep two étalons transmitting the same frequency, a constant pressure difference must be maintained between them. This enables each to be adjusted separately by a change of the difference, while a change of the total pressure scans them all together. For mechanical scanning, servo-control methods, such as those of Ramsay, are most useful, since they can be used not only to keep the pairs of interferometer surfaces parallel but also to hold the different spacings in constant ratios. Ramsay, Kobler & Mugridge [1970] have developed a filter consisting of three interferometers in series that can be tuned across the visible spectrum. The spacing of each interferometer is locked to that of a tunable reference interferometer.

The field-widened version of the Fabry–Perot interferometer is the *spherical Fabry–Perot* of Connes [1956], described in §9.4. The usable solid angle is no longer limited by the simple theory that has been given in (13.29) but there is a gain in solid angle by the factor

$$G=\pi\lambda^2R^2/8\mathscr{F}^2A \tag{13.31}$$

compared with a simple Fabry–Perot with the same resolving power R and with mirrors of area A; this gain has been given in an alternative form in (9.31). The spherical Fabry–Perot is therefore most useful at high resolving powers. It has the disadvantage that it has a fixed separation (although it can be scanned through an order without serious loss of performance) and thus a fixed resolution. It is commonly used to study the mode structure of lasers.

13.5. Birefringent filters

A single Michelson interferometer has a sinusoidal transmission function that is useless as a scanning function for direct spectroscopy. But several Michelson interferometers can be used in series to give a suitable function. If each interferometer has twice the delay of the preceding one, the transmission function for n interferometers is

$$k(\nu)=\tfrac{1}{2}\cos^2\pi\nu T\cos^2 2\pi\nu T\cos^2 2^2\pi\nu T\ldots\cos^2 2^{n-1}\pi\nu T, \tag{13.32}$$

where T is the delay of the first of the series. If this delay can be varied simultaneously for all interferometers, $k(\nu)$ acts as a scanning function. The function (13.32) represents a more efficient filter than given by the Airy function (13.24) that describes the Fabry–Perot interferometer. Thus the filtering efficiency, defined in §9.2, of a series of Michelson interferometers is

0.89 while it is 0.70 for a single Fabry–Perot, but again 0.89 for two in series of the same spacing.

Such an instrument was made by Lyot [1944], and independently by Öhman [1938], in its birefringent form. Each individual interferometer is a plate of quartz or calcite, cut parallel to its optic axis and set between polarizers. It has been named a *birefringent filter* and is tuned by changing the retardation of the plates. Lyot has tuned filters by changing their temperature and has introduced methods of field widening. The most commonly used 'element', or single interferometer, is shown in fig. 122. It is field widened by having the plate split into two halves which have their optic axes crossed. Between them is a tuning system comprising a rotatable half-wave plate between two fixed quarter-wave plates.

The peak transmittance of the filter is determined chiefly by the losses in the polarizers. These are usually dichroic polarizing sheets, as prism polarizers would greatly increase the length of the filter. Evans [1949] has shown how the number of polarizers used may be halved by placing two elements between each polarizer. A filter based on these principles has been made by Title & Rosenberg [1981] with a 6 GHz bandwidth, and by Beckers, Dickson & Joyce [1975] with a bandwidth of 8 GHz and tunable across the visible spectrum. These seem to represent the practical limit of resolution with available birefringent crystals. Evans [1949] has shown how Michelson interferometers with partially reflecting beam splitters may be used as elements to give finer resolution and this has been demonstrated by Title & Ramsey [1980].

A different type of birefringent filter has been made by Šolc [1965]. It consists of n plates of equal retardation set between two polarizers. Two forms are possible, and their theory is given by Evans [1958], and Fredga & Högbom [1971]. In the fan form, the plates have their optic axes at angles $\alpha/2, 3\alpha/2, \ldots, (2n-1)\alpha/2$ to the axes of the polarizers, which are parallel. In the folded form, the plate axes are at alternately $+\alpha/2$ and $-\alpha/2$ to the

Fig. 122 A field-widened, tunable element of a Lyot birefringent filter: C, crystal plate; P, polarizer.

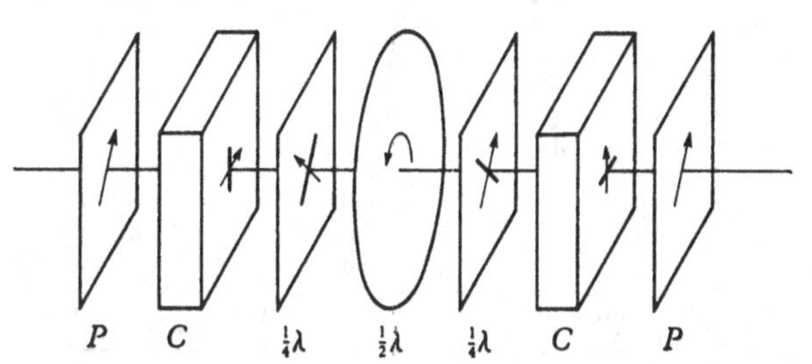

direction of the first polarizer and the second polarizer is crossed with this. The filter has a transmittance

$$k(\nu)=\frac{1}{2}\mathscr{T}\left[\frac{\sin n\chi}{\sin\chi}\sin\alpha\,\frac{\cos}{\sin}\,\pi\nu T\right]^2, \tag{13.33}$$

where

$$\cos\chi=\cos\alpha\cos\pi\nu T, \tag{13.34}$$

$\mathscr{T}$ being the transmittance of each plate and $T=\mu d/c$. In equation (13.33) the cosine applies for the fan form and the sine for the folded. For either form, the best value for the angle α is $\pi/2n$.

The transmission function for the Šolc filter is compared with that of the Lyot filter described above in fig. 123, and, without special correction, it is seen to have higher secondary transmission bands. However, they can be apodized by a modified design [Fredga & Högbom, 1971]. Šolc [1965] had apodized his filters experimentally.

A general theory covering both types of birefringent filters is given by Ammann [1971]. The main field of application of birefringent filters is the examination of the sun in a narrow spectral band. For manufacture they require large pieces of birefringent crystal, and in future the Fabry–Perot interferometer may supplant them, particularly as techniques improve for producing flat surfaces and uniform coatings.

Fig. 123 Transmittance of a Lyot filter (plain curve) compared with: (*a*) a simple Šolc filter; (*b*) an apodized Šolc filter [Beckers & Dunn, 1965].

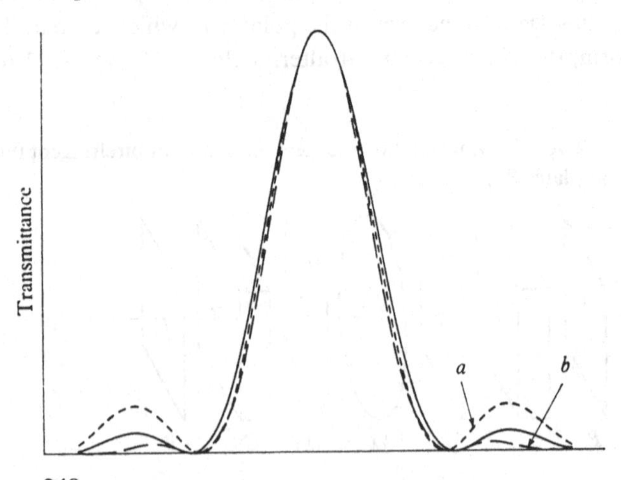

13.6. Sisam

A direct method of spectroscopy based on a two-beam interferometer was introduced by P. Connes [1959, 1960]. He used a collimated Michelson interferometer with the two mirrors replaced by identical diffraction gratings, as shown in fig. 124; prisms can also be used. The two gratings are rotated so that their images, as seen from outside the interferometer, rotated in opposite directions. The instrument is known by the acronym *Sisam* for '*spectromètre interférentiel à sélection par l'amplitude de modulation*' and it has been further described by Girard & Jacquinot [1967].

To give an alternating signal, the delay of the interferometer is changed linearly, for example by a saw-tooth oscillation of the compensating plate. As the gratings rotate relative to each other, the tilt vanishes in turn for each optical frequency that is diffracted directly back by the grating. A change of delay changes the output of this light, which is then fully modulated, but it has a decreasing effect on other radiation, for which there is an increasing tilt as it departs from this frequency. The interferometer acts as a *selective modulator* for this frequency, and this modulation is scanned across the spectrum. The instrument has the resolution of the gratings it uses: a sinc scanning function, if they are rectangular, apodized if they are suitably masked, and this is superimposed on a broad background. It has the solid angle of an interferometer,

$$\Omega = 2\pi/R, \tag{13.35}$$

but can be field widened [Egorova & Kiselev, 1978]. From signal-to-noise

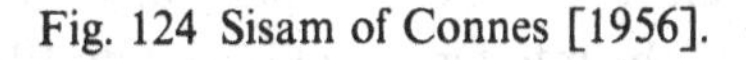
Fig. 124 Sisam of Connes [1956].

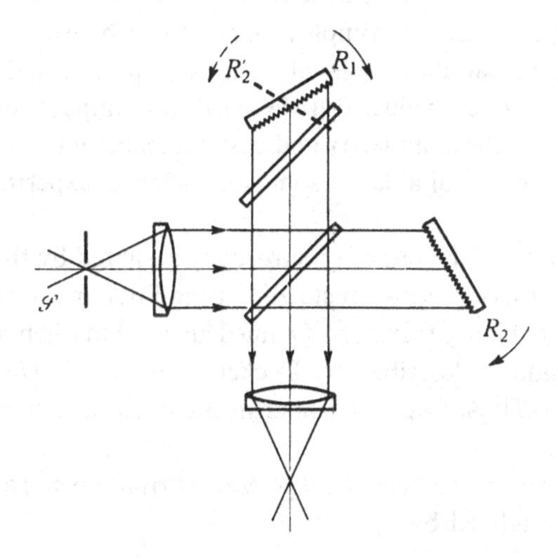

considerations, this instrument is equivalent to the Girard spectrometer. While the method gains by its increased étendue in regions where the limitation is detector noise, most of the gain is lost for photon noise since the large, unmodulated background to the scanning function gives a large increase in noise.

As discussed in §13.2.3, ruling errors in the gratings have a larger effect in the Sisam than in conventional spectrographs: they produce effects proportional to the amplitude rather than the intensity. But if the gratings are tilted slightly so that their directions of dispersion are inclined to the plane of the interferometer, these effects can be reduced practically to intensity effects with little reduction in performance of the instrument.

13.7. Theory of Fourier spectroscopy

Fourier-transform spectroscopy is of recent development and its techniques differ in many ways from those of direct spectroscopy. It is based on the result (6.1) that the interferogram produced by a two-beam interferometer and the spectral distribution of the radiation are Fourier cosine transforms. Its history is given by Loewenstein [1966], and Genzel & Sakai [1977], and it is described by Connes [1961], Mertz [1965], Vanasse & Sakai [1967], Bell [1972], and Vanasse [1977, 1981].

Fourier spectroscopy owes both its strength and its weakness to the need to compute a Fourier transform of the observed results. The operations of filtering out unwanted frequencies from the spectrum and apodizing the scanning function can be included in the mathematical computation, giving the method great versatility. Even the ultimate resolving power can be changed, either by the relatively simple procedure of changing the range of scan or by analytic continuation of the interferogram [Frieden, 1971]. As this is an analytic function in the complex plane, it can be extended by extrapolation, but with a sacrifice of signal-to-noise ratio that makes the possibility of limited practical value. But the need for computation of a transform adds greatly to the complexity and cost of the method and often means that spectra are not available for some time after an experiment is completed.

The multiplex advantage of Fourier spectroscopy is shared by the later technique of *Hadamard spectroscopy.* Instead of using Fourier transforms, this uses the orthogonal binary transforms named after the mathematician Hadamard. The method is described by Decker & Harwit [1974], and Harwit & Sloane [1979]. A dispersing instrument is used; it is not an interferometric method.

For an ideal Fourier interferometer, the spectral distribution $g(\nu)$ and the interferogram $G(\tau)$ are related by

$$G(\tau)=2\int_0^\infty g(\nu)\cos 2\pi\nu\tau\,d\nu, \tag{13.36}$$

$$g(\nu)=2\int_0^\infty G(\tau)\cos 2\pi\nu\tau\,d\tau; \tag{13.37}$$

as both are real, symmetrical functions, cosine transforms may be used. Alternative representations of (13.36) are in terms of the wavenumber σ and path difference p,

$$G'(p)=2\int_0^\infty g'(\sigma)\cos 2\pi\sigma p\,d\sigma, \tag{13.38}$$

or in terms of the reduced frequency $f=v\sigma=\nu v/c$ and recording time t,

$$G''(t)=2\int_0^\infty g''(f)\cos 2\pi ft\,df, \tag{13.39}$$

where v is the speed at which the path difference is being changed. The last expression represents Fourier spectroscopy as a means of reducing an optical frequency ν to a frequency f that can be handled by electronic equipment.

These are expressions for perfect systems that apply when the source and detector have negligible size. Otherwise, the radiation must be integrated over these, and the function $S(\nu\tau)$ of (7.25) and (7.26) included in the theory. As a function of $\nu\tau$, or p/λ_v, these depend on the geometry of the interferometer, as discussed in §7.5.2. A general expression for any interferogram that includes this, as well as other effects that depend on ν and τ separately, is

$$B(\tau)=2A(\tau)\int_0^\infty g(\nu)|S(\nu\tau)|\cos[2\pi\nu\tau+\arg S(\nu\tau)-\psi(\nu)]\,d\nu; \tag{13.40}$$

part of this has been given by Kiselyov & Parshin [1962]. This is no longer a symmetrical function of τ and a cosine transform is not sufficient to compute the spectrum. We can best express the interferogram as a full (exponential) transform by defining for negative ν,

$$g(-\nu)=g(\nu),\quad \psi(-\nu)=-\psi(\nu),\quad S(-\nu\tau)=S^*(\nu\tau).$$

Then

$$B(\tau)=A(\tau)\int_{-\infty}^{+\infty} g(\nu)S^*(\nu\tau)\,e^{i\psi(\nu)}\,e^{-2\pi i\nu\tau}\,d\nu. \tag{13.41}$$

In this expression $A(\tau)$ may represent the rectangular function that denotes that the interferogram is not recorded to infinity but to a finite value T of τ. The corresponding scanning function would then be

$$a(\nu)=\operatorname{sinc} 2\nu T. \tag{13.42}$$

If the scanning function is apodized, $A(\tau)$ denotes the tapering function used. It is a real function since $B(\tau)$ is real.

Intensity terms in ν only have already been included in $b(\nu)$ through the transmittance $\mathscr{T}(\nu)$ of the interferometer. The phase term $\psi(\nu)$ represents an important error because it is this that destroys the symmetry of the interferogram, so that a simple cosine transform is no longer sufficient to give the spectrum. Such an error occurs when the interferometer does not have zero path difference simultaneously for all frequencies. It can occur when the beam splitter and its compensator are not matched, while smaller errors are caused by the asymmetrical phase change on the two sides of a single-coating beam splitter, as discussed in §8.6.1, or by phase shifts in the electronic amplifier. The correction for this error is considered later.

The *source function* $S(\nu\tau)$ includes both amplitude and phase effects and depends on the parameters of the interferometer. Of these, shears and tilts can be regarded as errors of adjustment and can be eliminated by instrument design. As an indication of the precision required, a tilt giving n wavelengths variation of optical path across the plane of observation, or a shear giving the same variation across the source will, by the discussion in §8.1, both reduce the modulation of the interferogram by a factor of $2\mathbf{J}_1(n\pi)/n\pi$, or 0.72 for $n=\frac{1}{2}$. Shifts and leads can be eliminated by the compensation methods discussed in §8.3.1. For an uncompensated Michelson interferometer, the shift is a direct multiple of the delay, and the source functions discussed earlier in (8.8) can be represented as functions of $\nu\tau$ by the relation

$$\zeta=\Omega\nu\tau/\pi; \tag{13.43}$$

thus, for a collimated system

$$S(\nu\tau)=\frac{\sin\frac{1}{2}\Omega\nu\tau}{\frac{1}{2}\Omega\nu\tau}\,\mathrm{e}^{-(1/2)\mathrm{i}\Omega\nu\tau}. \tag{13.44}$$

The dependence of this function on $\nu\tau$ means that the interferometer is no longer a stationary system, since its scanning function varies with frequency. To minimize this variation, the source size, represented by the solid angle Ω, would need to be kept small so that $|S|$ did not fall greatly below unity; fig. 49 has shown the advantage of using collimated light with the interferometer since this represents the case for which $|S|$ decreases most slowly. Otherwise, the system can be treated as linear only over small ranges of ν. For each range, S may be written as a function of the mean frequency of the range, ν_0, and taken outside the integral so that (13.41) becomes

$$B(\tau) = A(\tau)|S(\nu_0\tau)| \int_{-\infty}^{+\infty} g(\nu)\,\mathrm{e}^{\mathrm{i}\psi(\nu)} \exp\left[-2\pi\mathrm{i}\nu\tau - \arg S(\nu_0\tau)\right] \mathrm{d}\nu. \tag{13.45}$$

When, however, the spectrum covers a large range of frequencies, this replacement of (13.41) by (13.45) is no longer valid. The variations of $S(\nu\tau)$ with ν become important: the function S acts as a further factor in the overall transmittance of the spectrometer, and must be taken into account in order to derive the correct spectral distribution $g(\nu)$ from the transform of the interferogram.

When (13.45) applies, the variations of S with τ are the more important. The modulus of S acts as a further taper, and the scanning function is a convolution of $a(\nu)$, due to the applied tapering $A(\tau)$, and $s(\nu/\nu_0)$, due to the effects of source size. For the case of (13.44)

$$s(\nu/\nu_0)\,\mathrm{rect}\left(\frac{2\pi\nu}{\Omega\nu_0}\right), \tag{13.46}$$

and J. Connes [1958] and Parshin [1962] have shown some examples of convolutions of this with unapodized and apodized scanning functions.

As the source function (13.44) becomes negative for $\frac{1}{2}\Omega\nu_0\tau > \pi$, the maximum modulation is obtained in the interferogram when

$$\Omega = 2\pi/\nu_0 T = 2\pi/R; \tag{13.47}$$

a Michelson interferometer has the same relation between solid angle and resolving power as a Fabry–Perot interferometer (13.29) or a Sisam (13.35). For a polarizing interferometer or for the field-widened systems discussed in Chapter 8,

$$\Omega = 2\pi G/R, \tag{13.48}$$

where G is the gain factor discussed there. These limits to the source size represent a source that extends from a central bright source fringe to the next bright fringe. This does not represent directly a phase variation from 0 to 2π, for the extra phase term $\arg S(\nu_0\tau)$ converts it to a variation from $-\pi$ to π.

This phase term is a second effect of source size and it acts as a distortion of the frequency scale of the spectrum. When the radiation is collimated, the distortion is linear and the exponential in (13.45) can be replaced by

$$\exp[-2\pi i\nu\tau(1-\Omega/4\pi)],$$

the scale of either ν or τ is multiplied by the constant $(1-\Omega/4\pi)$. If the monitoring fringes used to measure the path difference of the interferometer are derived from a beam of the same solid angle the error is automatically corrected, not only when it is linear, but also when it is the non-linear error of a non-collimated interferometer or an error of the mechanical drive.

Other geometrical effects that influence the interferogram are errors in the optical components and these can be included in the theory by the extension of the van Cittert–Zernike theorem, discussed in §6.4.1. These errors affect only the visibility of the interferogram and hence the efficiency of the instrument; they do not change the scanning function, as do similar errors in a dispersing spectrometer.

Two variants of the usual procedures of Fourier spectroscopy have been given by Mertz [1965]. The first is a *heterodyne* method, in which the interferogram is multiplied by a cosine function of some reference frequency ν_0. This gives

$$H(\tau)=2G(\tau)\cos 2\pi\nu_0\tau, \tag{13.49}$$

of which the transform is

$$h(\nu)=g(\nu+\nu_0)+g(\nu-\nu_0); \tag{13.50}$$

the required distribution g now appears as two sidebands at the sum and difference frequencies. The high frequencies can be removed by filtering $H(\tau)$, either electrically or mathematically, to leave a function whose transform is $g(\nu-\nu_0)$. This filtered function is much smoother than the original interferogram $G(\tau)$, since it has only low-frequency components.

In the second modification, the optical path, and hence the delay, has a sinusoidal modulation superimposed on a continuous drive:

$$p(t)=p+\lambda_0\cos\omega t, \quad \tau(t)=\tau+\frac{1}{\nu_0}\cos\omega t. \tag{13.51}$$

The interferogram obtained is then

$$G_m(\tau)=\int_0^\infty g(\nu)\cos 2\pi\nu\left[\tau+\frac{1}{\nu_0}\cos\omega t\right]\mathrm{d}\nu,$$

and the integrand can be expanded as a Jacobi series of Bessel functions. If the recorder accepts only the fundamental frequency ω, the result obtained is

$$G_m(\tau)=2\int_0^\infty \mathrm{J}_1(2\pi\nu/\nu_0)g(\nu)\sin 2\pi\nu\tau\,\mathrm{d}\nu. \tag{13.52}$$

This method involves sine transforms and gives the spectrum multiplied by a filtering function $J_1(2\pi\nu/\nu_0)$ that vanishes at $\nu=0$ and $\nu=0.61\nu_0$, or $\lambda=\infty$ and $\lambda=1.64\lambda_0$. The amplitude of $1/\nu_0$ of modulation should be chosen to suit the spectral range.

Mertz [1958] has used this method to correct for errors due to atmospheric turbulence and the Connes [1966] have further demonstrated its value for this. For, as they explain, the intensity fluctuations caused by turbulence are the present limiting factor when Fourier spectroscopy is applied to astronomy. Some of the fluctuations are those due directly to intensity variations in the image, the scintillation; other indirect effects follow from the image movement. Since the radiation is received by a detector, the sensitivity of which varies from place to place, image movement also causes fluctuations of the signal, even when the pupil and not the image itself is focused on the detector. Similar fluctuations can be caused by relative movements within the interferometer.

The fluctuations are proportional to the signal and, if allowed to become the main source of noise, the multiplex advantage would operate in reverse. Hence the importance of correcting for them. Since the background contains much more radiation than the interferogram, the first step is to eliminate the background, and its fluctuations, by recording the difference of the two complementary interferograms, the interferometer being designed to match these as closely as possible [Zehnpfennig *et al.*, 1979]. A next step is to record, not the interferogram itself, but a ratio, the interferogram divided by the background intensity. Finally the beam of radiation can be diffused to reduce the effects of image movement. An integrating sphere, although the best diffuser in theory, would cause a large loss of light and would greatly increase the étendue of the beam; the Connes [1966] have therefore used a light pipe. With these refinements, they have obtained spectra of Venus and Mars with a resolving power of 10^4. With a change to Mertz's technique of a modulated delay, they later obtained spectra of Jupiter with three times this resolving power.

Mertz [1967] has also described a method of reducing the effects of scintillation that makes use of a rapid scan in which the frequencies from the interferometer are much higher than those of the scintillation. To obtain a reasonable signal-to-noise ratio, a series of interferograms is taken and added together. Sakai [1977] has also discussed scintillation noise and how to reduce it, and Guelachvili [Vanasse, 1981] has given a review of causes of errors in Fourier spectroscopy.

Rapid scans are also needed when spectra are required that are changing with time. *Time-resolved Fourier spectroscopy* has been studied by Sakai & Murphy [1978].

13.8. Dispersive transforms and phase spectra

The spectral transmittance of a sample can be measured by Fourier spectroscopy by placing the sample between the source and the detector, but outside the interferometer. If it is placed inside, in one arm of the interferometer, the ratio of the Fourier transforms of the two interferograms obtained, with and without the sample, is the complex amplitude transmission factor. The two interferograms are either taken sequentially or at the same time by a double-beam method. The modulus of the result gives the absorption of the sample and hence the imaginary part of its refractive index, and the phase gives the real part. This is the method of measuring the variations of refractive index with frequency that was referred to in Chapter 10. The method is due to Chamberlain, Gibbs & Gebbie [1963], and Bell [1966], and is reviewed by Birch & Parker [1979], and Chamberlain [1979]. Although a Fourier spectrometer is usually used, Kolesov & Listvin [1979] have used a Sisam.

If a specimen of refractive index $n(\nu)$ and path length d, which is twice its thickness when double passage is used, is placed in one arm of an interferometer, the phase difference is changed by

$$\psi(\nu) = 2\pi\nu d[n(\nu) - 1]/c, \tag{13.53}$$

and the interferogram obtained is

$$B'(\tau) = \int_{-\infty}^{+\infty} g(\nu) \exp \mathrm{i}[-2\pi\nu\tau + \psi(\nu)] \, \mathrm{d}\nu.$$

Transformation of the interferogram gives

$$g(\nu)\, \mathrm{e}^{\mathrm{i}\psi(\nu)} = \int_{-\infty}^{+\infty} B'(\tau) \exp 2\pi \mathrm{i}\nu\tau \, \mathrm{d}\tau. \tag{13.54}$$

Both are double-sided exponential transforms. If, in addition, the specimen is absorbing, it has a complex refractive index $\hat{n} = n + \mathrm{i}k$ and the result of the transform (13.54) is then

$$t(\nu) g(\nu)\, \mathrm{e}^{\mathrm{i}\psi(\nu)} \tag{13.55}$$

where

$$t(\nu) = \exp 2\pi\nu d k(\nu)/c,$$

k being negative. The spectrum $g(\nu)$ is the transform of the interferogram obtained with no specimen in the interferometer.

In practice, the transform (13.54) is taken over a finite range of τ, centred about the interferogram maximum. When the specimen is inserted, the position of this maximum shifts, and it is convenient to take a new centre τ_2, which is different to the centre τ_1 for no specimen. Then the phase obtained from the ratio of the two transformed interferograms is no longer $\psi(\nu)$, given by (13.53) but is

$$2\pi\nu\{d[n(\nu)-1]/c+\tau_2-\tau_2\}. \tag{13.56}$$

From this phase it is possible to find $n(\nu)$, provided d and the difference between centres $\tau_2-\tau_1$ are known; it is not important that the centres chosen should be the exact centres of the interferograms.

This is the phase index. The group refractive index can also be obtained from the derivative of the results from the relation

$$\mathrm{d}\psi(\nu)/\mathrm{d}\nu=2\pi d n_g(\nu)/c. \tag{13.57}$$

For this it is no longer necessary to know the difference between the centres.

The preceding method has used interferograms obtained by varying the delay τ of an interferometer with a sample in one arm and a broad-band source of radiation. If monochromatic radiation is used instead and its frequency varied, for example, by a scanning monochromator, the resulting output still has the appearance of an interferogram [Kerl, 1979], but it now shows directly the variation of the amplitude transmission factor with frequency. The interferometer is again a convenient instrument to convert this to a variation of intensity. This technique may be regarded as a refinement of the use of channelled spectra.

13.8.1. Phase correction

When it is only the intensity spectrum $g(\nu)$ that is required, the phase term in (13.54) prevents the use of a simple cosine transform to obtain it. But it is very difficult to avoid all phase errors and, when one is present, full (double-sided) exponential transforms must be taken and the measured spectrum is the modulus, obtained as the square root of the sum of the squares of sine and cosine integrals of $B(\tau)$. This is not a linear process and the result cannot be expressed as the convolution of a single scanning function and the true spectrum. This method of overcoming a phase error is due to Connes [1961], who showed that the non-linearities can have a serious effect on the signal-to-noise ratio, the noise increasing in regions where the signal decreases. In addition it is no longer possible to identify negative regions of the spectrum, which occur when a thermal detector is radiating to the source. The non-linearity can be almost entirely eliminated by the use of the additional information that the phase error $\psi(\nu)$ is a smooth, slowly varying function, requiring a very small number of Fourier components to specify it. Then ψ can be computed from two-sided transforms of a short section of interferogram, either a small extension of the actual interferogram to negative values of τ or a short two-sided interferogram taken separately with a high-energy source. At such low resolving powers, the signal-to-noise ratio is very high and little extra noise is introduced.

The phase ψ' so calculated can be used to obtain the spectrum from (13.54), either by multiplication [Mertz, 1965], as

$$g(\nu) = g(\nu)\, e^{i\psi(\nu)}\, e^{-i\psi'(\nu)}, \tag{13.58}$$

or by convolution of its transform with the interferogram [Forman, Steel & Vanasse, 1966], as

$$g(\nu) = \mathscr{F}_c^{-1}\{B(\tau) * \mathscr{F}[e^{-i\psi'(\nu)}]\}, \tag{13.59}$$

the spectrum being obtained by a simple cosine transform of the corrected interferogram. In both cases, no non-linear operation is used. Before the introduction of fast Fourier transforms, the second method reduced computing time, but the gain is now less significant.

In all cases where full transforms are required, Mertz [1965], and Sakai & Vanasse [1966] have shown that these should be taken for interferograms that are centred. But these interferograms are not symmetrical, since there is no value of τ for which all frequencies are in phase. The centre is taken, therefore, at the value of τ for which the phase difference between the arms is stationary for frequencies at the middle of the spectral range [Mertz, 1965; Sakai & Vanasse, 1966]. In the extreme case of an interferogram that is not centred, that of a one-sided exponential transform, the relation (2.41) shows that the result obtained would be the convolution of a single scanning function with a complex function, the real part of which is the spectrum and the imaginary part the Hilbert transform of the spectrum. Some of the Hilbert transform of the spectrum will be included in any Fourier transform calculated over an unsymmetrical range of τ, but now the two parts, the spectrum and its Hilbert transform, are each convolved with a different scanning function. Examples have been given by Steel & Forman [1966]. There is also a phase error associated with the Hilbert transform, discussed by Mertz [1965], and this increases linearly as the centre is displaced.

When phase errors of the interferometer are present, the centre of the interferogram changes with frequency and the range of τ cannot be symmetrical for all frequencies. The use of only a small section of the interferogram to compute the error ψ increases the effect of these asymmetries. The method of phase correction is therefore accurate only for small errors. If the error is large the computed value of ψ' will include an additional phase term due to the Hilbert transform and one convolution will not yield a completely symmetrical interferogram. The technique can be repeated, however, until the interferogram is symmetrical.

Under some circumstances, the deliberate introduction of a known phase error into an interferometer can be advantageous. An interferogram may cover an entirely different range of intensities to the corresponding spectrum. Its maximum value, at $\tau = 0$, is the integral of the whole spectrum and, particularly for absorption spectra, this may be very large in comparison with all other values. The detector therefore may be required to cover a larger dynamic range than is necessary in direct systems, although the ultimate sensitivities required may be similar.

The dynamic range can be reduced by filters that limit the spectral range, or by *chirping* the interferogram by a deliberate introduction of a phase error [Mertz, 1965; Sheahen, 1974, 1975]. The central maximum and hence the dynamic range are considerably reduced, as shown in fig. 125. The phase error is produced by having two dispersive media in the interferometer, so that both the shift and the delay at one frequency can be made zero, but a change of delay with frequency remains. To analyse a chirped interferogram, the same technique of phase correction is used, only now the phase is a function that is known from the construction of the interferometer.

13.9. Methods of computation

The spectrum is obtained from the recorded interferogram by either analogue or digital computation. Analogue methods record the interferogram as a continuous function and give the integral (13.37)

$$g(\nu) = 2\int_0^\infty G(\tau)\cos 2\pi\nu\tau \, d\tau$$

directly for a range of frequencies ν. For digital computation the interferogram is sampled at a finite interval $\delta\tau$, given by (13.11), and the

Fig. 125 An interferogram (*a*) with a deliberate phase error (chirping), and (*b*) with no phase error [Mertz, 1965].

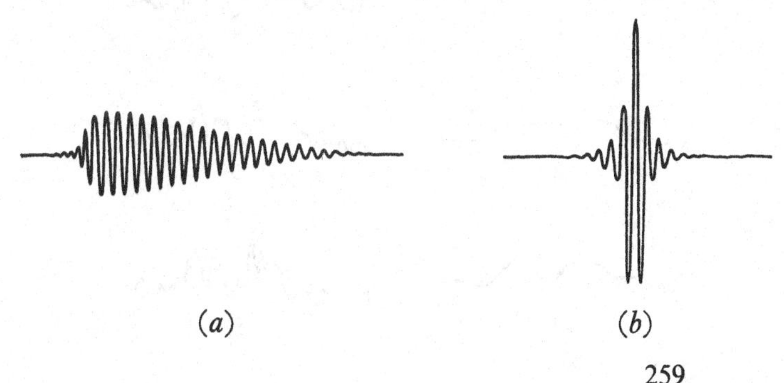

integral is replaced by the series

$$g(\nu) = G(o) + 2 \sum_{n=1}^{\infty} G(n\,\delta\tau) \cos 2\pi\nu n\,\delta\tau. \tag{13.60}$$

Two distinct methods of carrying out either computation are possible. The first is Fourier *analysis* where the interferogram is recorded and stored. Each value $g(\nu)$ of the spectral distribution is then found in turn. The second method is that of Fourier *synthesis* where again $G(\tau)$ is sampled, each value taken in turn and multiplied by the appropriate cosine factors, and the result stored as a function of frequency. As more of the interferogram is taken, the function in the store approaches progressively the spectrum $g(\nu)$: an example is shown in fig. 126, taken from the digital computer of Yoshinga [1965], while Hoffman & Vanasse [1966] have developed an analogue method of synthesis. Both analysis and synthesis require the same series of multiplications by cosines, although in a different order; they differ in the function that is stored. Although analysis is more commonly used, synthesis has the advantage that it can compute the result in real time as the interferogram is being recorded.

When the interferogram is not recorded continuously but sampled, it is desirable that each sample should represent an integrated value over as

Fig. 126 Progressive development of the spectrum as more of the interferogram is used [from Yoshinaga, 1965].

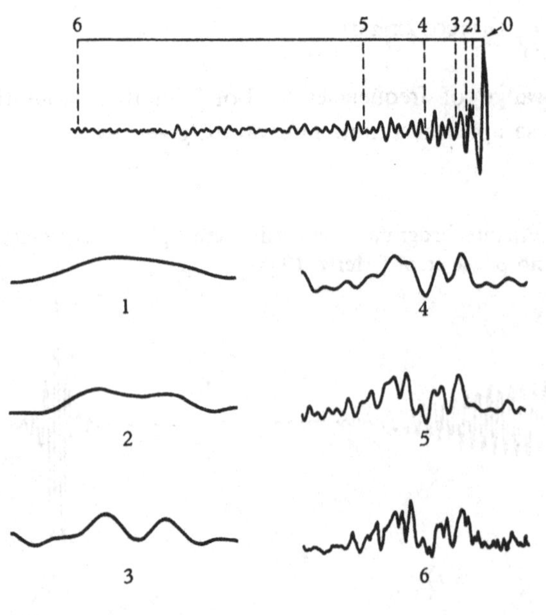

long a time of observation as possible. The Connes [1966] have used a *stepped record* in which the delay is increased by the amount $\delta\tau$ as rapidly as possible, since this represents wasted time, and then kept constant for a finite observing time over which the signal is integrated. As detector noise increases considerably at low frequencies, it is usual to record the interferogram at audio frequencies, either by chopping the incident radiation, or by changing the delay very rapidly, so that the frequencies of the cosines in (13.39) lie in this range. For such a rapid scan [Mertz, 1965], the time of observation is usually too short, so several such interferograms are taken in succession and added together.

The simplest analogue method of Fourier analysis is that of Mertz [1965] who recorded the interferogram on a magnetic wire or tape and transformed it by a wave analyser. A more precise analogue analyser, suitable for $R \approx 10^4$, has been made by Pritchard *et al.* [1967].

Although there is no sampling of the interferogram in analogue analysis, the spectrum need be computed only at the intervals $\delta\nu = 1/2T$ given by the sampling theorem. To reduce noise, the results should be filtered to reject any optical frequencies lying outside the range of interest $\Delta\nu$. For a rapid scan, this is equivalent to using an electrical filter that transmits a band Δf where, as in (13.39), $f = \nu v/c$, optical frequencies ν having been effectively changed to electronic frequencies f. In digital computation, such filtering is usually done numerically.

When the transform is computed by digital methods, the interferogram also is sampled. If the detector is sensitive up to a maximum frequency ν_d, the interferogram contains noise frequencies up to this value and should initially be sampled at intervals not larger than $\delta\tau = 1/2\nu_d$. After this noise is removed by mathematical filtering, the results need be retained only for intervals $\delta\tau = N/2\nu_m$, as in (13.12). Mathematical filtering of the noise [Connes & Nozal, 1961] is given by convolution of the interferogram with the transform of the filter function desired, for example a rectangular function covering the range $\Delta\nu$; Vanasse & Sakai [1967] give practical details. In the same operation, the phase correction can be included and, if desired, any correction for the transmittance $\mathscr{T}(\nu)$ and the detector sensitivity $d(\nu)$. Then the interferogram would be convolved with the transform of

$$f(\nu)\,\mathrm{e}^{-\mathrm{i}\psi'(\nu)}/[\mathscr{T}(\nu)d(\nu)], \tag{13.61}$$

where $f(\nu)$ is the desired filter function.

A cosine transform then completes the computation. Initially the required cosines were computed by the Chebyshev formulae. For m spectral elements, this required of the order of m^2 operations, and computing time became a major limitation. For resolving powers of 10^6, times of 100 years were estimated. This has now been reduced to well under an hour by the introduction of the fast Fourier transform. This has been developed by Cooley & Tukey [1965], and introduced into Fourier spectroscopy by Forman [1966]. It reduces the number of operations required to the order of $m \log_2 m$. Sakai [1977] discusses the changes required to adapt their algorism to suit this application.

13.10. Interferometers for Fourier spectroscopy

The design of a Fourier spectrometer calls on several of the results of earlier theory. A systematic approach has also been given by Breckinridge & Schindler [Vanasse, 1981]. It requires an interferometer with a variable delay, in which this variation introduces no accidental shears and tilts, since the full étendue obtainable is usually needed to collect enough radiation. The spectrometer is commonly used in the infrared, where optical surfaces need not be as good as for visible-light interferometers, but where fewer transparent materials are available for beam splitters. Refractive indices of coating materials are high and a single coating on a beam splitter is usual. For longer wavelengths, stretched plastic sheets of Melinex or Mylar [Gebbie, 1961] are used, and these are symmetrical. Metal grids and meshes are also used, the former acting as a polarizing beam splitter [Costley *et al.*, 1977; Challener *et al.*, 1980].

13.10.1. Michelson interferometers

The interferometer most commonly used for Fourier spectroscopy is the Michelson interferometer with the effective source at infinity. It corresponds then to the Twyman–Green interferometer used for optical testing although it differs greatly in mechanical design. While the Twyman–Green interferometer is traditionally made as a series of components spread out on a baseplate, Fourier spectrometers are made as compact, rigid cubes [Mertz, 1965]. For the shorter infrared wavelengths, the beam splitter is a coating, usually of germanium or silicon, on a plate of calcium fluoride, barium fluoride, or rocksalt. As cements transparent for this region are not available, and it is difficult (though not impossible) to make surfaces on these materials that are flat enough for optical contact, the compensator is usually separate. This gives the asymmetries discussed in Chapter 8 and hence a phase error.

The light reflected from the second surface of the beam splitter support can interfere with the main beam to produce, superimposed on the spectrum obtained, the sinusoidal variations of a channelled spectrum. To avoid this error, the plate should be wedged so that the unwanted reflexion does not reach the detector, or it should be made thick compared with the maximum path difference of the interferometer. In the latter case the interferometer, acting as an interference spectrometer, will not resolve the channelled spectrum. Weaker channelled spectra may also be formed by interference between a pair of unwanted reflexions and these may be avoided by the same methods.

The movement of the component that changes the optical path is usually by a magnetic motor, a solenoid moving in a permanent magnetic field against viscous friction. A screw drive is also used. Any irregularities in the drive act in the same way as errors in a grating-ruling machine, but with the enhanced effect common to interferometers. For any but very low resolving powers, a separate measurement of the delay in the interferometer is necessary. This can be made by recording, as well as the interferogram, the fringes given by a separate monochromatic source or by using these fringes to control the delay. The fringes may come from the same interferometer, with chromatic beam splitters used to separate this light from the radiation being studied, or with the aperture of the interferometer shared by the two beams. Alternatively, a separate interferometer may be used, rigidly connected to the first. For the far infrared, where less precision is necessary, the mirror movement is usually recorded by moiré fringes.

To avoid tilts and shears, reliance is usually placed on the quality of the slide carrying the moving component and Gebbie, Habell & Middleton [1962] have used optically worked glass slides. Parallelogram mountings have also been used.

An alternative approach is through servo control of the moving mirror. Parsche & Luchner [1973] have controlled the tilt by monitoring, on a position-sensitive detector, the angle at which a beam from a laser is reflected. The method used by Ramsay [1962] for Fabry–Perot interferometers can be adapted to a Michelson. A beam of white light passes in turn through both sides of the interferometer and the fringes of superposition used to keep these paths equal.

When a monochromatic source is used to monitor the delay, three beams can be used to control tilt as well. Three sets of fringes are kept in phase. Such a method combines well with the step method of recording, suitable for digital computers, and the Connes [1966] have incorporated servo control of the position at each step. The time for which the interferometer rests at each step is controlled by the total radiation received so that the instrument acts as a ratio recorder, giving the ratio of the interferogram to the background radiation.

Since it is not usually possible to limit the size of the source itself to the required value, the limiting aperture follows the interferometer, being placed at the focus of a lens or mirror if a collimated interferometer is desired. A second optical system is placed near this aperture to focus the exit pupil of the interferometer on the surface of the detector with the maximum useful solid angle. As this is effectively increased by a factor of n^2 by immersing the detector in a medium of refractive index n, detectors are sometimes made as a coating on the back of a lens of high index.

The tilt-compensated interferometers of §8.3.2 are also used. The combination of a retroreflector and plane mirror becomes inconveniently long when a cat's-eye is used, and the use of both sides of a moving mirror and the geometry of a Möbius band also becomes complicated when corrected for polarization, as shown in fig. 127. This last instrument is freer from the effects of vibration than uncompensated interferometers, since the moving mirror is generally the component most susceptible to vibration, but it has proved to be unduly sensitive to variations in temperature.

13.10.2. Michelson with retroreflectors

Retroreflectors are less sensitive than plane mirrors to errors of the drive that changes the delay, and the arrangement of fig. 128 is often used, particularly for high resolving powers [Connes & Michel, 1975; Sakai, 1977]. If the beam enters the retroreflectors off centre, a phase error due to the beam splitter can be avoided, and the two emergent beams can be taken to separate detectors [Fellgett, 1958]. As the two interferograms can be

Fig. 127 A Möbius-band interferometer, corrected for polarization [Pritchard *et al.*, 1967].

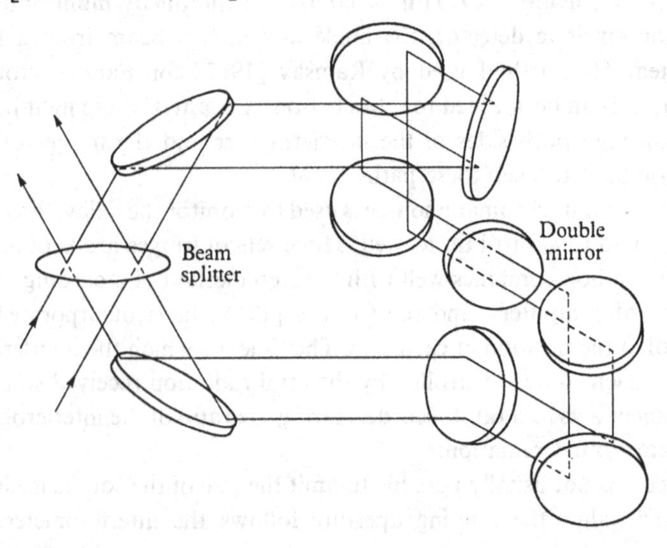

made complementary, their difference has double the modulation, while the background is eliminated. Roland [1965] has used such an instrument with trihedral mirrors as retroreflectors, but the Connes [1966] have used cat's eyes to avoid polarization effects.

The sensitivity of the retroreflectors to lateral displacement can be overcome by connecting the two together, now with no twist of the beams. The retroreflectors can be mounted back to back, as suggested by Erikson [Steel, 1971].

13.10.3. Field-widened systems

To give delays without shifts, the compensation methods of Chapter 8 are used to produce field-widened Michelson systems; these were developed initially for this application. They become important for resolving powers greater than about 2000. Baker [1977] has given a review of the systems used. The method using telescopes is not commonly used, since, in the infrared, mirror telescopes would be desirable, and the arms of the interferometer become very long. In general, it is the size of the beam splitter that is the practical limitation, although the size of the end mirrors determines the étendue of the interferometer. If the arms are too long, a theoretical gain in étendue due to an increased solid angle may be nullified by the reduction in the effective area of the end mirrors. Of the wedge methods, that of Bouchareine & Connes [1963], shown in fig. 52(*a*), is the more convenient and, for a prism angle θ, the gain in solid angle is, by (8.29),

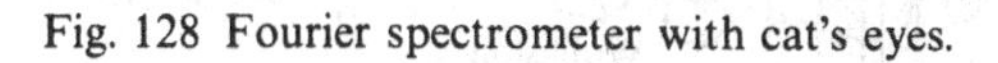

$$G = 2\cot^2\theta.$$

The optical path introduced depends on the refractive index of the prisms, being proportional to $n - 1/n$. Since the refractive index depends on frequency, the spectrum is obtained as a function not of ν but of $(n^2 - 1)\nu/n$.

Fig. 128 Fourier spectrometer with cat's eyes.

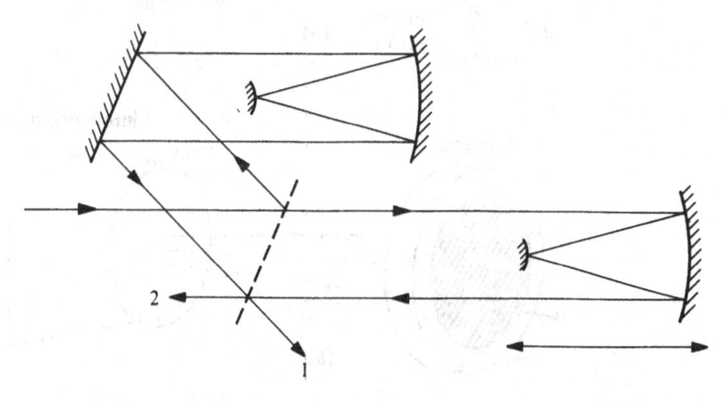

For interferometers with cat's eyes, the curvature of the convex mirror can be changed as the cat's eye is moved [Cuisenier & Pinard, 1967; Adelman, 1977].

13.10.4. Polarizing and diffraction forms

The birefringent equivalent of a Michelson interferometer is the Soleil compensator. This has been used for Fourier spectroscopy by Mertz [1958], and Sinton [1963] has used the more complex form shown in fig. 129 in which both interferograms are collected in both polarizations so that the usual losses in polarization interferometers are avoided. The interferometer is made of rutile and can be used up to $\nu = 75$ THz or $\lambda = 4\ \mu$m. Polarizing interferometers for the far infrared are made with metal-grid beam splitters. In the form due to Martin and Puplett [Lambert & Richards, 1978], shown in fig. 130, the direction of polarization is rotated through $\frac{1}{2}\pi$ by a pair of roof mirrors in each arm, set at $\pi/4$ to the plane of the instrument. These are coated so as to produce little phase difference between orthogonal linear polarizations.

Diffraction forms have been described by Lohmann [1962*b*] and Fonck *et al.* [1978]. Another is the 'mock interferometer' of Mertz, Young &

Fig. 129 Other forms of Fourier spectrometers: (*a*) polarization form of Sinton [1963]; (*b*) diffraction form, the mock interferometer of Mertz, Young & Armitage [1962].

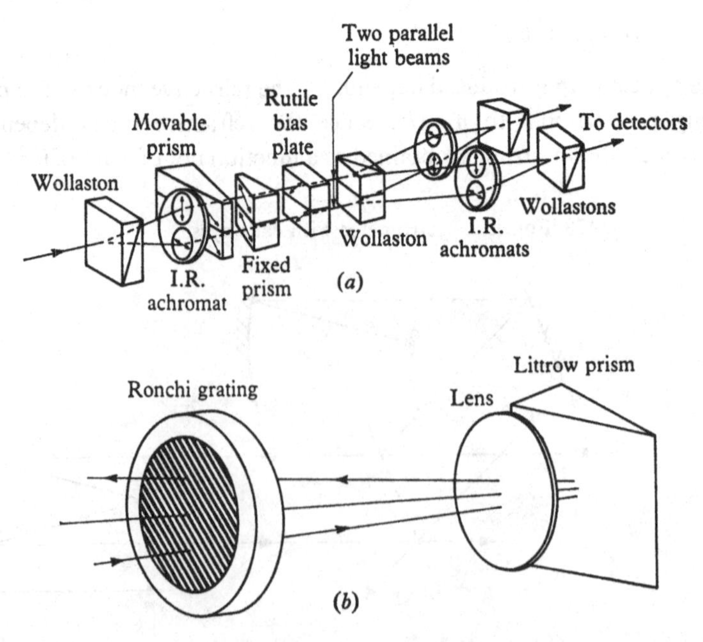

Armitage [1962], also shown in fig. 129. This latter is closely related to the Girard spectrograph only the grilles now used, equally spaced clear and opaque straight bands, have a sinusoidal rather than a peaked autocorrelation function and thus produce a Fourier spectrometer rather than a direct one. The two grilles may be on the same disk, which is rotated to record an interferogram. As the delay of the interferogram depends on the sine of the angle, a non-uniform rate of rotation is required. This instrument has been studied further by Selby [1966], and Ring & Selby [1966].

Other non-interferometric multiplex systems such as Hadamard spectrometers, are similar in principle but use different grilles.

13.10.5. Lamellar grating

All these interferometers have employed division of amplitude. One of the earliest Fourier spectrometers, the lamellar grating of Strong & Vanasse [1958, 1960; Henry & Tanner, 1979], uses division of wavefront. Essentially it consists of two plane mirrors, each broken up into strips, so that one mirror can pass through the plane of the other, as shown schematically in fig. 131. It has not the radial symmetry of the other instruments and must be used with a slit as effective source, narrow enough so that adjacent strips, one on each mirror, are coherently illuminated; this requires a change of slit width as the wavelength is changed. The length of the slit is equal to the diameter of the circular source of an equivalent Michelson interferometer.

Fig. 130 The Martin–Puplett polarizing interferometer for the far infrared: a view looking down at 45° into the roof reflectors.

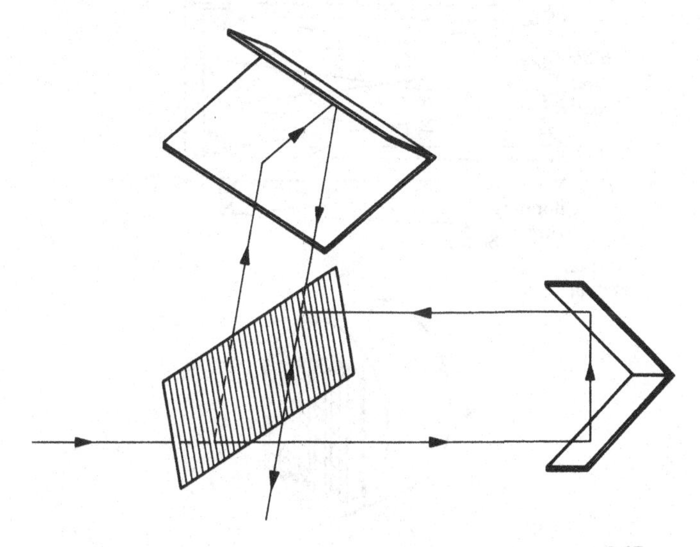

The loss in étendue is partially offset by the higher transmittance, for there is no beam splitter to cause losses. The theory given by Vanasse & Sakai [1967], and an experimental comparison by Richards [1964] have shown that the lamellar grating is more efficient than a Michelson interferometer in the far infrared ($\nu < 3$ THz, $\lambda > 0.1$ mm approximately). Its range of usefulness is much larger than that of a Michelson, whose beam splitters need changing to suit each region. The range of the lamellar grating is limited at high frequencies by the optical quality of the mirrors, at low frequencies by the width of the facets. When the grating is made of metal and the facets become narrow compared with the wavelength of the radiation, the delay depends not on the velocity in free space but on the cavity velocity in the slots, which act as waveguides.

13.10.6. Sims

The lamellar grating sacrifices étendue, but retains the multiplex advantage. The converse applies to an instrument, suggested by Prat, and developed by Fortunato & Maréchal [Esplin, 1977], known as *Sims*, '*Spectromètre interférentiel à modulation sélective*'. This is an interferometer with a tilt, which, with a polychromatic source, produces an interferogram of fine fringes. These have high contrast only at the central white-light fringe.

Fig. 131 The lamellar grating of Strong and Vanasse [1958].

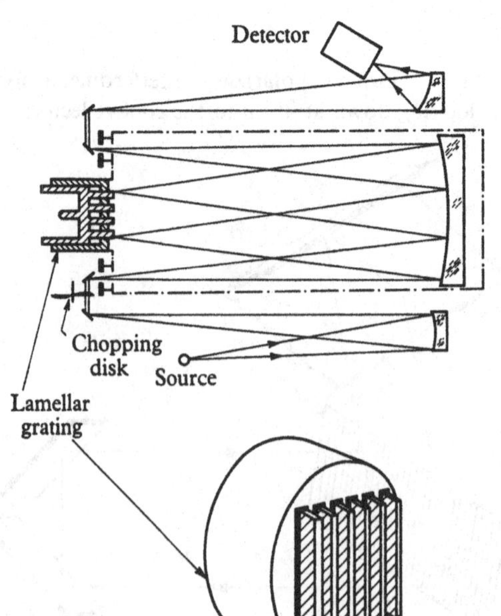

If this interferogram is scanned by a sinusoid or a grid, the transmitted radiation is modulated by an amount proportional to the intensity of the spectral component that gives a fringe spacing equal to that of the grid. Thus the whole spectrum can be covered by changing the period of the grid, or, more conveniently, changing the tilt of the interferometer.

The instrument has the wide field of an interferometer and so has a performance equivalent to a Girard spectrometer.

13.11. Fringe visibility and correlation interferometry

Although Michelson used the interferometer now named after him for interference spectroscopy, he did not measure the full interferogram but only its envelope, the visibility. Provided the spectral distribution $g(\nu)$ that is being studied covers only a narrow spectral range, fringes can be observed over a reasonable range of τ. If ν is written as ν_0+v, (6.18) and (6.24) give

$$\gamma_{11}(\tau)=\mathrm{e}^{-2\pi i\nu_0\tau}\int_{-\infty}^{+\infty}\tilde{\mu}(v)\,\mathrm{e}^{-2\pi iv\tau}\,\mathrm{d}v, \tag{13.62}$$

where $\tilde{\mu}(v)=g(\nu)/G(o)$, the normalized spectral distribution, now treated as a function of v.

If the original spectral distribution is symmetrical about $\nu=\nu_0$, $\tilde{\mu}(v)$ is symmetrical about $v=0$ and the integral in (13.62) is real. When the interfering beams have equal intensity, the visibility is given by $m=|\gamma|$ and

$$m(\tau)=\pm\int_{-\infty}^{+\infty}\tilde{\mu}(\nu)\,\mathrm{e}^{-2\pi iv\tau}\,\mathrm{d}v, \tag{13.63}$$

and the inverse transform gives the spectrum. Only the difference frequencies $v=\nu-\nu_0$ are involved and it was for this reason that Michelson used the method: his facilities for performing Fourier transforms could cope with these but not with the higher frequencies required to transform the interferogram. He successfully used the method to select the red cadmium line as the most useful line as a length standard, and this line was not bettered until artificial isotopes became readily available. The method of visibility is still used in this field [Terrien, 1960], while Michel [1967] has constructed an interferometer with six fixed delays to find, from the visibility, the profiles of lines emitted by plasmas.

The method is applicable, however, only when it is known that the spectral line is symmetrical, otherwise (13.63) no longer holds and the phase as well as the modulus of $\gamma_{11}(\tau)$ is needed to find the spectral distribution. Separate measurement of the phase has been used by Rowley & Hamon [1963] to study the orange line of krypton 86. In essence, the amount of information obtained is that given by a correctly sampled interferogram.

Fig. 132 Method of measuring spectral profiles by correlation interferometry, proposed by Mandel [1963*a*].

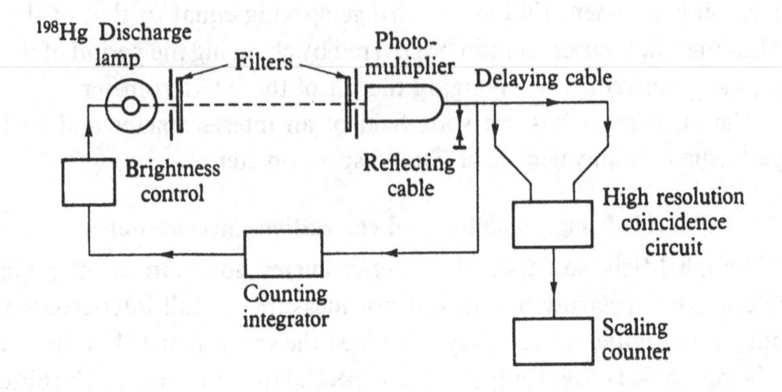

When the phase is not measured, the discussion of §6.9 shows the possibility of obtaining the phase from the modulus. The practical applications of this possibility are less important in simple interferometry, where the interferogram rather than its visibility can usually be recorded, than they are in correlation interferometry. There the result obtained is proportional to the square of the visibility, with no phase information.

In correlation interferometry, the delay may be in the propagation either of the radiation or of the electrical signals derived from it. In the first case, two detectors are used at different distances from the source of radiation and the delay arises in the same way as in a classical interferometer. In the second case, a single photo-electric detector is used and the autocorrelations are taken of the fluctuating signal with different delay lines in the equipment. The method, as proposed by Mandel [1963*a*], is shown in fig. 132, and Martienssen & Spiller [1964] have reported an experiment of this type. In either case, correlation interferometry is limited to very high resolving powers. The relaxation time for the emission of photo-electrons from present photocathodes is of the order of 10^{-8} s, and to this must be added the variations in transit time of the electrons from the cathode, via the amplifying dynodes of a photo-multiplier, to the anode. A photomultiplier will therefore not distinguish time intervals shorter than $\delta\tau \sim 10^{-8}$ s, and any autocorrelation function would be sampled at intervals greater than this. This would give a free spectral range $\Delta\nu = 1/2\delta\tau$ of 50 MHz or $\Delta\lambda = 5$ pm at $\nu = 545$ THz ($\lambda = 550$ nm), while the resolution limit would be small compared with this. Correlation methods are reviewed by Pike [1976], and Gebbie & Twiss [1966].

14

Interference imagery

The study of the spatial distribution of radiation implies the formation of some form of image of the distribution. When an interferometer is used, we may speak of *interference imagery*, and of *Fourier imagery* for two-beam interferometers, where the 'image' is the Fourier transform of the distribution. This is a two-dimensional transform, the reciprocal coordinate being the vector shear.

14.1. Survey of applications

As a technique, interference imagery has not the advantages over classical methods that give interference spectroscopy its great importance. Optical image-forming systems are now highly developed instruments, combining high resolving power and a large field of view. The change from the one dimension of spectroscopy to two dimensions means that, to compete with conventional systems, an interferometer must handle a very large amount of information: the number of picture elements involved is very much greater than the number of spectral elements encountered in spectroscopy. This has meant that computing capacity has been a greater limitation to the development of Fourier imagery than it was to Fourier spectroscopy. But with the continuing improvement of the speed and capacity of computers and of computing methods, and with the introduction of specially designed computers [Frater & Skellern, 1978], digital image processing is an established technique, and Fourier imagery is now a practical alternative to conventional image-forming systems.

In many problems, however, the large amount of information in a conventional image is not needed, and interference methods of measurement can often be designed to give only the desired information. These interference methods may be more efficient than a classical image-forming system, sending all the energy into the results desired, or they may be simpler. A radio interferometer with a dilute aperture, across which relatively few receptors are spread in a cross or circular array, has the same resolving power as a huge parabolic mirror that fills the whole aperture, out to the edge of the array. Interferometers are used in radio astronomy because it is impossible to build paraboloids large enough to give the desired resolution. Similar practical reasons dictate the use of interferometers at visible frequencies to measure stellar diameters.

Other applications of interference imagery follow from particular properties of interferometers. Since a two-beam interferometer measures Fourier transforms, this instrument is used when it is not the intensity distribution but its transform that is of interest.

14.2. Radio astronomy

Interference imagery has developed much more slowly with visible radiation than it has with radio waves, and radio astronomy provides the best illustration of the technique. Because the radio wavelengths used are about one million times those of visible radiation, the largest radio telescopes, although considerably larger than those for the visible, have a much lower resolving power. This has led to the early adoption of interference methods, by which resolving powers can be increased without requiring telescopes of a mechanically impossible size.

Radio astronomy studies the outside universe with radio waves that penetrate the earth's atmosphere through the atmospheric window at 10 MHz to 30 GHz (30 m to 1 cm), the second important window after that around the visible region. The early history is given by Pawsey & Bracewell [1955], and methods and instruments are described by Christiansen & Högbom [1969], and Cole [1977].

A radio telescope does not produce a two-dimensional image directly. It is essentially a radiometer that measures the mean radiance within a certain band of frequency, coming from a certain direction, and averaged over some acceptance angle. As stated in §3.3, the radiance is often expressed as the *brightness temperature*, the temperature of a black body which, within the same frequency band, would have the same radiance. A picture can be built up, however, by scanning the radiometer across the object to give a sequence of measurements. A typical radio image of the sun is shown in fig. 133.

The essential parts of a radio telescope are illustrated in fig. 134. An aerial accepts radiation from a particular direction and delivers the power obtained to the receiver. The same Fourier relation holds between the distribution of field across the aperture of an aerial and the distribution of field at infinity. In terms of radiant power, the aerial is characterized by an instrumental function called the *reception pattern*. By a reciprocity relation, this is the same as the *radiation pattern* or angular distribution of radiation produced when the same aerial is used to radiate power. As the telescope is a linear system, the final output is the convolution of the distribution of radiance at the object and the reception pattern; this convolution is called *aerial smoothing* of the object.

To overcome the limitations imposed in practice on the resolving power of a single aerial, radio interferometers are used. In radio astronomy an interferometer means an aerial consisting of two or more widely separated parts and corresponds directly to the optical interferometer with division of wavefront. When such an aerial is used to transmit radiation, the far-field pattern consists of a series of interference fringes, and the reception pattern has the same form. The separate parts of the aerial may be linear arrays of dipoles or, more commonly, paraboloidal reflectors with either a dipole or a horn at the focus. A two-element interferometer will record separate Fourier components of the object distribution. Multi-element inter-

Fig. 133 Radio image of the sun at 3 GHz ($\lambda = 9$ cm) [Bracewell, 1962].

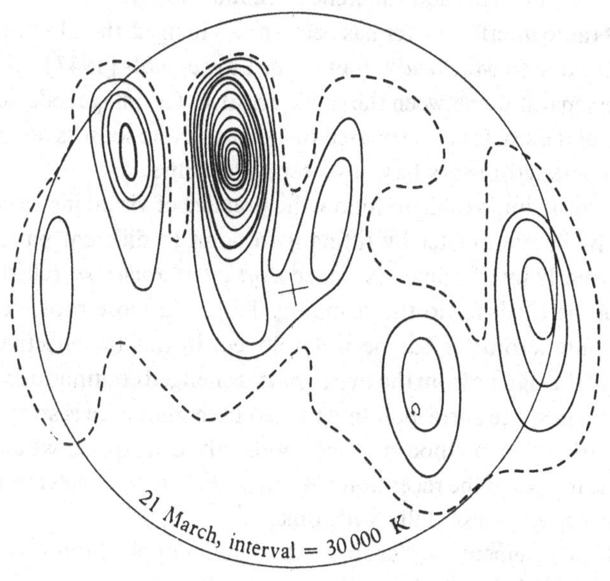

Fig. 134 The components of a radio telescope [Pawsey & Bracewell].

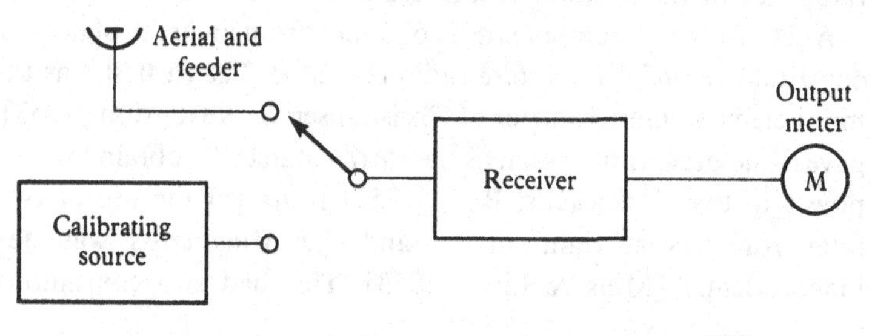

ferometers receive many components simultaneously and thus approach a direct record of the object, but with the repetitions inherent in the image given by a grating system.

The signals from each separate element are proportional to the radiation field. These can be added and squared to give a power distribution that corresponds directly to the intensity in a two-beam interferometer, an interferogram superimposed on a uniform background. This is an addition interferometer. But there are often advantages in eliminating this background, particularly when it has fluctuations which swamp the interferogram. It is done by a multiplication interferometer, discussed in § 6.2. In one method of achieving this, the interferometer is switched between the sum and difference of the two signals so that the interferogram is the alternating component of the output power, having a frequency equal to that of the switching. Another method is based on a correlator that integrates both the sum and difference simultaneously.

The first radio interferometer has been shown in fig. 2, the Lloyd's mirror arrangement due to McCready, Pawsey & Payne-Scott [1947], where the shear is the separation between the single aerial and its image reflected from the surface of the sea. It has been used to locate discrete sources on the solar disk. Later interferometers have two separate elements.

It is obviously impossible to follow the practice of visual instruments of scanning the interferometer by tilting it to point to different parts of the object. The early interferometers, called *drift interferometers*, relied on the rotation of the earth to do the scanning. To give a more rapid scan, the *swept-lobe* interferometer has been developed. In this the effective phase difference of the signals from the two aerials is changed continuously so that the direction in space corresponding to zero path difference is swept across the sky. With either method a point source, fixed in space, would move through the fringes of the reception pattern of the interferometer and give a signal that varied sinusoidally with time.

A simple two-element interferometer has similar applications to its visual analogue, the Michelson stellar interferometer. It is most useful when only a small amount of information is required, for example, the diameter of a radio star or the location of a discrete source of radio emission.

A set of Fourier components is obtained from an array of aerials, which constitute an *unfilled-aperture* radio telescope. The earliest was the linear multi-element interferometer of Christiansen & Warburton [1953], which gave a one-dimensional scan as the earth rotated. To obtain high resolving power in two dimensions, Ryle [1952] proposed the use of two linear interferometers at right angles, and the *Mills cross* was developed independently [Mills & Little, 1953]. This had two continuous linear arrays at right angles. The form with discrete sampling of spatial frequencies, is the cross of Christiansen & Mathewson [1958], shown in fig. 5. This is a typical example of a dilute-aperture aerial and its transfer function has been indicated in fig. 81. It is, however, not the simplest array for sampling all these spatial frequencies, for, if one arm is removed, the same spatial frequencies are still present in the transfer function.

Another example of a dilute aperture is the ring array of Wild [1961, 1965] of 96 parabolic aerials equally spaced around a circle of 3 km diameter. As this has no longer a rectangular form, the reception pattern is not repeated in a rectangular lattice but it has enhanced ring lobes, widely spaced from the centre. Inside the innermost of these, the pattern is similar to the reception pattern of an annular aperture of negligible width and has the form Λ_0^2, but this can be corrected to be approximately the Λ_1^2 function of the image obtained from a full circular aperture.

When the object studied is not changing, a two-element interferometer can be used instead of a multi-element array. The different Fourier components are measured in succession, the vector spacing of the two aerials being varied both in distance and direction. From this a complete picture can be obtained by Fourier transformation. This is the method of *aperture synthesis* of Ryle & Hewish [1960].

For moderate separations of the aerials, the two signals can be brought together and correlated directly. At larger separations, a radio link is used, or both signals are beaten against the same continuous reference wave. At extreme separations, up to the diameter of the earth, this common reference signal is replaced by two signals from very stable oscillators, that is, two accurate clocks that remain in synchronism. Alternatively, the signals are first rectified and their intensities correlated; a method known as *post-detector correlation*, which is the equivalent of the intensity interferometer described later. These techniques are treated by Christiansen & Högbom [1969], and Wohlleben & Mattes [1973].

14.2.1. Measurement of polarization

A two-element interferometer has another use in radio astronomy as one of the methods of measuring the degree of polarization of the radiation. Although this application of interferometry is not a study of spatial distributions, it is, as stated in §6.6 one of the few applications to polarimetry (unless an optical quarter-wave plate is regarded as an interferometer). It is therefore arbitrarily included with the other aspects of radio astronomy.

Aerials are polarized and have different sensitivities to different states of polarization of the incident radiation. A dipole, for example, is insensitive to linearly polarized radiation in which the electric vector is normal to it. If the two elements of an interferometer are both linearly polarized and are set close together, two interferograms can be obtained, one with the aerials parallel, one with them crossed. These interferograms are a function of the phase difference between the two signals. For the two aerials parallel and in the x-direction, the intensity in the interferogram is

$$\begin{aligned} I_{\parallel} &= \langle |E_x \mathrm{e}^{\mathrm{i}\psi} + E_x|^2 \rangle \\ &= 2J_{xx}(1+\cos\psi), \end{aligned} \tag{14.1}$$

where J_{xx} is the matrix element defined in (6.36). The crossed intensity is

$$\begin{aligned} I_{\perp} &= \langle |E_x \mathrm{e}^{\mathrm{i}\psi} + E_y|^2 \rangle \\ &= J_{xx} + J_{yy} + 2|J_{xy}|\cos(\arg J_{xy} + \psi), \end{aligned} \tag{14.2}$$

since $J_{yx} = J_{xy}^*$. From the second interferogram (14.2), the modulation gives $|J_{xy}|$ and the phase difference between it and (14.1) gives $\arg J_{xy}$. The modulation of (14.1) gives J_{xx} and the background of (14.2) $J_{xx}+J_{yy}$. Hence the coherency matrix $\mathbf{J}$ is fully specified, and from it any alternative description of the polarization can be derived, such as the degree of polarization or the Stokes parameters.

14.3. Stellar diameters

Astronomical telescopes for visible light have much higher resolving powers than radio telescopes and there have been fewer needs to use interference imagery. These needs can be included under the title of measurements of stellar diameters, which, as well as these simple measurements, will be taken to cover variations of diameter with position angle or time, the separations of binary stars, and limb darkening. The construction of suitable interferometers is much more difficult, since optical paths must match to a much smaller wavelength tolerance. Two difficulties stand out, that of making the interferometer mechanically stable, and that of overcoming the effects of atmospheric turbulence, or *seeing*.

The air above a telescope is neither uniform nor stationary, but has continuous variations of refractive index, which lead to random fluctuations of the optical path from a star to the telescope. The severity of the fluctuations depends on the observatory site and on weather, but they can be characterized by a cell size over which the variation is small, up to 100 mm for good conditions, and a lifetime of 1 to 10^{-3} s. If the telescope has an aperture less than the cell size, the image is little degraded by the turbulence, although it will move around as the slope of the path variations changes. A larger telescope, with many cells across its aperture, will give a blurred image.

This is illustrated by the transfer functions shown in fig. 135. A telescope with no aberration follows curve (*a*), which is the autocorrelation function of the telescope aperture. When turbulence is present, its autocorrelation function has a form like curve (*b*), it drops to zero at the spatial frequency of the cell size and then oscillates about zero up to the limit of resolution of the telescope aperture. But this applies only to an instantaneous image. The turbulence is varying and the image that is seen has the transfer function (*c*), effectively that of a telescope with an aperture equal to the cell size. A larger telescope collects more light but gives no greater resolving power.

14.3.1. Michelson stellar interferometer

When a dilute aperture is used to study a distribution of radiation, the number of different shears possible is chosen to sample the Fourier transform at just sufficient spatial frequencies for the amount of information desired. Thus, if the object is known to be radially symmetrical, the sampling need be in one dimension only of spatial frequency, and a shear that is variable in magnitude but not direction is sufficient. Further, since the transform of a symmetrical function is real and has zero phase, measurements of the visibility only are sufficient. If, in addition, the further simplification is made of assuming the star to be a disk of either uniform radiance or of a radiance that follows some prescribed law of limb darkening, measurements at one or two shears are sufficient to determine the diameter of such a model.

Fig. 135 Transfer functions of an astronomical telescope: (*a*) with no aberration; (*b*) short exposure through the atmosphere; (*c*) time-average of (*b*) by long exposure; (*d*) average of squared modulus of (*b*) from many short exposures.

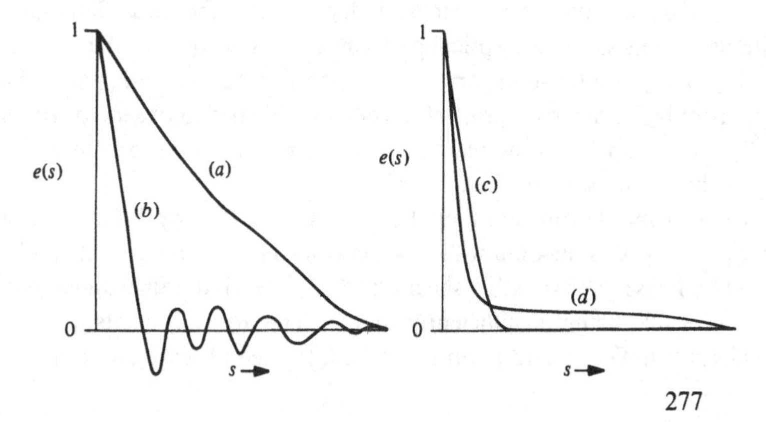

Interference imagery

The theory given earlier in (7.20) and (7.21) has shown that a uniform source of angular diameter 2α and of mean wavelength λ, with an interferometer having a shear s, gives fringes of visibility

$$\mathscr{V} = \Lambda_1(2\pi\alpha s/\lambda), \tag{14.3}$$

where, as in (2.28), $\Lambda_1(x)$ is the normalized Bessel function, $2J_1(x)/x$. In general two measurements of the visibility at different shears are needed to fix the scale of this function (or the corresponding function for a model with limb darkening) and hence give α, but when photo-electric measurements are made, it is usual to take more readings and choose the scale for Λ_1 to give the best fit to these. One single measurement suffices if this is the location of the first zero of Λ_1, found as the value s_1 of the shear at which the visibility vanishes. Then

$$2\alpha = 1.22\lambda/s, \tag{14.4}$$

the occurrence of zero visibility being judged visually.

This method of measuring stellar diameters was suggested by Fizeau. The telescope objective is covered except for two small holes of equal size, the separation of which can be varied to vary the shear. The fringes, formed in the focal plane, have a spacing (spatial period) of $\lambda f/s$, where f is the focal length. The shear, however, is limited to the diameter of the objective, but now the full resolving power of the aperture is attainable, provided the holes are not larger than the cell size of the turbulence. The instrument is as stable as the telescope itself.

To increase the resolving power beyond that of the telescope, Michelson [1920] has introduced the *stellar interferometer* shown in fig. 136. Four mirrors are mounted on a rigid girder in front of the telescope, the outer mirrors having a variable separation that gives the variable shear. The inner mirrors are fixed in front of the two apertures at the objective and the fringes in the focal plane have a fixed spacing, independent of the shear. To obtain white-light fringes, the two optical paths must be kept equal as the shear is altered, and a wedge compensator is incorporated in one path, the adjustment being made in terms of the channelled spectra observed with a small spectroscope. A tilting plate in the other beam is used to ensure that the two beams overlap in the focal plane.

Two such interferometers have been made, the first by Michelson & Pease [1921] with a maximum shear of 6 m and a resolution of 0.02″, and a second by Pease [1930] with a shear up to 15 m; the latter instrument was not successful, being insufficiently stable. The measurements present considerable difficulty and require a skilled, trained observer, and further development of this interferometer had to wait until photomultipliers and servo systems made a new approach possible.

14.3.2. Stellar intensity interferometer

The few stars that have been measured with a Michelson stellar interferometer are all cool, nearby stars. Any major increase in the resolving power of the interferometer by an increase of the possible shear presents considerable practical difficulties. It will be necessary to control the two optical paths sufficiently well to keep the central fringe reasonably stationary and to measure the maximum visibility of fringes that are changing rapidly both in visibility and position. These difficulties can be largely overcome by the use of the intensity interferometer of Hanbury Brown & Twiss [1958]. This measures the correlation between the fluctuations of the intensities at two points, rather than that of the radiation field, and is the first practical example of an interferometer based on fourth-order correlations. More precisely, the measurement made is the correlation between the fluctuations in the outputs of two photomultipliers that have been fed through two bandlimited amplifiers. As shown in §6.7, the result obtained is proportional to $|\gamma_{12}|^2$, the square of the visibility given in (14.3). The stellar diameter can be derived in the same way as from the results from a Michelson interferometer, if some (symmetrical) model is chosen for the radiance distribution of the star.

The correlation is reduced when there is a delay in the interferometer. A delay represents the time difference for signals, from the star, following the two paths to the correlator and it may occur either in the optical path or in

Fig. 136 Michelson's stellar interferometer.

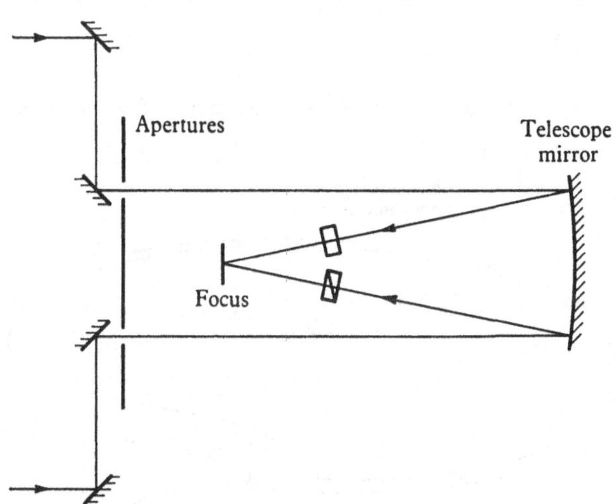

the electronic system. Delays in the optical path can arise from atmospheric turbulence or errors in tracking the star. Electronic delays can be due to errors in matching the transmission lines or to differences, within the photomultipliers, of the intervals between the arrival of a photon at the photocathode and the arrival of the electrons at the final collector. Whatever the cause, the delay should be small compared with $1/\Delta f$, where Δf is the electrical bandwidth of the power spectrum transmitted by the amplifiers. This is typically 120 MHz, so that delays should be small compared with 10^{-8} s or path differences small compared with 3 m. The great advantage of a correlation interferometer is now apparent, for variations between the two optical paths in the atmosphere are unlikely to reach even a small fraction of 3 m, while it is easy to match transmission lines to this accuracy. In a classical interferometer, variations of path difference must be small compared with the wavelength of the radiation, so there is a large gain in the tolerance on stability.

The limitations of an intensity interferometer are those of energy rather than resolving power, since it is much less efficient than a Michelson interferometer [Twiss, 1969]. If a filter is used in front of the photomultipliers to limit the bandwidth of the radiation and hence increase the correlation by increasing the degeneracy parameter δ in (6.39), the available energy is reduced, and the final correlation remains the same; it depends therefore only on the energy per unit spectral bandwidth. In practice, such filters were used, but to limit the photocurrent to a suitable value and to define the wavelength for which the result (14.3) was obtained. To increase the energy received at the photomultipliers, the radiation from the star was focused by two large reflectors, but integration times of several hours were required to give a reasonable signal-to-noise ratio. The instrument is illustrated in fig. 137.

Fig. 137 The stellar intensity interferometer [from Hanbury Brown, 1974].

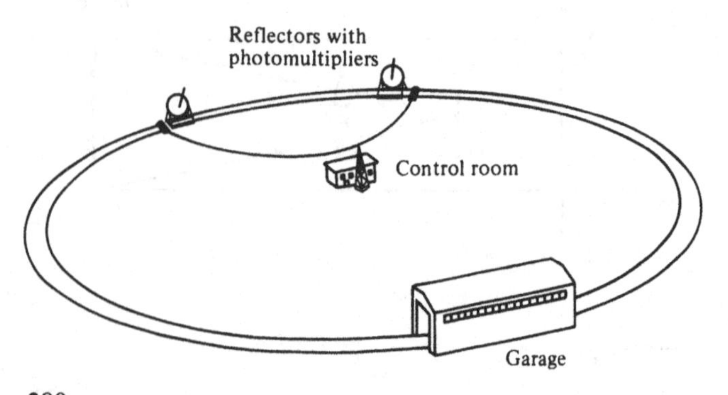

The prototype interferometer, described by Hanbury Brown & Twiss [1958] was successfully tested on Sirius. A large instrument was then set up at Narrabri, Australia [Hanbury Brown, 1974]. This had collectors of 6.7 m diameter, each made up of 250 separate mirrors, mounted on carriages that rode on a circular track of 188 m diameter. The two carriages moved around the track at a constant chordal distance, the shear, following a star for the period of integration.

The first measurements made, of the diameter of Vega, are reproduced in fig. 138. Since then, the interferometer has measured the 32 bright stars it is capable of measuring and then closed down. While Hanbury Brown [1974] has described the design of a larger instrument, it now appears that the instruments described later will supplant it.

14.3.3. Modern stellar interferometers

The low efficiency of the intensity interferometer has led to a further look at the Michelson interferometer and at other methods obtaining high resolving power. These are covered in the conference report by Davis & Tango [1979]. Several of these are photo-electric versions of the Michelson stellar interferometer, such as that at Monteporzio, described by Tango & Twiss [1980]. The instrument described by Davis [Davis & Tango, 1979] may be taken as an example; it is illustrated in fig. 139.

The interferometer obtains stability by being fixed to the ground, the

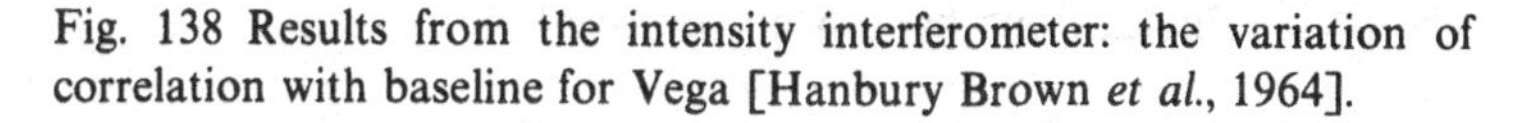

Fig. 138 Results from the intensity interferometer: the variation of correlation with baseline for Vega [Hanbury Brown *et al.*, 1964].

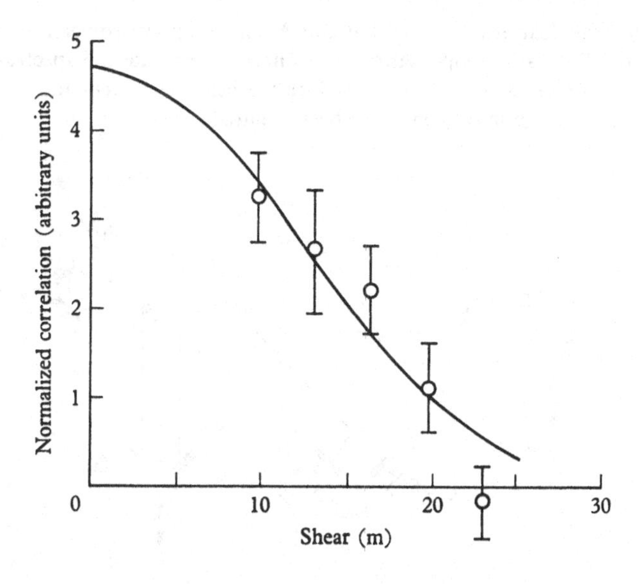

light being collected by two siderostats. It is necessary to add retroreflectors (or, to avoid polarization problems, roof mirrors) moving along tracks as path compensators to correct for the changing paths from the star to the siderostat mirrors. Some small path fluctuations remain, but these are smaller than the coherence length of the spectral band of the light that is used. In order not to waste light, it is split into several such spectral bands. Another approach would be servo correction of these remaining path differences.

Image movement is corrected by servo systems so that the two beams arrive together always with no tilt. The two complementary interferograms are detected separately in each spectral band and the differences are proportional to

$$(I_1 I_2)^{1/2} |\gamma_{12}| \cos \psi, \tag{14.5}$$

where ψ is now the fluctuating phase term due to turbulence. If the phase difference between the arms of the interferometer is changed by $\frac{1}{2}\pi$, either by switching with time or by splitting the light again to form another pair of interferograms, the difference becomes

$$(I_1 I_2)^{1/2} |\gamma_{12}| \sin \psi, \tag{14.6}$$

and the sum of the squares of these is

$$I_1 I_2 |\gamma_{12}|^2. \tag{14.7}$$

The phase fluctuations have been removed.

Fig. 139 Main features of a fixed-baseline Michelson interferometer: C – coelostats, P – path compensators, F – filters to separate the spectral bands, D – detectors. Telescopes to reduce the beam diameters and the correction of image movement have been omitted.

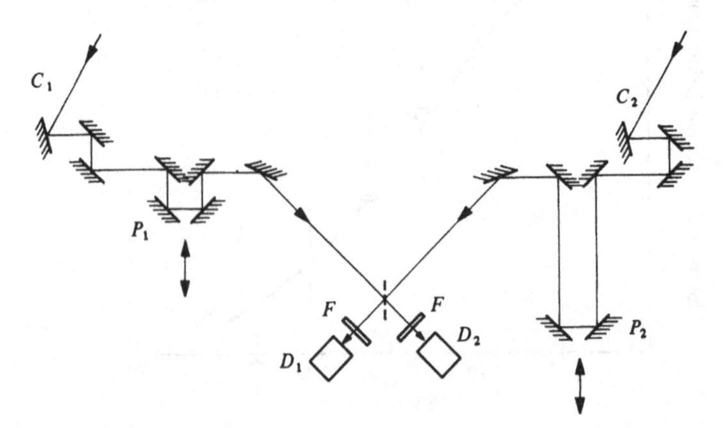

14.3.4. Speckle interferometry

This technique, also called *speckle imagery* to distinguish it from the metrological technique of §12.4, is due to Labeyrie [1974, 1976]. Further theory has been given by Dainty [1973, 1975]. While the image from a large telescope appears as a uniform blur, with high magnification and short exposures it is seen to be a rapidly changing speckle pattern. For a point source, each speckle has roughly the size of the Airy disk of the full aperture of the telescope.

To obtain stellar images with high resolving power, Labeyrie made a series of photographs of these speckle images with short exposures and high magnification. Their intensity can be expressed as

$$I(\mathbf{x}) = O(\mathbf{x}) * K(\mathbf{x}), \tag{14.8}$$

where $O(\mathbf{x})$ is the intensity distribution of the star, and $K(\mathbf{x})$ is the spread function of the atmosphere and telescope. He formed the two-dimensional Fourier transforms $i(\mathbf{u})$ of these in the coherent optical system of fig. 15. This is obtained as an amplitude, the corresponding intensity being $|i(\mathbf{u})|^2$. These intensities are all added to give a result proportional to

$$|i(\mathbf{u})|^2 = |o(\mathbf{u})|^2 \langle |k(\mathbf{u})|^2 \rangle. \tag{14.9}$$

The transfer function for this process, $\langle |k(\mathbf{u})|^2 \rangle$, is shown in fig. 135, curve (*d*). Because the negative regions have been inverted, this mean does not drop to zero like $\langle k(\mathbf{u}) \rangle$ of curve (*c*), but remains finite out to the limit of resolution of the full aperture. It can be measured by repeating the process on an unresolved star or estimated from the known statistics of the seeing, and then used to derive the object power spectrum $|o(\mathbf{u})|^2$ from the measured $|i(\mathbf{u})|^2$. The transform of this is the autocorrelation function of the object.

Other experimental procedures have been used. The images $I(\mathbf{x})$ can be recorded, digitized, and their autocorrelation functions computed and added, to give finally the transform of equation (14.9). Both methods yield the autocorrelation of the object. To obtain the object itself, there is the problem of phase retrieval, discussed in §6.9. When there is a nearby unresolved star, the solution by adding a reference point can be used; it is known as *speckle holography*. Other methods of obtaining the object itself, such as averaging the individual bright speckles, have been given by Labeyrie [1976] and Dainty [1975], and Bates [1982].

By this technique, images have been obtained from large telescopes with their full diffraction-limited resolving power, and Labeyrie is extending the technique to combine an array of separate telescopes. This will have the advantage, over the Michelson interferometer, of larger apertures and more light, but this may make it more difficult to meet the stability requirements and to separate beams of larger étendue into several spectral bands. Which is the more useful is not yet clear.

14.3.5. Heterodyne interferometry

Another approach for obtaining high resolving powers is through heterodyne interferometry. This is a direct transfer to higher frequencies of the same technique in radio astronomy. Its chief use has been in the infrared atmospheric window at 27 THz (11 μm) with the two receiving stations using a CO_2 laser as the local oscillator, and is described, for example, by Sutton [Davis & Tango, 1979]. Other work has been reviewed by Labeyrie [1976].

14.4. Optical transfer function

In image-forming systems, the instrumental function is the *spread function*, the image of a point through the system. Its two-dimensional Fourier transform is the *optical transfer function.* In both the theory of image formation and in the testing of optical systems, there are often advantages in using the transfer function rather than the spread function, for, when the optical instrument is only one stage of a series of linear systems, convolutions are replaced by products.

Since the optical transfer function is the Fourier transform of the image of a point, a two-beam interferometer will measure it directly. The spatial frequency for which the transfer function is measured is, to a constant factor, the shear of the interferometer used. As stated in §7.5, a radial-shear interferometer and a rotational-shear interferometer with a small shear both have a shear that is a constant times the position vector in the plane of observation of the interferogram. They can therefore give an intensity distribution that provides a complete picture of the real part of the transfer function; the imaginary part can be obtained if the interferometer has added a phase difference of $\frac{1}{2}\pi$ between the two beams. But for photo-electric measurement of the transfer function it is more convenient to use an interferometer in which the shear is constant over the field but variable in time. Then all the energy is concentrated in turn into the Fourier component being measured. This is a simple lateral-shearing interferometer in which the effective source is the image being studied and the angular shear is centred on a point at this image. For an image at infinity the shear should be a linear separation of the beams.

In this, the usual method of measurement, the interferometer acts as a *wave analyser* to give a Fourier analysis of the image of an illuminated pinhole or slit (the image of a line normal to the direction of shear has the same Fourier transform but gives more light than a point image). A photomultiplier receives all the light through the interferometer and measures the transfer function at successive values of the shear. However, the lamp that illuminates the slit and the photomultiplier can be interchanged to reverse the direction of the light. The interferometer then acts as a *signal generator*, forming a sinusoidal object of variable spatial frequency. The image of this through the system under test is analysed by the slit and photomultiplier. While the wave-analyser method has the light going in the more convenient direction for setting up the instrument, the reverse method has the advantage that the photomultiplier behind the slit is better protected from stray light.

As the optical transfer function is complex, a full measurement involves both its modulus and phase. If the phase difference between the two beams of the interferometer is varied continuously, the phase of the transfer function is the difference between this phase and that of the resulting alternating signal. The signal-generator method has then the further advantage here that the object fringes can be detected separately to give a phase reference.

Hopkins [1955] proposed the use of shearing interferometers for measuring transfer functions, and instruments using partially reflecting beam splitters have been reviewed by H.H. Hopkins [1962]. The interferometer used by Montgomery [1964] is shown in fig. 140. The lens being tested forms an image of an illuminated slit at infinity and this collimated light goes to the interferometer. The variable shear between the two beams is produced by rotating together two plane parallel plates of glass, one in each beam. The phase between the beams is changed by moving backwards and forwards a small-angle prism in one beam. As the photomultiplier used for measuring the transfer function collects radiation over some finite area, there should be no variations of phase difference, and hence intensity, over this area: the interferometer should have no tilt. The design of fig. 140(*a*) is compensated so that an inclination of either one of the mirrors or the beam splitter does not introduce a tilt.

Birefringent interferometers have also been used and, if these are made without tilt, no tilt is introduced as the shear is altered. Tsuruta [1963*a*] has made an interferometer in which a two-component calcite lens acts as a Wollaston prism of variable shear when it is displaced laterally. The image to be analysed is formed at this lens. For analysing an image at infinity, a lateral separation between the two beams is given either by a plate of calcite, cut obliquely to the optic axis, or by two such plates cemented together to form a Savart polariscope. Either of these has a fixed shear. To obtain a variable shear, Tsuruta [1963*b*] has used two wedged polariscopes, as shown in fig. 140(*b*). These are crossed with a fixed Savart polariscope to give results down to zero shear, and a Soleil–Babinet compensator produces a variation in phase between the two terms. The variable-shear systems shown in fig. 68 have also been used [Steel, 1964*b*].

Interference methods are not those most commonly used to measure transfer functions. For routine testing, an interferometer is rather too delicate an instrument, requiring careful setting up and protection from vibration. The polarization interferometers are less sensitive than those with partially reflecting beam splitters and separated paths for the two beams, but those with wave plates are limited to the frequency for which these have been made. Other methods of obtaining the transfer function are by computation from the measured wave aberration, as a computed Fourier transform of the image of a slit or, most commonly, by measurements on the images of specially prepared sinusoidal objects. This last method has not the versatility of an interferometric measurement, but as a consequence is less susceptible to maladjustment. It has also the advantage that it can measure integrated transfer functions for white light. For an interferometer to make such measurements, it would require a shear proportional to wavelength, which is given only by interferometers with division by diffraction (of which a sine-wave test object can be regarded as a trivial case). This condition is satisfied if moiré fringes are used as test objects, as proposed by Lohmann [1959] and used by Sapper & Pulvermacher [1965].

14.4.1. Precision focusing

When an optical system has aberration, this reduces the transfer function most rapidly at some spatial frequency near one half that of the limit of

Fig. 140 Lateral-shearing interferometer for measuring optical transfer functions.

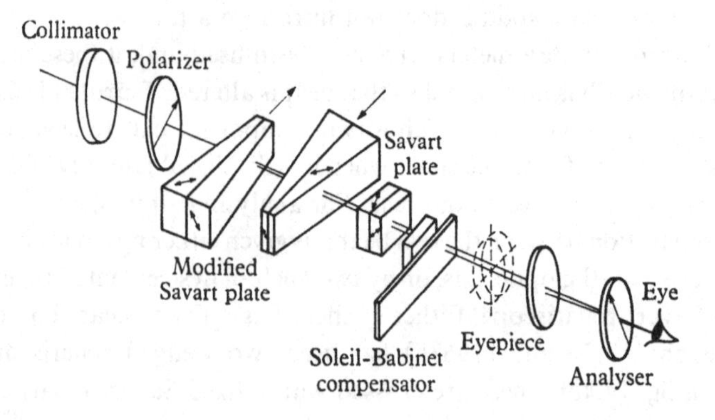

resolution. This property can be used to increase the precision with which an optical system can be focused on a target object.

The precision to which an ordinary imaging system can be focused is some fraction of the focal range, which is, in the notation of fig. 12,

$$\Delta z = \frac{\lambda}{n' \sin^2 \alpha'}. \tag{14.10}$$

For visual settings, this fraction is about 0.1. Simon [1961] has shown two ways in which this can be reduced so that, in combination, a factor of about 10^{-3} is possible. The first gain is obtained if, instead of a target such as a line, which contains all spatial frequencies, a sine-wave target is used of the spatial frequency most affected by a defocusing. The second is obtained if setting to maximum sharpness is replaced by a null setting. The sine-wave target can be obtained by interference [Steel, 1962]. If the fringes are imaged through a calcite plate, two sets of fringes are produced, separated longitudinally and out of register, so that a dark fringe of one is in line with a bright fringe of the other. When the focal setting is midway between these, no variation of intensity is obtained.

An interferometer can also be used to scan focus [Steel, 1980]. If the shear is varied, the position of fringe localization moves towards or away from the interferometer. A periodically varying shear can be obtained in an interferometer by rotating components, for example, calcite plates as in fig. 68. This is more convenient than the reciprocating motion needed to scan the focus of an image-forming system.

Bibliography

The *italic figures* in square brackets indicate the pages of the text in which the reference is quoted.

Abramson, N. (1970) *Appl. Opt.*, **9**, 2311–20. [*224*]
Abramson, N. (1971) *Appl. Opt.*, **10**, 2155–61. [*224*]
Abramson, N. (1972) *Appl. Opt.*, **11**, 1143–7. [*224*]
Abramson, N. (1976) *The Engineering Uses of Coherent Optics*, ed. E. R. Robertson, pp. 647–63. Cambridge University Press. [*216*]
Abramson, N. and Bjelkhagen, H. (1979) *Appl. Opt.*, **18**, 3870–80. [*216*]
Adelman, N.T. (1977) *Appl. Opt.*, **16**, 2075–7. [*266*]
Ammann, E.O. (1971) *Progress in Optics*, ed. E. Wolf, **9**, 123–77. Amsterdam: North-Holland. [*248*]
Aplet, L.J. and Carson, J.W. (1964) *Appl. Opt.*, **3**, 544–5. [*59*]
Archbold, E., Burch, J.M. and Ennos, A.E. (1967) *J. Sci. Instrum.*, **44**, 489–94. [*195, 218*]
Archbold, E., Burch, J.M., Ennos, A.E. and Taylor, P.A. (1969) *Nature, (GB)*, **222**, 263–5. [*225*]
Armitage, J.D. and Lohmann, A. (1965) *Opt. Acta*, **12**, 185–92. [*115*]
Aronowitz, F. (1971) *Laser Applications I*, ed. M. Ross. pp. 133–200. New York: Academic Press. [*184*]
Babcock, H.W. (1962) *Appl. Opt.*, **1**, 415–20. [*177*]
Baer, T., Kowalski, F.V. and Hall, J.L. (1980) *Appl. Opt.*, **19**, 3173–7. [*57*]
Baird, K.M. (1963) *Appl. Opt.*, **2**, 471–9. [*171, 174*]
Baird, K.M., Evenson, K.M., Hanes, G.R., Jennings, D.A. and Peterson, F.R. (1979) *Opt. Lett.*, **4**, 263–4. [*30*]
Baird, K.M. and Hanes, G.R. (1974) *Rep. Prog. Phys.*, **37**, 927–50. [*57*]
Baird, K.M. and Howlett, L.E. (1963) *Appl. Opt.*, **2**, 455–62. [*170*]
Baker, D. (1977) *Spectrometric Techniques*, ed. G.A. Vanasse, **1**, 71–106. New York: Academic Press. [*265*]
Baker, D., Steed, A. and Stair, A.T. (1981) *Appl. Opt.*, **20**, 1734–46. [*240*]
Balhorn, R., Kunzmann, H. and Lebowsky, F. (1972) *Appl. Opt.*, **11**, 742–8. [*56*]
Baltes, H.P. ed. (1978) *Inverse Source Problems in Optics*. Berlin: Springer-Verlag. [*207*]
Banning, M. (1947). *J. Opt. Soc. Am.*, **37**, 792–7. [*133*]
Bateman, H. (1954) *Tables of Integral Transforms*, ed. A. Erdélyi *et al.*, vol. 1, New York: McGraw-Hill. [*13*]
Bates, R.H.T. (1982) *Optik*, **61**, 247–62. [*87, 283*]
Bates, W.J. (1947) *Proc. Phys. Soc.*, **59**, 940–50. [*5, 6, 51, 197*]

Beckers, J.M., Dickson, L. and Joyce, R.S. (1975) *Appl. Opt.*, **14**, 2061–6. [*247*]
Beckers, J.M. and Dunn, R.B. (1965) *A Ray-Tracing Program for Birefringent Filters.* AFCRL-65-605, US Air Force, Office of Aerospace Research. [*248*]
Bell, E.E. (1966) *Infrared Phys.*, **6**, 57–74; Russell, E.E. and Bell, E.E. *ibid.* 75–84. [*256*]
Bell, E.E. and Sanderson, R.B. (1972) *Appl. Opt.*, **11**, 688–9. [*241*]
Bell, R.J. (1972) *Introductory Fourier Transform Spectroscopy.* New York: Academic Press. [*250*]
Bellevue Conference (1958) *Les Progrés Récents en Spectroscopie Interférentielle.* Paris: CNRS; *or see J. Phys. Radium.*, **19**, No. 3, 185–436. [*154*]
Bellevue Conference (1966) *Les Méthodes Nouvelles de Spectroscopie Instrumentale.* Paris: CNRS; *or see J. Phys. (Fr).*, **28**, suppl. C2. [*154*]
Bennett, F.D. and Kahl, G.D. (1953) *J. Opt. Soc. Am.*, **43**, 71–8. [*124*]
Bennett, S.J. and Gill, P. (1980) *J. Phys. E: Sci. Instrum.*, **13**, 174–7. [*172*]
Bennett, S.J., Ward, R.E. and Wilson, D.C. (1973) *Appl. Opt.*, **12**, 1406. [*56*]
Benton, S.A. (1977) *Applications of Holography and Optical Data Processing*, ed. E. Maron, A.A. Friesem and E. Wiener-Avnear, pp. 401–9. Oxford: Pergamon. [*67*]
Beran, M.J. and Parrent, G.B. (1964) *Theory of Partial Coherence.* Englewood Cliffs, NJ: Prentice-Hall. [*72, 78*]
Bergstrand, E. (1956) *Handbuch der Physik*, ed. Flügge, vol. **24**, 1–43. Berlin: Springer-Verlag. [*182*]
Bien, F., Camac, M., Caulfield, H.J. and Ezekiel, S. (1981) *Appl. Opt.*, **20**, 400–3. [*170*]
Birch, J.R. and Parker, J.J. (1979) *Infrared and Millemeter Waves*, ed. K.J. Button, vol. **2**, 137–271. New York: Academic Press. [*256*]
Bloom, A.L. (1968) *Gas Lasers.* New York: John Wiley. [*54*]
Bonse, U. and Rauch, H. ed. (1979) *Neutron Interferometry.* Oxford: Clarendon Press. [*1*]
Boone, P.M. (1975) *Opt. Acta*, **22**, 579–89. [*223*]
Boone, P.M. and de Backer, L.C. (1976) *Optik.*, **44**, 343–55. [*229*]
Born, M. and Wolf, E. (1959) *Principles of Optics.* London: Pergamon. [*100, 107, 127, 243*]
Bottema, M. (1960) *Optics in Metrology*, ed. P. Mollet, pp. 42–9. Oxford: Pergamon. [*165*]
Bottema, M. (1972) *J. Opt. Soc. Am.*, **62**, 1438–43. [*113*]
Bouchareine, P. and Connes, P. (1963) *J. Phys. (Fr)*, **24**, 134–8. [*118, 119, 265*]
Bouchareine, P. and Janest, A. (1978) *First European Congress on Optics Applied to Metrology, Proc. SPIE* (Soc. Photo-opt. Instrum. Eng.), **136**, 38–42. [*172*]
Boulouch, R. (1893) *J. Phys. (Fr) 3ᵉ sér.*, **2**, 316–20. [*8*]
Boyd, G.D. and Kogelnik, H. (1962) *Bell Syst. Tech. J.*, **41**, 1347–69. [*156*]
Bracewell, R.N. (1962) *Handbuch der Physik*, ed. S. Flügge, vol. **54**, 42–129. Berlin: Springer-Verlag. [*273*]
Bracewell, R.N. (1965) *The Fourier Transform and Its Applications.* New York: McGraw-Hill. (2nd ed. 1978). [*12*]
Bradley, D.J., Bates, B., Juulman, C.O.L. and Majumdar, S. (1965) *Jpn. J. Appl. Phys.*, **4**, suppl. 1, 467–72. [*149*]
Bradley, D.J. and Mitchell, C.J. (1969) *Appl. Opt.*, **8**, 707–9, 710. [*154*]

Brames, B.J. and Dainty, J.C. (1981) *J. Opt. Soc. Am.*, **71**, 1542–5. [*88*]
Briers, J.D. (1976) *Opt. Quantum Electron*, **8**, 469–501. [*221*, *223*]
Brillet, A. and Hall, J.L. (1979) *Phys. Rev. Lett.*, **42**, 549–52. [*182*]
Brouwer, W. (1964) *Matrix Methods in Optical Instrument Design*. New York: W.A. Benjamin. [*27*, *29*]
Brown, D.S. (1962) *J. Sci. Instrum.*, **39**, 71–2. [*201*, *202*]
Bruce, C.F. (1966) *Appl. Opt.*, **5**, 1447–52. [*154*]
Bruce, C.F. and Ciddor, P.E. (1967) *Metrologia*, **3**, 109–18. [*171*, *175*]
Bruce, C.F. and Duffy, R.M. (1975) *Rev. Sci. Instrum.*, **46**, 379–82. [*168*]
Bruce, C.F. and Sharples, F.P. (1975) *Appl. Opt.*, **16**, 3082–5. [*188*]
Bryngdahl, O. (1965) *Progress in Optics*, ed. E. Wolf, **4**, 37–83. Amsterdam: North-Holland. [*209*, *210*]
Bryngdahl, O. and Lee, W.H. (1974) *J. Opt. Soc. Am.*, **64**, 1606–15. [*204*]
Bryngdahl, O. and Lohmann, A.W. (1968) *J. Opt. Soc. Am.*, **58**, 141–2. [*60*]
Bünnagel, R. (1965) *Z. Instrumenkd.*, **73**, 214–15. [*188*]
Burch, C.R. (1940) *Mon. Not. R. Astron. Soc.*, **100**, 488–90. [*195*]
Burch, J.M. (1952) *The Screw-Testing Interferometer*. Thesis, Bristol. [*195*]
Burch, J.M. (1953) *Nature* (*GB*), **171**, 889–90. [*200*]
Burch, J.M. (1960) *Research* (*GB*), **13**, 2–7. [*69*, *178*]
Burch, J.M. (1963) *Progress in Optics*, ed. E. Wolf, **2**, 73–108. Amsterdam: North-Holland. [*177*]
Burch, J.M. and Forno, C. (1975) *Opt. Engng.*, **14**, 178–85. [*229*]
Burch, J.M. and Tokarski, J.M. (1968) *Opt. Acta*, **15**, 101–11. [*228*]
Burge, R.E., Fiddy, M.A., Greenway, A.H. and Ross, G. (1976) *Proc. R. Soc. Lond.*, A**350**, 191–212. [*87*]
Butters, J.N. and Leendertz, J.A. (1971) *Opt. Laser Tech.*, **3**, 26–30. [*226*]
Cagnet, M. (1954) *Rev. Opt.* (*Fr*), **33**, 1–25, 113–24, 229–41, 552. [*161*, *180*]
Candler, C. (1951) *Modern Interferometers*. London: Hilger and Watts. [*3*, *108*, *169*, *176*, *178*, *192*, *197*, *243*]
Cha, S. and Vest, C.M. (1981) *Appl. Opt.*, **20**, 2787–94. [*207*]
Chabbal, R. (1957) *J. Rech. Cent. Natl. Rech. Sci.*, **8**, 77–106. [*245*]
Chabbal, R. (1958) *Rev. Opt.* (*Fr*), **37**, 49–103, 336–70, 501–50, 608. [*147*, *159*]
Chabbal, R. and Pelletier, R. (1965) *Photographic and Spectroscopic Optics. Jpn. J. Appl. Phys.*, **4**, suppl. 1, 445–7. [*245*]
Challener, W.A., Richards, P.L., Zilio, S.C. and Garvin, H.L. (1980) *Infrared Phys.*, **20**, 215–22. [*262*]
Chamberlain, J. (1979) *The Principles of Interferometric Spectroscopy*. Chichester: John Wiley. [*256*]
Chamberlain, J.E., Gibbs, J.E. and Gebbie, H.A. (1963) *Nature* (*GB*). **198**, 874–5. [*256*]
Chang, B.J., Alferness, R. and Leith, E.N. (1975) *Appl. Opt.*, **14**, 1592–600. [*137*]
Chantry, G.W. (1982) *J. Phys. E.: Sci. Instrum.*, **15**, 3–8. [*150*]
Chebotayev, V.P. (1980) *Metrology and Fundamental Constants*, ed. A.F. Milone, F. Giacomo and S. Leschiutta, pp. 623–85. Amsterdam: North-Holland. [*57*, *171*]
Christiansen, W.N. and Högbom, J.A. (1969) *Radiotelescopes*. Cambridge University Press. [*161*, *272*, *275*]
Christiansen, W.N., Labrum, N.R., McAlister, K.R. and Mathewson, D.S. (1961) *Proc. Inst. Electr. Eng.*, **108**B, 48–58. [*8*]

Christiansen, W.N. and Mathewson, D.S. (1958) *Proc. Inst. Radio Eng.*, **46**, 127–31. [*274*]
Christiansen, W.N. and Warburton, J.A. (1953) *Aust. J. Phys.*, **6**, 190–202, 262–71. [*274*]
Ciddor, P.E. (1973) *Aust. J. Phys.*, **26**, 783–96. [*173*]
Ciddor, P.E. (1982) *Advances in Optical and Electron Microscopy*, **8**, ed. R. Barer, London: Academic Press. [*174*]
CIE (1957) *International Lighting Vocabulary*. Paris: Commission Internationale d'Eclairage. [*30*]
Clapham, P.B. (1971) *Opt. Acta*, **18**, 563–75. [*131*]
Clapham, P.B., Downs, M.J. and King, R.J. (1969) *Appl. Opt.*, **8**, 1965–74. [*133*]
Clothier, W.K. (1965) *Metrologia*, **1**, 36–56. [*173*]
Clothier, W.K., Sloggett, G.J. and Bairnsfather, H. (1980) *Opt. Engng*, **19**, 834–42. [*154, 173, 188*]
Cole, T.W. (1977) *Progress in Optics*, ed. E. Wolf, **15**, 187–244. Amsterdam: North-Holland. [*72, 272*]
Cole, T.W. (1979) *J. Opt. Soc. Am.*, **69**, 554–7. [*72*]
Collier, R.J., Burckhardt, C.B. and Lin, L.H. (1971) *Optical Holography*. New York: Academic Press. [*60*]
Connes, J. (1958) *J. Phys. Radium*, **19**, 197–208. [*251*]
Connes, J. (1961) *Rev. Opt. (Fr)*, **40**, 45–79, 116–40, 171–90, 231–65. [*145, 235, 238, 250, 257*]
Connes, J. and Connes, P. (1966) *J. Opt. Soc. Am.*, **56**, 896–910. [*255, 261, 263, 265*]
Connes, J., Delouis, H., Connes, P., Guelachvili, G., Maillard, J.-P. and Michel, G. (1970) *Nouv. Rev. Opt.*, **1**, 3–22. [*241*]
Connes, J. and Nozal, V. (1961) *J. Phys. Radium*, **22**, 359–66. [*261*]
Connes, P. (1956) *Rev. Opt. (Fr)*, **35**, 37–43. [*121, 154, 155, 246*]
Connes, P. (1959) *Rev. Opt. (Fr)*, **38**, 157–201, 416–41; (1960) *ibid.*, **39**, 402–36. [*249*]
Connes, P. and Michel, G. (1975) *Appl. Opt.*, **14**, 2067–84. [*240, 264*]
Cooley, J.W. and Tookey, J.W. (1965) *Math. Comput.*, **19**, 297–301. [*262*]
Corno, J., Lamare, M. and Simon, J. (1977) *J. Opt. Paris*, **8**, 33–8. [*196*]
Costley, A.E., Hursey, K.H., Neill, G.F. and Ward, J.M. (1977) *J. Opt. Soc. Am.*, **67**, 979–81. [*262*]
Cuisenier, M. and Pinard, J. (1967) *J. Phys. (Fr)* **28**, suppl. C2, 97–104. [*121, 131, 266*]
Culshaw, W. (1959) *Inst. Radio Eng. Trans.*, MTT-**7**, 221–8. [*138, 150*]
Culshaw, W. (1960) *Inst. Radio Eng. Trans.*, MTT-**8**, 182–9. [*153*]
Culshaw, W. (1961) *Inst. Radio Eng. Trans.*, MTT-**9**, 135–44. [*155*]
Culshaw, W., Richardson, J.M. and Kerns, D.M. (1960) *Interferometry: NPL Symposium No. 11*, 329–53. London HMSO. [*139, 150*]
Cuny, B. (1955) *Rev. Opt. (Fr)*, **34**, 460–4. [*129*]
Dainty, J.C. (1973) *Opt. Commun.*, **7**, 129–34. [*283*]
Dainty, J.C. ed. (1975) *Laser Speckle and Related Phenomena*. Berlin: Springer-Verlag. [*283*]
Dändliker, R. (1980) *Progress in Optics*, ed. E. Wolf, **17**, 1–84. Amsterdam: North-Holland. [*57, 223*]

Davis, J. and Tango, W. ed. (1979) *High Angular Resolution Stellar Interferometry*, IAU Symposium 50. Univ. Sydney: Astron. Dep. [*281*, *284*]
Decker, J. and Harwit, M. (1974) *Progress in Optics*, ed. E. Wolf, **12**, 101–62. Amsterdam: North-Holland. [*250*]
Denby, D., Quintanilla, G.E. and Butters, J.N. (1976) *The Engineering Uses of Coherent Optics*, ed. E.R. Robertson, pp. 171–97. Cambridge University Press. [*231*]
Denisyuk, Yu.N. (1962) *Dokl. Akad. Nauk SSSR*, **144**, 1275–8; Engl. transl. *Sov. Phys. Dokl.*, **7**, 543–5. [*66*]
Denisyuk, Yu.N. (1963) *Opt. Spektrosk.*, **15**, 522–3; Engl. transl. *Opt. Spectrosc.*, **15**, 279–84. [*66*]
Denisyuk, Yu.N. (1980) *Zh. Prikl. Spektrosk.*, **33**(3), 397–414, Engl. transl. *J. Appl. Spectrosc.*, **33**, 901–15. [*66*]
Deslattes, R.D. (1980) *Metrology and Fundamental Constants*, ed. A.F. Milone *et al.* pp. 38–113. Amsterdam: North-Holland. [*171*, *173*]
Dew, G.D. (1966) *J. Sci. Instrum.*, **43**, 409–15. [*187*, *188*]
Dew, G.D. (1974) *Opt. Acta*, **21**, 609–14. [*189*]
Doherty, V.J. and Shafer, D. (1980) *International Lens Design Conference*, ed. R.E. Fischer. *Proc. SPIE* (Soc. Photo-opt. Instrum. Eng.), **237**, 195–200. [*129*]
Dörband, B. (1982) *Optik*, **60**, 161–74. [*223*]
Dorenwendt, K. (1975) *Proc. 5th Int. Conf. Atomic Masses and Fundamental Constants*, ed. J.H. Sanders and A.H. Wapstra, pp. 403–9. New York: Plenum Press. [*172*]
Dorenwendt, K. and Bönsch, G. (1976) *Metrologia*, **12**, 57–60. [*139*, *172*]
Duffieux, P.M. (1946) *L'Intégrale de Fourier et ses Applications à l'Optique*. Besançon: privately printed; 2nd ed. (1970), Paris: Masson. [*12*]
Duffieux, P.M. (1969) *Appl. Opt.*, **8**, 329–32. [*243*]
Dufour, C. and Picca, R. (1945) *Rev. Opt. (Fr)*, **24**, 19–34. [*150*]
Durst, F., Melling, A. and Whitelaw, J.H. (1976) *Principles and Practice of Laser-Doppler Anemometry*. London: Academic Press; 2nd ed. (1981). [*54*]
Dyson, J. (1949) *Nature (GB)*, **164**, 229. [*213*]
Dyson, J. (1953) *Proc. R. Soc. Lond.*, **A216**, 493–501. [*210*]
Dyson, J. (1957) *J. Opt. Soc. Am.*, **47**, 386–90. [*201*]
Dyson, J. (1959) *J. Opt. Soc. Am.*, **49**, 713–16. [*195*]
Dyson, J. (1963*a*) *Opt. Acta*, **10**, 171–7. [*166*]
Dyson, J. (1963*b*) *J. Opt. Soc. Am.*, **53**, 690–4. [*178*]
Dyson, J. (1970) *Interferometry as a Measuring Tool*. Brighton: Machinery Publishing Co. [*3*]
Dyson, J., Williams, R.V. and Young, K.M. (1962) *Nature (GB)*, **195**, 1291–2. [*209*]
Egorova, L.V. and Kiselev, B.A. (1979) *Opt. Spektrosk.*, **45**, 196–9; Eng. transl. *Opt. Spectrosc.*, **45**, 107–9. [*249*]
Elssner, K.-E., Grzanna, J.V. and Schulz, G. (1980) *Opt. Acta*, **27**, 563–80. [*189*]
Endo, J., Matsuda, T. and Tonomura, A. (1979) *Jpn. J. Appl. Phys.*, **18**, 2291–4. [*1*]
Engelhard, E. (1957) *NBS Circular 581: Metrology of Gage Glocks*, pp. 1–20. US Dept. Commerce. [*175*, *176*, *180*]
Ennos, A.E. (1968) *J. Sci. Instrum. 2nd ser. (J. Phys. E.)*, **1**, 731–4. [*221*]
Ennos, A.E. (1978) *Progress in Optics*, ed. E. Wolf, **16**, 231–88. Amsterdam: North-Holland. [*225*, *226*]

Ennos, A.E. (1980) *Opt. Commun.*, **33**, 9–12. [*229*]
Erf, R.A. ed. (1978) *Speckle Metrology.* New York: Academic Press. [*225, 227*]
Esplin, R.W. (1977) *Opt. Engng*, **17**, 73–81, [*268*]
Evans, J.W. (1949) *J. Opt. Soc. Am.*, **39**, 229–42, 412. [*135, 247*]
Evans, J.W. (1958) *J. Opt. Soc. Am.*, **48**, 142–5. [*247*]
Evenson, K.M., Day, G.W., Wells, J.S. and Mullen, L.O. (1972) *Appl. Phys. Lett.* (*US*), **20**, 133–4. [*30*]
Fabry, Ch. and Perot, A. (1897) *Ann. Chim. Phys.*, **12**, 459–501. [*243*]
Fellgett, P. (1951) *On the Theory of Infrared Sensitivities and Its Application to the Investigation of Stellar Radiation in the Near Infrared.* Thesis, Cambridge. [*240*]
Fellgett, P. (1958) *J. Phys. Radium*, **19**, 187–91, 237–40. [*240, 264*]
Fienup, J.R. (1978) *Opt. Lett.*, **3**, 27–9. [*88*]
Filler, A.S. (1964) *J. Opt. Soc. Am.*, **54**, 762–7. [*235, 236*]
Fisher, L.R., Parker, N.S. and Sharples, F. (1980) *Opt. Engng*, **19**, 798–800. [*178*]
Fleming, J.W. (1977) *Infrared Phys.*, **17**, 263–9. [*240*]
Fonck, R.J., Huppler, D.A., Roesler, F.L., Tracy, D.H. and Daehler, M. (1978) *Appl. Opt.*, **17**, 1739–47. [*266*]
Forman, M.L. (1966) *J. Opt. Soc. Am.*, **56**, 978–9. [*262*]
Forman, M.L., Steel, W.H. and Vanasse, G.A. (1966) *J. Opt. Soc. Am.*, **56**, 59–63. [*258*]
Forman, P.F. (1979) *Interferometry*, ed. G.W. Hopkins, *Proc. SPIE* (Soc. Photo-opt. Instrum. Eng.), **192**, 41–8. [*193*]
Fox, A. G. and Li, T. (1961) *Bell Syst. Tech. J.*, **40**, 453–88. [*155*]
Françon, M. (1956) *Handbuch der Physik*, ed. S. Flügge, **24**, 171–460. Berlin: Springer-Verlag. [*3*]
Françon, M. (1957) *J. Opt. Soc. Am.*, **47**, 528–35. [*136*]
Françon, M. (1961) *Progress in Microscopy.* Oxford: Pergamon. [*210, 213*]
Françon, M. (1966) *Optical Interferometry.* London: Academic Press [*3*]
Françon, M. (1979) *Optical Image Formation and Processing.* New York: Academic Press. [*36*]
Françon, M. and Sergent, B. (1955) *Opt. Acta*, **2**, 182–4. *136*]
Frater, R.H. and Skellern, D.J. (1978) *Astrophys.*, **68**, 391–6, 397–403. [*271*]
Fredga, K. and Högbom, J.A. (1971) *Solar Phys.*, **20**, 204–27. [*247, 248*]
Frieden, B.R. (1971) *Progress in Optics*, ed. E. Wolf, **9**, 311–407. Amsterdam: North-Holland. [*250*]
Froome, K.D. (1954) *Proc. R. Soc. Lond.*, A**223**, 195–215. [*139*]
Froome, K.D. (1958) *Proc. R. Soc. Lond.*, A**249**, 109–22. [*139*]
Froome, K.D. and Essen, L. (1969) *The Velocity of Light and Radio Waves.* London: Academic Press. [*182*]
Fymat, A.L. (1972) *Appl. Opt.*, **11**, 160–73. [*85*]
Fymat, A.L. (1981) *Opt. Engng*, **20**, 25–30. [*105*]
Gabor, D. (1949) *Proc. R. Soc. Lond.*, A**197**, 454–86. [*2, 60*]
Gamo, H. (1964) *Progress in Optics*, ed. E. Wolf, **3**, 187–332. Amsterdam: North-Holland. [*87*]
Gates, J.W.C. and Bennett, S.J. (1968) *J. Sci. Instrum. ser. 2* (*J. Phys. E.*), **1**, 1171–4. [*214*]

Gebbie, H.A. (1961) *Advances in Quantum Electronics*, ed. J.R. Singer, pp. 155–63. New York: Columbia University Press. [*262*]
Gebbie, H.A., Habell, K.J. and Middleton, S.P. (1962) *Optical Instruments and Techniques*, ed. K.J. Habell, pp. 43–50. London: Chapman and Hall. [*263*]
Gebbie, H.A. and Twiss, R.Q. (1966) *Rep. Progr. Phys.*, **29**, 729–54. [*270*]
Genzel, L. (1965) *Photographic and Spectroscopic Optics*, *Jpn. J. Appl. Phys.*, **4**, suppl. 1, 353–7. [*150*]
Genzel, L. and Sakai, K. (1977) *J. Opt. Soc. Am.*, **67**, 871–9. [*250*]
Giacomo, P. (1980) *Metrology and Fundamental Constants*, ed. A.F. Milone *et al.*, pp. 114–48. Amsterdam: North-Holland. [*173*]
Giallorenzi, T. (1981) *Opt. Laser Technol.*, **13**, 73–8. [*140*]
Gingell, D., Todd, I. and Heavens, O.S. (1982) *Opt. Acta*, **29**, 901–8. [*213*]
Girard, A. (1965) *Photographic and Spectroscopic Optics*, *Jpn. J. Appl. Phys.*, **4**, suppl. 1, 379–84. [*239*]
Girard, A. and Jacquinot, P. (1967) *Advanced Optical Techniques*, ed. A.C.S. van Heel, pp. 71–121. Amsterdam: North-Holland. [*249*]
Glauber, R.J. (1963) *Phys. Rev.*, **13**, 2529–39. [*85*]
Golay, M.J.E. (1973) *J. Opt. Soc. Am.*, **63**, 1217–21. [*120*]
Goldsmith, J.E.M., Weber, E.W., Kowalski, F.V. and Schawlow, A.L. (1979) *Appl. Opt.*, **18**, 1983–7. [*170*]
Goodman, J.W. (1968) *Introduction to Fourier Optics*. San Francisco: McGraw-Hill. [*36*, *69*]
Guelachvili, G. (1972) *Nouv. Rev. Opt.*, **3**, 317–36. [*241*]
Guild, J. (1956) *The Interference Systems of Crossed Diffraction Gratings*. Oxford: Clarendon Press. [*39*]
Guild, J. (1960) *Diffraction Gratings as Measuring Scales*. London: Oxford University Press. [*5*, *177*, *178*]
Guillard, M. (1963–4) *Rev. Opt.* (*Fr*), **42**, 463–82; **43**, 21–32, 64–88, 349–73. [*213*]
Hall, J.L. and Lee, S.A. (1976) *Appl. Phys. Lett.* (*US*), **29**, 367–9. [*172*]
Hanbury Brown, R. (1974) *The Intensity Interferometer, Its Application to Astronomy*. London: Taylor & Francis. [*280*, *281*]
Hanbury Brown, R., Hazard, C., Davis, J. and Allen, L.R. (1964) *Nature* (*GB*), **201**, 1111–12. [*281*]
Hanbury Brown, R. and Twiss, R.Q. (1954) *Philos. Mag.*, (7) **45**, 663–82. [*6*]
Hanbury Brown, R. and Twiss, R.Q. (1956) *Nature* (*GB*), **177**, 27–9. [*6*]
Hanbury Brown, R. and Twiss, R.Q. (1957) *Proc. R. Soc. Lond.*, A**242**, 300–24; **243**, 291–319. [*85*]
Hanbury Brown, R. and Twiss, R.Q. (1958) *Proc. R. Soc. Lond.*, A**248**, 199–221, 222–37. [*279*, *281*]
Hanes, G.R. (1959) *Can. J. Phys.*, **37**, 1283–92. [*142*]
Hanes, G.R. (1963) *Appl. Opt.*, **2**, 465–70. [*142*, *167*]
Hansen, G. (1942) *Zeiss Nachr.*, **4**, 109–22. [*99*, *104*, *110*, *194*]
Hansen, G. (1955) *Optik*, **12**, 5–16. [*110*, *194*]
Hariharan, P. (1969) *Appl. Opt.*, **8**, 1925–6. [*8*, *9*]
Hariharan, P. (1974) *Opt. Engng*, **1**, 257–8. [*189*]
Hariharan, P. (1975*a*) *Appl. Opt.*, **14**, 1056–7. [*204*]
Hariharan, P. (1975*b*) *Appl. Opt.*, **14**, 2319–21. [*8*]
Hariharan, P. (1982*a*) *Opt. Lett.*, **7**, 274–6. [*56*]

Hariharan, P. (1982*b*) *Progress in Optics*, ed. E. Wolf, **20**, to be published, Amsterdam: North-Holland. [*67*]
Hariharan, P. (1983) *Holography*. Cambridge University Press, to be published. [*60*]
Hariharan, P. and Hegedus, Z. S. (1973) *Opt. Commun.*, **9**, 152–5. [*216*]
Hariharan, P. and Hegedus, Z.S. (1975) *Opt. Commun.*, **14**, 148–51. [*198*]
Hariharan, P., Hegedus, Z.S. and Steel, W.H. (1979) *Opt. Acta*, **26**, 289–91. [*67*]
Hariharan, P., Oreb, B.F. and Brown, N. (1982) *Opt. Commun.*, **41**, 393–6. [*223*]
Hariharan, P. and Sen, D. (1960) *J. Opt. Soc. Am.*, **50**, 1026–7. [*165*]
Hariharan, P. and Sen, D. (1961*a*) *Proc. Phys. Soc.*, **77**, 328–34. [*203*]
Hariharan, P. and Sen, D. (1961*b*) *J. Sci. Instrum.*, **38**, 428–32. [*202*]
Hariharan, P. and Sen, D. (1962) *Opt. Acta*, **9**, 159–75. [*202*]
Hariharan, P. and Steel, W.H. (1974) *Appl. Opt.*, **13**, 721. [*190*]
Harress, F. (1912) *Die Geschwindigkeit des Lichtes Beweglichen Körpen*. Diss. Jena. [*183*]
Harrison, G.R. (1949) *J. Opt. Soc. Am.*, **39**, 522–8. [*242*]
Harrison, G.R., Lord, R.C. and Loofbourow, J.R. (1948) *Practical Spectroscopy*, New York: Prentice-Hall. [*244*]
Harrison, G.R., Sturgis, N., Baker, S.C. and Stroke, G.W. (1957) *J. Opt. Soc. Am.*, **47**, 15–22, [*176*]
Harvey, A.F. (1963) *Microwave Engineering*. London: Academic Press. [*139*]
Harwit, M. and Sloane, N.J.A. (1979) *Hadamard Transform Optics*. New York: Academic Press. [*250*]
Heavens, O.S. (1955) *Optical Properties of Thin Solid Films*. London: Butterworths. [*127*]
Heavens, O.S. (1960) *Rep. Progr. Phys.*, **23**, 1–65. [*127*]
Henry, R.L. and Tanner, D.B. (1979) *Infrared Phys.*, **19**, 163–74. [*267*]
Hercher, M. (1968) *Appl. Opt.*, **7**, 951–66; *ibid.*, **8**, 709. [*154*]
Herriott, D.R. (1961) *J. Opt. Soc. Am.*, **51**, 1142–5. [*187*]
Herriott, D.R. (1967) *Progress in Optics*, ed. E. Wolf, **6**, 171–209. Amsterdam: North-Holland. [*141*]
Hildebrand, B.P. (1976) *The Engineering Uses of Coherent Optics*, ed. E.R. Robertson, pp. 647–63. Cambridge University Press. [*219*]
Hill, R.M. (1963) *Opt. Acta*, **10**, 141–52. [*150*]
Hill, R.M. and Bruce, C.F. (1962) *Aust. J. Phys.*, **15**, 194–222; (1963) *ibid.*, **16**, 282. [*142, 167*]
Hirschberg, J.G. and Platz, P. (1965) *Appl. Opt.*, **4**, 1375–81. [*244*]
Hoffman, J.E. and Vanasse, G.A. (1966) *Appl. Opt.*, **5**, 1167–9. [*260*]
Holden, J. (1949) *Proc. Phys. Soc.*, **62B**, 405–17. [*144*]
Hopf, F.A. (1980) *J. Opt. Soc. Am.*, **70**, 1320–3. [*28*]
Hopf, F.A. and Cervantes, M. (1982) *Appl. Opt.*, **21**, 668–77. [*190*]
Hopf, F.A., Tomita, A., Al-Jumaily, G., Cervantes, M. and Leipmann, T. (1981) *Opt. Commun.*, **36**, 487–90. [*198*]
Hopkins, H.H. (1951) *Proc. R. Soc. Lond.*, **A208**, 263–77. [*81, 82*]
Hopkins, H.H. (1955) *Opt. Acta*, **2**, 23–9. [*285*]
Hopkins, H.H. (1962) *Proc. Phys. Soc.*, **79**, 889–919. [*285*]
Hopkins, R.E. (1962) *J. Opt. Soc. Am.*, **52**, 1218–22. [*194*]
Hopkinson, G.R. (1978) *J. Opt. Paris*, **9**, 151–5. [*166*]
Horn, W. (1958) *Jahrb. Opt. Feinmech.*, **5**, 85–106. [*210*]

Houston, J.B., Buccini, C.J. and O'Neill, P.K. (1967) *Appl. Opt.*, **6**, 1237–42. [*197*]
Hutcheson, E.T., Hass, G. and Coulter, J.K. (1971) *Opt. Commun.*, **3**, 213–16. [*149*]
Hutley, M. (1976) *J. Phys. E.: Sci. Instrum.*, **2**, 513–20. [*69*]
Ichioka, Y. and Lohmann, A.W. (1972) *Appl. Opt.*, **11**, 2597–602. [*219*]
Imai, M., Ohashi, T. and Ohtsuka, Y. (1981) *Opt. Commun.*, **39**, 7–10. [*140*]
Ineichen, B., Eglin, P. and Dändliker, R. (1980) *Appl. Opt.*, **19**, 2191–5. [*228*]
Ingelstam, E. (1954) *Ark. Fys.*, **7**, 309–32. [*185*]
Ingelstam, E. and Johansson, L.P. (1958) *J. Sci. Instrum.*, **35**, 15–17. [*213*]
Israelachvili, J.N. (1973) *J. Colloid Interface Sci.*, **44**, 259–72. [*178*]
Jacquinot, P. (1958) *J. Phys. Radium*, **19**, 223–9. [*238*]
Jacquinot, P. (1960) *Rep. Prog. Phys.*, **23**, 267–312. [*154*]
Jacquinot, P. and Dufour, C. (1948) *J. Rech. Cent. Natl. Rech. Sci.*, **2**, 91–103. [*239, 244*]
Jacquinot, P. and Roizen-Dossier, B. (1964) *Progress in Optics*, ed. E. Wolf, **3**, 29–186. Amsterdam: North-Holland. [*234, 236*]
Johnson, E.R. and Scholes, J.F.M. (1948) *Aust. J. Sci. Res.*, **A1**, 464–71. [*102, 208, 209*]
Jonathan, J.M. and May, M. (1980) *Appl. Opt.*, **19**, 624–30. [*105*]
Jones, F.E. (1981) *J. Res. Natl. Bur. Stand.*, **86**, 27–32. [*173*]
Jones, R.C. (1941) *J. Opt. Soc. Am.*, **31**, 488–503. [*39*]
Kerl, K. (1979) *Opt. Acta*, **26**, 1209–24. [*257*]
Khanna, S.M. and Leonard, D.G.B. (1982) *Optics in Biomedical Sciences*, ed. P. Greguss and G. von Bally, pp. 88–91. [*166*]
Kimmit, M.F. (1977) *Infrared Phys.*, **17**, 459–66. [*237*]
King, R.J. and Raine, K.W. (1981) *Opt. Engng*, **20**, 39–43. [*166*]
Kinosita, K. (1953). *J. Phys. Soc. Jpn.*, **8**, 219–25. [*147, 148*]
Kiselyov, B.A. and Parshin, P.F. (1964) *Opt. Spektrosk.*, **17**, 940–3, Engl. transl. *Opt. Spectrosc.*, **17**, 511–13. [*238, 251*]
Knox, K.T. (1976) *J. Opt. Soc. Am.*, **66**, 1236–9. [*88*]
Kohler, D. and Mandel, L. (1973) *J. Opt. Soc. Am.*, **63**, 126–34. [*87*]
Kolesov, Yu.I. and Listvin, V.N. (1979) *Opt. Spektrosk.*, **46**, 1010–12. Engl. transl. *Opt. Spectrosc.*, **46**, 568–9. [*256*]
Koppelmann, G., Rudolph, H. and Schrech, K. (1975) *Optik*, **43**, 35–52. [*187*]
Korotaev, V.V. and Pankov, E.D. (1981) *Opt. Mekh. Promst.*, **48**(1), 9–12; Engl. transl. *Sov. J. Opt. Technol.*, **48**, 9–13. [*106*]
Korpel, A. (1981) *Proc. IEEE* (Inst. Elect. Electron. Eng.), **69**, 48–53. [*58*]
Krug, W., Rienitz, J. and Schulz, G. (1964) *Contributions to Interference Microscopy*. London: Hilger & Watts. [*211*]
Kubota, H. (1961) *Progress in Optics*, ed. E. Wolf, **1**, 211–51. Amsterdam: North-Holland. [*165*]
Kuhn, H. (1951) *Rep. Progr. Phys.*, **14**, 64–94. [*180*]
Kuhn, H. and Wilson, B.A. (1950) *Proc. Phys. Soc.*, **B63**, 745–55. [*149*]
Kwaaitaal, Th., Luymes, B.J. and van der Pijll, G.A. (1980) *J. Phys. D.: Appl. Phys.*, **13**, 1005–15. [*167*]
Kwon, O., Wyant, J.C. and Hayslett, C.R. (1980) *Appl. Opt.*, **19**, 1862–9. [*190*]
Labeyrie, A. (1974) *Nouv. Rev. Opt.*, **5**, 141–51. [*283*]

Labeyrie, A. (1976) *Progress in Optics*, ed. E. Wolf, **14**, 47–87. Amsterdam: North-Holland. [*283, 284*]

Lambert, D.K. and Richards, P.L. (1978) *J. Opt. Soc. Am.*, **68**, 1124–9. [*91, 266*]

Layer, H.P., Deslattes, R.D. and Schweitzer, W.G. (1976) *Appl. Opt.*, **15**, 734–43. [*172*]

Lee, W.H. (1978) *Progress in Optics*, ed. E. Wolf, **16**, 119–232. Amsterdam: North-Holland. [*68*]

Leeb, W.R., Schiffner, G. and Scheiterer, E. (1979) *Appl. Opt.*, **18**, 1293–5. [*184*]

Leendertz, J.A. (1970) *J. Phys. E.: Sci. Instrum.*, **3**, 214–18. [*227*]

Leistner, A.J. (1975) *Appl. Opt.*, **15**, 293–8. [*151*]

Leith, E.N. and Swanson, G.J. (1980) *Appl. Opt.*, **19**, 638–44. [*6*]

Leith, E.N. and Upatnieks, J. (1962) *J. Opt. Soc. Am.*, **52**, 1123–30. [*61*]

Leith, E.N. and Upatnieks, J. (1963) *J. Opt. Soc. Am.*, **53**, 1377–81. [*61*]

Leith, E.N. and Upatnieks, J. (1964) *J. Opt. Soc. Am.*, **54**, 1295–301. [*61, 67*]

Lenhardt, K. and Burckhardt, Ch. (1977) *Optik*, **47**, 215–22. [*106*]

Lenouvel, L. and Lenouvel, F. (1938) *Rev. Opt. (Fr)*, **17**, 350–61. [*197*]

Lighthill, M.J. (1958) *An Introduction to Fourier Analysis and Generalised Functions.* Cambridge University Press. [*16*]

Linnik, V.P. (1933*a*) *Akad. Nauk. SSSR Dokl.*, **1**, 18–23. [*89, 210*]

Linnik, V.P. (1933*b*) *Akad. Nauk. SSSR Dokl.*, **1**, 208–10. [*205*]

Liu, L.S. and Klinger, J.H. (1979) *Interferometry*, ed. G.W. Hopkins. *Proc. SPIE* (Soc. Photo-Opt. Instrum. Eng.), **192**, 17–26. [*57, 169*]

Liu, C.Y.C. and Lohmann, A.W. (1973) *Opt. Commun.*, **8**, 372–7. [*87*]

Loewenstein, E.V. (1966) *Appl. Opt.*, **5**, 845–54. [*250*]

Lohmann, A. (1959) *Opt. Acta*, **6**, 37–41. [*286*]

Lohmann, A.W. (1962*a*) *Opt. Acta*, **9**, 1–12. [*5*]

Lohmann, A.W. (1962*b*) *Optical Instruments and Techniques*, ed. K.J. Habell, pp. 58–61. London: Chapman & Hall. [*266*]

Lohmann, A.W. and Paris, D.P. (1967) *Appl. Opt.*, **6**, 1739–48. [*68*]

Luc, P. and Gerstenkorn, S. (1978) *Appl. Opt.*, **17**, 1327–31. [*240*]

Luhmann, N.C. (1979) *Instrumentation and Techniques for Plasma Diagnostics*, ed. K.J. Button, vol. 2, New York: Academic Press. [*181*]

Lyot, B. (1944) *Ann. Astrophys.*, **7**, 31–79. [*134, 247*]

Lyubimov, V.V., Shur, V.L. and Etsin, I.Sh. (1978) *Opt. Spektrosk.*, **45**, 368–73, Engl. transl. *Opt. Spectrosc.*, **45**, 204–7. [*139, 172*]

Macek, W.M. and Davis, D.T.M. (1963) *Appl. Phys. Lett. (US)*, **2**, 67–8. [*183, 184*]

Mack, J.E., McNutt, D.P., Roesler, F.L. and Chabbal, R. (1963) *Appl. Opt.*, **2**, 873–85. [*154, 159, 246*]

Macleod, H.A. (1969) *Thin-Film Optical Filters.* London: Adam Hilger. [*42, 150*]

Magome, N., Imamura, T., Ueha, S. and Tsujiuchi, J. (1981) *Opt. Commun.*, **36**, 347–50. [*1*]

Mahé, C. and Marioge, J.P. (1978) *J. Opt. Paris*, **9**, 127–30. [*197*]

Maitland, A. and Dunn, M.H. (1969) *Laser Physics.* Amsterdam: North-Holland. [*155*]

Malacara, D. ed. (1978) *Optical Shop Testing.* New York: John Wiley. [*185, 194, 203, 204, 219*]

Mandel, L. (1963*a*) *Electromagnetic Theory and Antennas*, ed. E.C. Jordan, part 2, pp. 811–17. New York: Pergamon. [*270*]

Mandel, L. (1963*b*) *Progress in Optics*, ed. E. Wolf, **2**, 181–248. Amsterdam: North-Holland. [*86*]
Mandel, L. and Wolf, E. (1965) *Rev. Mod. Phys.*, **37**, 231–87. [*72, 85, 87*]
Maréchal, A. (1947) *Rev. Opt. (Fr)*, **26**, 257–77. [*38, 151, 152*]
Marioge, J.P., Bonino, B. and Mullot, M. (1975) *Appl. Opt.*, **14**, 2283–5. [*188*]
Martienssen, W. and Spiller, E. (1964) *Am. J. Phys.*, **32**, 919–26. [*270*]
Matsuda, K. (1980) *Appl. Opt.*, **19**, 2643–6. [*68, 204*]
McCready, L.L., Pawsey, J.L. and Payne-Scott, R. (1947) *Proc. R. Soc. Lond. A*, **190**, 357–75. [*274*]
McNutt, D.P. (1965) *J. Opt. Soc. Am.*, **55**, 288–92. [*159, 245*]
Mercier, R. and Lowenthal, S. (1980) *Opt. Commun.*, **33**, 251–6. [*61, 219*]
Mertz, L. (1958) *J. Phys. Radium*, **19**, 233–6. [*255, 266*]
Mertz, L. (1965) *Transformations in Optics*, New York: John Wiley. [*97, 117, 123, 250, 254, 258, 259, 261, 262*]
Mertz, L. (1967) *J. Phys. (Fr)*, **28**, suppl. C2, 11–13. [*255*]
Mertz, L., Young, N.O. and Armitage, J. (1962) *Optical Instruments and Techniques*, ed. K.J. Habell, pp. 51–7. London: Chapman & Hall. [*266*]
Michel, J.J. (1967) *J. Phys. Fr)*, **28**, suppl. C2, 109–12. [*269*]
Michelson, A.A. (1902) *Light Waves and Their Uses*. University of Chicago Press. [*1*]
Michelson, A.A. (1920) *Astrophys. J.*, **51**, 257–62. [*278*]
Michelson, A.A. (1927) *Studies in Optics*. University of Chicago Press. [*1*]
Michelson, A.A. and Benoît, J.R. (1895) *Trav. Mém. Bur. Int. Poids Mes.*, **11**, 1–85. [*169*]
Michelson, A.A. and Gale, H.G. (1925) *Astrophys. J.*, **61**, 137–9, 140–5. [*183*]
Michelson, A.A. and Morley, E.W. (1886) *Am. J. Sci. ser. 3*, **31**, 377–86. [*182*]
Michelson, A.A. and Morley, E.W. (1887) *Philos. Mag.*, (5) **24**, 449–63. [*182*]
Michelson, A.A. and Pease, F.G. (1921) *Astrophys. J.*, **53**, 249–59. [*102, 279*]
Mills, B.Y. and Little, A.G. (1953) *Aust. J. Phys.*, **6**, 272–8. [*75, 274*]
Minkwitz, G. and Schulz, G. (1964) *Opt. Acta*, **11**, 89–99. [*100*]
Monchalin, J.-P., Kelly, M.J., Thomas, J.E., Kurnit, N.A., Szöke, A., Zernike, F., Lee, P.H. and Javan, A. (1981) *Appl. Opt.*, **20**, 736–57. [*172*]
Montgomery, A.J. (1964) *J. Opt. Soc. Am.*, **54**, 191–8. [*285, 286*]
Mooney, C.F. and Barlow, B.L. (1965) *J. Opt. Soc. Am.*, **55**, 1178–9. [*177*]
Mori, S. (1978) *Opt. Acta*, **25**, 219–31. [*206*]
Murty, M.V.R.K. (1960) *J. Opt. Soc. Am.*, **50**, 83–4. [*122*]
Murty, M.V.R.K. (1964) *Appl. Opt.*, **3**, 531–4. [*204*]
Netterfield, R.P. (1977) *Opt. Acta*, **24**, 69–79. [*133*]
Nomarski, G. (1955) *J. Phys. Radium*, **16**, 9S–13S. [*136*]
Norton, R.H. and Beer, R. (1976) *J. Opt. Soc. Am.*, **66**, 259–64. [*235*]
Öhman, Y. (1938) *Nature (GB)*, **141**, 157–8. [*247*]
O'Neill, E.L. (1963) *Introduction to Statistical Optics*, Reading, Mass: Addison-Wesley. [*15, 29*]
Ostrovsky, Yu.I., Butusov, M.M. and Ostrovskaya, G.V. (1980) *Interference by Holography*. Berlin: Springer-Verlag. [*215, 218*]
Otte, G. (1969) *J. Sci. Instrum. ser. 2 (J. Phys. E)*, **2**, 622–3. [*151*]
Pancharatnam, S. (1963) *Proc. Indian Acad. Sci. A*, **57**, 231–43. [*95*]

Papoulis, A. (1968) *Systems and Transforms with Applications in Optics.* New York: McGraw Hill. [*12*]
Parsche, H. and Luchner, K. (1973) *Optik*, **38**, 298–310. [*263*]
Parshin, P.F. (1962) *Opt. Spektrosk.*, **13**, 740–5; Engl. transl. *Opt. Spectrosc.*, **13**, 418–21. [*253*]
Pawsey, J.L. and Bracewell, R.N. (1955) *Radio Astronomy.* Oxford: Clarendon Press. [*272, 273*]
Pease, F.G. (1930). *Sci. American*, **143**, 290–3. [*279*]
Peck, E.R. (1957) *J. Opt. Soc. Am.*, **47**, 250–2. [*113*]
Peřina, J. (1971) *Coherent Light.* London: van Nostrand Reinhold. [*72*]
Petit, R. ed. (1980) *Electromagnetic Theory of Gratings.* Berlin: Springer-Verlag. [*242*]
Philpot, J.StL. (1952) *Contraste de Phase et Contraste par Interférences*, ed. M. Françon, pp. 42–7. Paris: Revue d'Optique. [*213*]
Pike, E.R. (1976) *Very High Resolution Spectroscopy*, ed. R.A. Smith, pp. 51–73. London: Academic Press. [*270*]
Post, E.J. (1967) *Rev. Mod. Phys.*, **39**, 475–93. [*183*]
Poulter, K.F. and Nash, P.J. (1979) *J. Phys. E.: Sci. Instrum.*, **12**, 931–6. [*173*]
Pritchard, J.L., Bullard, A., Sakai, H. and Vanasse, G.A. (1967) *J. Phys. (Fr)*, **28**, suppl. 2, 67–72. [*261*]
Pritchard, J.L., Sakai, H., Steel, W.H. and Vanasse, G.A. (1967) *J. Phys. (Fr)*, **28**, suppl. 2, 91–6. [*264*]
Pryputniewicz, R. and Stetson, K.A. (1976) *Appl. Opt.*, **15**, 725–8. [*221*]
Pryputniewicz, R.J. and Stetson, K.A. (1980) *Appl. Opt.*, **19**, 2201–5. [*220*]
Puntambeker, P.N. (1973) *Proc. Indian Natl. Sci. Acad.*, A**39**, 257–68. [*161*]
Raine, K.W. and Downs, M.J. (1978) *Opt. Acta*, **25**, 549–58. [*128*]
Ramsay, J.V. (1962) *Appl. Opt.*, **1**, 411–13. [*154, 263*]
Ramsay, J.V. (1969) *Appl. Opt.*, **8**, 569–75. [*152*]
Ramsay, J.V., Kobler, H. and Mugridge, E.G.V. (1970) *Solar Phys.*, **12**, 492–501. [*246*]
Ramsay, J.V., Netterfield, R.P. and Mugridge, E.G.V. (1974) *Vacuum*, **24**, 337–40. [*151*]
Richards, P.L. (1964) *J. Opt. Soc. Am.*, **54**, 1474–84. [*268*]
Rienitz, J., Minor, U. and Urbach, E. (1962) *Optical Instruments and Techniques*, ed. K.J. Habell, pp. 163–72. London: Chapman & Hall. [*213*]
Ring, J. and Schofield, J.W. (1972) *Appl. Opt.*, **11**, 507–16. [*120*]
Roberts, R.B. (1975) *J. Phys. E: Sci. Instrum.*, **8**, 600–2. [*178*]
Robertson, E.R. ed. (1976) *The Engineering Uses of Coherent Optics.* Cambridge University Press. [*215*]
Roddier, F., Roddier, C. and Demarq, J. (1978) *J. Opt. Paris*, **9**, 145–9. [*106*]
Roesler, F.L. and Traub, W. (1966) *Appl. Opt.*, **5**, 463–8. [*188*]
Roland, G. (1965) *Spectroscopie par Transformation de Fourier.* Thèse, Liège. [*128, 265*]
Ronchi, V. (1962) *Atti Fond. Ronchi*, **17**, 93–143, 240–51. [*198*]
Rosendahl, G.R. (1960) *J. Opt. Soc. Am.*, **50**, 287–9, 859–61. [*27*]
Rosenthal, A.H. (1962) *J. Opt. Soc. Am.*, **52**, 1141–8. [*183*]
Rowland, J.T. and Agrawal, G.P. (1981) *Opt. Laser Technol.*, **13**, 239–44. [*184*]
Rowley, W.R.C. and Hamon, J. (1963) *Rev. Opt. (Fr)*, **42**, 519–31. [*269*]

Rozhdestvenskii, D.S. (1951) *Papers on Anomalous Dispersion in Metal Vapours.* Moscow: Akad. Nauk SSSR. [*180*]
Rubin, L. (1980) *Opt. Engng*, **19**, 815–24. [*200*]
Ryle, M. (1952) *Proc. R. Soc. Lond.*, A**211**, 351–75. [*75*, *274*]
Ryle, M. and Hewish, A. (1960) *Mon. Not. R. Astron. Soc.*, **120**, 220–30. [*275*]
Sagnac, G. (1914) *J. Phys. Radium*, **4**, 177–95. [*183*]
Saito, T., Imamura, T., Honda, T. and Tsujiuchi, J. (1980) *J. Opt. Paris*, **11**, 285–92. [*61*]
Sakai, H. (1971) *Aspen International Conference on Fourier Spectroscopy*, 1970, ed. G.A. Vanasse, A.T. Stair and D.J. Baker, pp. 19–41. AFCRL-71-0019. US Air Force. [*241*]
Sakai, H. (1977) *Spectrometric Techniques*, ed. G.A. Vanasse, **1**, 1–70. New York: Academic Press. [*255*, *262*, *264*]
Sakai, H. and Murphy, R.E. (1978) *Appl. Opt.*, **17**, 1342–6. [*255*]
Sakai, H. and Vanasse, G.A. (1966) *J. Opt. Soc. Am.*, **56**, 131–2. [*258*]
Sakurai, Y. and Tanaka, K. (1965) *Trans. Soc. Instrum. Control Eng. Jpn*, **1**, 25–31. [*177*]
Sapper, G.R. and Pulvermacher, H. (1965) *Optik*, **23**, 101–8. [*286*]
Schmahl, G. and Rudolph, D. (1976) *Progress in Optics*, ed. E. Wolf, **14**, 195–244. Amsterdam: North-Holland. [*69*]
Schnurr, A.D. and Mann, A. (1981) *Opt. Engng*, **20**, 412–16. [*197*]
Schulz, G. (1964) *Opt. Acta*, **11**, 43–60, 131–43. [*53*]
Schulz, G. (1976) *Opt. Acta*, **23**, 1029–38. [*218*]
Schulz, G. and Schwider, J. (1967) *Appl. Opt.*, **6**, 1077–84. [*187*]
Schulz, G. and Schwider, J. (1976) *Progress in Optics*, ed. E. Wolf, **13**, 95–167. Amsterdam: North-Holland. [*186*, *188*, *219*]
Schumann, W. and Dubas, M. (1979) *Holographic Interferometry.* Berlin: Springer-Verlag. [*217*]
Schwider, J. (1965) *Opt. Acta*, **12**, 65–79. [*158*]
Schwider, J. (1968) *Opt. Acta*, **15**, 351–72. [*141*]
Schwider, J. (1979) *Appl. Opt.*, **18**, 2364–7. [*161*]
Schwider, J., Grzanna, J., Spolaczyk, R. and Burow, R. (1980) *Opt. Acta*, **27**, 683–98. [*219*]
Selby, M.J. (1966) *Infrared Phys.*, **6**, 21–32; Ring, J. and Selby, M.J. *ibid.*, 33–43. [*267*]
Shankland, R.S. (1973) *Appl. Opt.*, **12**, 2280–7. [*1*]
Sheahen, T.P. (1974, 1975) *Appl. Opt.*, **13**, 2907–11; **14**, 1004–12. [*259*]
Sheem, S.K. and Giallorenzi, T.G. (1979) *Opt. Lett.*, **4**, 23–31. [*140*]
Shiotake, N., Tsuruta, T., Itoh, Y., Tsujiuchi, J., Takeya, N. and Matsuda, K. (1968) *Jpn J. Appl. Phys.*, **7**, 904–9. [*231*]
Shurchiff, W.A. (1962) *Polarized Light.* Harvard University Press. [*39*]
Simon, J. (1961) *Rev. Opt. (Fr)*, **40**, 213–30. [*287*]
Sinton, W.M. (1963) *J. Quantum Spectrosc. Radiat. Transfer*, **3**, 551–8. [*266*]
Sivtsov, G.P. (1980) *Opt. Mekh. Promst.*, **47**(7), 15–17; Engl. transl. *Soc. J. Opt. Technol.*, **47**, 392–4. [*27*]
Slater, P.N., Betz, H.T. and Henderson, G. (1965) *Photographic and Spectroscopic Optics*, *Jpn J. Appl. Phys.*, **4**, suppl. 1, 440–4. [*154*]
Smartt, R.N. and Ramsay, J.V. (1964) *J. Sci. Instrum.*, **41**, 514. [*131*, *133*]

Smartt, R.N. and Steel, W.H. (1975) *Jpn J. Appl. Phys.* **14**, suppl. 14-1, 351–6. [*205, 206, 207*]
Smith, R.H. (1956) *Modern Methods of Microscopy*, ed. A.E.J. Vickers, pp. 76–86. London: Butterworths. [*210, 212*]
Smith, P.W. (1965) *IEEE J. Quantum Electron.*, **1**, 343–8. [*56*]
Snyder, J.J. (1978) *Kvantovaya Elektron.*, **5**, 1682–4; Engl. transl. *Sov. J. Quantum Electron.*, **8**, 959–60. [*172*]
Šolc, I. (1965) *J. Opt. Soc. Am.*, **55**, 621–5. [*247, 248*]
Solomakha, D.A. and Toropov, A.K. (1977) *Kvantovaya Elektron.*, **4**, 1637–60; Engl. transl. *Sov. J. Quantum Electron.*, **7**, 929–42. [*172*]
Som, S.C. (1963) *Opt. Acta*, **10**, 179–86. [*82*]
Speer, R.J., Chrisp, M., Turner, D., Mrowka, S. and Tregidgo, K. (1979) *Appl. Opt.*, **18**, 2002–12. [*206*]
Spornik, N.M. and Yanichkin, V.I. (1971) *Opt. Mekh. Promst.*, **38**(8), 38–40; Engl. transl. *Sov. J. Opt. Technol.*, **38**, 487–9. [*198*]
Steel, W.H. (1962) *J. Opt. Soc. Am.*, **52**, 1153–5. [*287*]
Steel, W.H. (1964*a*) *J. Opt. Soc. Am.*, **54**, 151–6. [*113*]
Steel, W.H. (1964*b*) *Opt. Acta*, **11**, 9–19. [*286*]
Steel, W.H. (1964*c*) *Opt. Acta*, **11**, 211–17. [*123*]
Steel, W.H. (1965*a*) *J. Sci. Instrum.*, **42**, 102–4. [*202*]
Steel, W.H. (1966) *Progress in Optics*, ed. E. Wolf, **5**, 145–97. Amsterdam: North-Holland. [*105, 201, 202*]
Steel, W.H. (1971) *Aspen International Conference on Fourier Spectroscopy*, 1970, ed. G.A. Vanasse, A.T. Stair and D.J. Baker, pp. 43–57. AFCRL-71-0019. US Air Force. [*265*]
Steel, W.H. (1974) *Opt. Acta*, **21**, 599–608. [*113*]
Steel, W.H. (1975*a*) *Opt. Commun.*, **14**, 108–9. [*204*]
Steel, W.H. (1975*b*) *Opt. Acta*, **22**, 563–7. [*116*]
Steel, W.H. (1980) *Opt. Commun.*, **32**, 214–16. [*98, 287*]
Steel, W.H. and Forman, M.L. (1966) *J. Opt. Soc. Am.*, **56**, 982. [*258*]
Steel, W.H., Smartt, R.N. and Giovanelli, R.G. (1961) *Aust. J. Phys.*, **14**, 201–11. [*132*]
Sternberg, R.S. and James, J.F. (1964) *J. Sci. Instrum.*, **41**, 225–7. [*123*]
Stetson, K.A. (1974) *J. Opt. Soc. Am.*, **64**, 1–10. [*221*]
Stetson, K.A. (1976) *J. Opt. Soc. Am.*, **66**, 1267–71. [*229*]
Stetson, K.A. (1977) *J. Opt. Soc. Am.*, **67**, 1587–90. [*229*]
Stetson, K.A. (1978) *Speckle Metrology*, ed. R.K. Erf, pp. 295–320. New York: Academic Press. [*225*]
Stetson, K.A. (1979) *J. Opt. Soc. Am.*, **69**, 1705–10. [*217*]
Stevenson, W.H. (1970) *Appl. Opt.*, **9**, 649–52. [*58*]
Strong, J.D. (1958) *Concepts of Classical Optics*. San Francisco: Freeman. [*131, 214*]
Strong, J.D. and Vanasse, G.A. (1958) *J. Phys. Radium*, **19**, 192–6. [*267, 268*]
Strong, J. and Vanasse, G.A. (1960) *J. Opt. Soc. Am.*, **50**, 113–18. [*267*]
Tai, M.H. and Harwit, M. (1976–7) *Appl. Opt.*, **15**, 2664–6; **16**, 3071–2. [*238*]
Takasaki, H. (1966) *Appl. Opt.*, **5**, 759–64. [*168*]
Takasaki, H. (1979) *Opt. Acta*, **26**, 1009–19. [*5, 231*]
Tanaka, K. and Ohtsuka, Y. (1977) *J. Opt. Paris*, **8**, 37–40. [*167*]

Tango, W.J. and Twiss, R.Q. (1974) *Appl. Opt.*, **13**, 1814–19. [*139*]
Tango, W.J. and Twiss, R.Q. (1980) *Progress in Optics*, ed. E. Wolf, **17**, 239–77. Amsterdam: North-Holland. [*281*]
Taylor, W.G.A. (1957) *J. Sci. Instrum.*, **34**, 399–402. [*186*]
Terrien, J. (1959) *Rev. Opt. (Fr)*, **38**, 29–37. [*28, 122, 173*]
Terrien, J. (1960) *Interferometry: NPL Symposium No. 11*, 435–56. London: HMSO. [*269*]
Thelen, A. (1976) *Appl. Opt.*, **15**, 2983–5. [*131*]
Thompson, S.P. (1905) *Proc. Opt. Conv.* (1st), 216–40. London: Williams & Norgate. [*5, 132*]
Title, A.M. and Ramsey, H.E. (1980) *Appl. Opt.*, **19**, 2046–58. [*247*]
Title, A.M. and Rosenberg, W.J. (1981) *Opt. Engng*, **20**, 815–23. [*247*]
Tolansky, S. (1947) *High Resolution Spectroscopy*. London: Methuen. [*243*]
Tolansky, S. (1955) *An Introduction to Interferometry*. London: Longmans. [*3, 148, 191, 192, 243*]
Tolansky, S. (1960) *Surface Microtopography*. London: Longmans. [*191, 192*]
Tolansky, S. (1970) *Multiple-Beam Interference Microscopy of Metals*. London: Academic Press. [*190*]
Tremblay, R. and Boivin, A. (1966) *Appl. Opt.*, **5**, 249–78. [*137*]
Tsujiuchi, J., Takeya, N. and Matsuda, K. (1969) *Opt. Acta*, **16**, 709–22. [*222*]
Tsuruta, T. (1963*a*) *Appl. Opt.*, **2**, 371–8. [*285*]
Tsuruta, T. (1963*b*) *J. Opt. Soc. Am.*, **53**, 1156–61. [*285*]
Tsuruta, T. and Itoh, Y. (1969) *Opt. Commun.*, **1**, 34–6. [*229*]
Twiss, R.Q. (1969) *Opt. Acta*, **16**, 423–51. [*86, 280*]
Twiss, R.Q. and Little, A.G. (1959) *Aust. J. Phys.*, **12**, 77–93. [*86*]
Tyler, G.A. and Thompson, B.J. (1976) *Opt. Acta*, **23**, 685–700. [*60*]
Väisälä, Y. (1954) *P-V. Scéance Commun. Int. Poids Mes.*, **24**, M60–6. [*169*]
Vali, V. and Shorthill, R.W. (1977) *Appl. Opt.*, **16**, 290–1. [*140*]
Vanasse, G.A. ed. (1977) *Spectrometric Techniques*, vol. **1**. New York: Academic Press. [*250*]
Vanasse, G.A. ed. (1981) *Spectrometric Techniques*, vol. **2**, New York: Academic Press. [*250, 255, 262*]
Vanasse, G.A. and Sakai, H. (1967) *Progress in Optics*, ed. E. Wolf, **6**, 259–77. Amsterdam: North-Holland. [*235, 250, 261, 268*]
van Heel, A.C.S. (1961) *Progress in Optics*, ed. E. Wolf, **1**, 289–329. Amsterdam: North-Holland. [*213*]
van Heel, A.C.S. and Simons, C.A.J. (1967) *Appl. Opt.*, **6**, 803–6. [*197*]
Véron, D. (1979) *Infrared and Millimeter Waves*, ed. K.J. Button, **2**, 69–135. New York: Academic Press. [*181*]
Vest, C.M. (1979) *Holographic Interferometry*. New York: John Wiley. [*207, 215, 223, 225*]
Walker, J. (1904) *Analytical Theory of Light*. Cambridge University Press. [*102*]
Wallard, A.J. (1973) *J. Phys. E: Sci. Instrum.*, **6**, 793–807. [*57*]
Walles, S. (1969) *Ark Fys.*, **40**, 299–403. [*83*]
Walles, S. (1970) *Opt. Acta*, **17**, 899–913. [*84*]
Warde, C. (1976) *Appl. Opt.*, **15**, 2730–5. [*243*]
Watson, G.N. (1944) *A Treatise on the Theory of Bessel Functions*, 2nd ed. Cambridge University Press. [*99*]

Weinberg, F.J. (1963) *Optics of Flames*. London: Butterworths. [*208*, *209*]
Weinberg, F.J. and Wood, N.B. (1959) *J. Sci. Instrum.*, **36**, 227–30. [*137*]
Weingärtner, I. and Rosenbruch, K.-J. (1982) *Opt. Acta*, **29**, 519–29. [*68*]
Welford, W.T. (1969*a*) *Opt. Acta*, **16**, 371–6. [*229*]
Welford, W.T. (1969*b*–70) *Opt. Commun.*, **1**, 123–5, 311–14. [*97*]
Weller, R. and Shepard, B.M. (1948) *Proc. Soc. Exp. Stress Anal.*, **6**, 35–8. [*231*]
Wild, J.P. (1961) *Proc. R. Soc. Lond*, *A*, **262**, 84–99. [*163*, *275*]
Wild, J.P. (1965) *Proc. R. Soc. Lond. A*, **286**, 449–509. [*163*, *275*]
Williams, W.E. (1930) *Applications of Interferometry*. London: Methuen. [*3*]
Wohlleben, R. and Mattes, R. (1973) *Interferometrie in Radioastronomie und Radartechnik*. Würzburg: Vogel-Verlag. [*275*]
Wolfe, W.L. (1978) *Handbook of Optics*, ed. W.G. Driscol, 7–1 to 7–157. New York: McGraw-Hill. [*133*]
Woods, P.T., Shotton, K.C. and Rowley, W.R.C. (1978) *Appl. Opt.*, **17**, 1048–54. [*172*]
Wyant, J.C. and Smith, F.D. (1975) *Appl. Opt.*, **14**, 1607–12. [*204*]
Yoshinaga, H. (1965) *Photographic and Spectroscopic Optics*, *Jpn J. Appl. Phys.*, **4**, suppl. 1, 420–8. [*260*]
Zakharov, V.P., Tychinskii, V.P., Snezhko, Yu.A., Evtikhiev, N.N. and Vanetsian, R.A. (1976) *Izmer. Tekh.*, **19**(10), 33–6; Engl. transl. *Meas. Tech.*, **19**, 1433–7. [*167*]
Zehnpfennig, T.F., Shepherd, O., Rappaport, S., Reidy, W.P. and Vanasse, G. (1979) *Appl. Opt.*, **18**, 1996–2002. [*255*]
Zernike, F. (1934) *Physica*, **1**, 689–704. [*35*]
Zernike, F. (1947) *Physica*, **13**, 279–88. [*182*]
Zernike, F. (1948) *Proc. Phys. Soc.*, **6**, 158–64. [*50*, *185*]
Zhang, G.Y. (1981) *Opt. Acta*, **28**, 585–8. [*157*]

Index

Index